International Law and the Global South

Perspectives from the Rest of the World

Series editor

Leïla Choukroune, Director, University of Portsmouth Research and Innovation Theme in Democratic Citizenship, University of Portsmouth, Portsmouth, UK

This book series aims to promote a complex vision of contemporary legal developments from the perspective of emerging or developing countries and/or authors integrating these elements into their approach. While focusing on today's law and international economic law in particular, it brings together contributions from, or influenced by, other social sciences disciplines. Written in both technical and non-technical language and addressing topics of contemporary importance to a general audience, the series will be of interest to legal researchers as well as non-lawyers.
In referring to the "rest of the world", the book series puts forward new and alternative visions of today's law not only from emerging and developing countries, but also from authors who deliberately integrate this perspective into their thinking. The series approach is not only comparative, post-colonial or critical, but also truly universal in the sense that it places a plurality of well-informed visions at its center.

The Series

- Provides a truly global coverage of the world in reflecting cutting-edge developments and thinking in law and international law
- Focuses on the transformations of international and comparative law with an emphasis on international economic law (investment, trade and development)
- Welcomes contributions on comparative and/or domestic legal evolutions

More information about this series at http://www.springer.com/series/13447

Stellina Jolly · Nafees Ahmad

Climate Refugees in South Asia

Protection Under International Legal Standards and State Practices in South Asia

Springer

Stellina Jolly
Faculty of Legal Studies
South Asian University
New Delhi, India

Nafees Ahmad
Faculty of Legal Studies
South Asian University
New Delhi, India

ISSN 2510-1420 ISSN 2510-1439 (electronic)
International Law and the Global South
ISBN 978-981-13-3136-7 ISBN 978-981-13-3137-4 (eBook)
https://doi.org/10.1007/978-981-13-3137-4

Library of Congress Control Number: 2018911908

This Springer imprint is published by the registered company Springer Nature Singapore Pte Ltd.
The registered company address is: 152 Beach Road, #21-01/04 Gateway East, Singapore 189721, Singapore

Acknowledgements

The concept of this book emanated and emerged out of our experience of teaching the courses of International Environmental Law and International Refugee Law at the Faculty of Legal Studies (FLS), South Asian University (SAU), New Delhi, India. The curious questions and the vibrant deliberation with the dynamic minds at FLS, in and around the classroom, gestated the concept for this book. After this initial stage of fruition, many rendered their assistance in putting together the current version. We would like to place on record our gratitude to all these people who offered their support and encouragement at all stages of the book.

Reviewers of this volume deserve a special mention for their valuable suggestions and comments. We would also like to thank the entire Springer team, especially Sagarika Ghosh and Nupoor Singh, for their encouraging response to our project, the constant cooperation, and for ensuring an accelerated publication process.

Special thanks are extended to Dr. Kavita Sharma, President, South Asian University, for her enthusiastic response and kind encouragement for this project. We acknowledge Prof. Sasanka Perera, Vice President, South Asian University, and Prof. Santosh C. Panda, Vice President, South Asian University, for fostering South Asian focus-oriented research. This book project is motivated by this endeavour of SAU to promote the South Asian-focused research. We also express our gratitude to our colleagues at FLS-SAU for their constant encouragement.

We place on record our gratitude to our teachers—Prof. Veer Singh, Former Vice Chancellor, NALSAR, Hyderabad; Prof. P. S. Jaswal, Vice Chancellor, Rajiv Gandhi National Law University, Patiala—for their sustained interest and unbridled encouragement of our academic endeavours.

We take this opportunity to acknowledge the research assistance rendered by Amit Kumar, Ravneet Sandhu, Tanaya Thakur, Manini Syali and Sumedha Chaudhury, our students of LLM at FLS-SAU. We also acknowledge the research assistance rendered by Kanika Jamwal, a student at RGNUL, Patiala.

The past year has been exhausting and enervating, and this book would not have been achievable without the incessant help of our family. We want to thank our respective spouses Amit and Aaisha Khan, children Agastya and Shezan and the rest of our families, who supported and encouraged us in this academic journey all through the time it took us away from them.

Contents

About the Authors

Dr. Stellina Jolly is an assistant professor at the Faculty of Legal Studies, South Asian University (SAU), New Delhi, India. Her teaching interests include international environmental law, conflict of laws and bio-ethics. She has published an edited collection titled 'Private International Law: South Asian States' Practice', published by Springer in 2017. She has undertaken projects and consultancies with various organizations including the Rajiv Gandhi Foundation, European Union etc. She was awarded an educational grant on Civil Society Law from ICNL and USAID. Her publications include articles in the *International Journal of Family Law and Policy* (Oxford), *European-Asian Journal of Law and Governance* (CPG), *UNESCO Journal of Bio-Ethics, International Journal of Public Law and Policy* (Inderscience), and *ISIL Year Book on International Humanitarian Law and Refugee Law* (ISIL) etc. She has served on the editorial boards of several peer-reviewed journals, including the *Indian Journal of Human Rights, the International Journal of Bioethics* and the *NUJS International Journal of Legal Studies and Research*. Stellina has been a Resource Person and External Reviewer for the Ministry of Law, Government of India sponsored Research Project on Judicial Reforms since June 2016 at the Indian Institute of Management (IIM), Kashipur. Further, she has been a visiting professor at the Indian Institute of Public Administration (IIPA), Indian Society of International Law (ISIL), and National Law University Delhi.

Dr. Nafees Ahmad is an assistant professor at the Faculty of Legal Studies, South Asian University (SAU), New Delhi. Holding a doctorate in international refugee law and human rights, his work focuses on global forced displacement and migrations, climate refugees & climate change human displacement, policy, asylum, durable solutions and SAARC issues. He has introduced a new Program at the SAU called Comparative Constitutional Law of SAARC Nations for LLM along with international human rights and international refugee law. His publications include papers in the *Asia-Pacific Journal of Human Rights and Law* (Brill), *Kings' Student Law Review* (King's College London), *Groningen Journal of International Law* (University of Groningen), *ISIL Year Book on International Humanitarian*

Law and Refugee Law etc. His book on international refugee law and human rights is in press. Dr. Ahmad is also an active blogger, writer and Op-Ed contributor to many international sites in the fields of forced migration, refugee research, human rights, international relations and diplomacy etc. Dr. Ahmad has been a Resource Person and External Reviewer for the Ministry of Law, Government of India sponsored Research Project on Judicial Reforms since June 2016 at the Indian Institute of Management (IIM), Kashipur.

Abbreviations

AALCO	Asian-African Legal Consultative Organization
ABC-CLIO	American Bibliographic Centre–Clio Press
ACCSAP	National Climate Change Strategy, and Action Plan for Afghanistan
ADB	Asian Development Bank
ADR	Alternative Dispute Resolution
AHCR	American Convention on Human Rights
AJNE	Asian Judges Network on the Environment
AOSIS	Alliance of Small Island States
ASEAN	Association of Southeast Asian Nations
AU	African Union
CBDR	Common but Differentiated Responsibilities
CCA	Climate Change Adaptation
CCDESO	Climate Change Displacement Environment and Sciences Organisation
CCDO	Climate Change Displacement Organisation
CCDPs	Climate Change Displaced Persons
CCVI	Climate Change Vulnerability Index
CDM	Clean Development Mechanism
CED	UN Convention for the Protection of All Persons from Enforced Disappearance
CEDAW	UN Convention on the Elimination of All Forms of Discrimination Against Women
CERD	UN Convention on the Elimination of All Forms of Racial Discrimination
CIGI	Centre for International Governance Innovation
CIL	Customary International Law
CMW	International Convention on the Protection of the Rights of All Migrant Workers and Members of Their Families
COP	Conference of Parties

CRC	Convention on the Rights of the Child, 1989
CRED	Centre for Research on the Epidemiology of Disasters
CRPD	Convention on the Rights of Persons with Disabilities
CSOs	Civil Society Organisations
DDMA	District Disaster Management Authorities
DRR	Disaster Risk Reduction
EC	Environment Courts
ECA	Environment Court Act
ECHR	Convention for the Protection of Human Rights and Fundamental Freedoms (European Convention on Human Rights)
ECOSOC	Economic and Social Council
ECOWAS	Economic Community of West African States
EPAs	Environmental Protection Agencies
EPG	Eminent Persons Group
ESIL	European Society of International Law
FMR	Forced Migration Review
GEO	Global Environment Outlook
GJIE	Global Judicial Institute for the Environment
GLACC	Global Legal Action on Climate Change
GLOF	Glacial Lake Outburst
HRBA	Human Rights-Based Approach
HRP	Humanitarian Response Plan
Ibid	*Ibidem*
ICCL	International Climate Change Law
ICCPR	International Covenant on Civil and Political Rights
ICELA	Indian Council for Enviro-legal Action
ICESCR	International Covenant on Economic, Social and Cultural Rights
ICJ	International Court of Justice
IDMC	Internal Displacement Monitoring Centre
IDP	Internally Displaced Person
IHL	International Humanitarian Law
IHRL	International Human Rights Law
INDC	Intended Nationally Determined Commitments
IOM	International Organisation for Migration
IPCC	Intergovernmental Panel on Climate Change
IRL	International Refugee Law
ISAJI	Intra-South Asian Judicial Interactions
IUCN	International Union for the Conservation of Nature and Natural Resources
LDC	Least Developed Countries
MAHAGENCO	Maharashtra State Power Generation Company Limited
MCCPF	Maldives Climate Change Policy Framework
MoEF	Ministry of Forests and Environment
MOU	Memorandum of Understanding

MPCB	Maharashtra Pollution Control Board
NAP	National Action Plan
NAPA	National Adaptation Plan
NAPCC	National Action Plan on Climate Change
NAPCC	National Advisory Committee on Climate Change
NATCOMs	National Communication on Climate Change
NDMA	National Disaster Management Authority
NDMP	National Disaster Management Plan
NEC	National Environment Commission
NEPA	National Environmental Protection Agency
NGT	National Green Tribunal
NHRC	National Commission of Human Rights
NSC	National Conservation Strategy
OAS	Organisation of American States
OAU	Organisation of African Unity
OHCHR	Office of the United Nations High Commissioner for Human Rights
PAK	Pakistan-administered Kashmir
PDMA	Provincial Disaster Management Authorities
PEPC	Pakistan Environmental Protection Council
PIL	Public Interest Litigation
PNG	Papua New Guinea
PSNR	Permanent Sovereignty over Natural Resources
R2P	Responsibility to Protect Framework
REDD	Reducing Emission from Deforestation and Degradation
SAARC CFDM	SAARC Comprehensive Disaster Management Framework
SAARC	South Asian Association for Regional Cooperation
SADKN	South Asian Disaster Knowledge Network
SAJIE	South Asian Judicial Institute for the Environment
SAJNE	South Asian Judges Network on Environment
SAPCCs	State Action Plans on Climate Change
SDF	SAARC Development Fund
SIDS	Small Island Developing States
SLP	Special Leave Petition
SRD	SAARC Regional Declaration
TWAIL	Third World Approaches of International Law
UDHR	Universal Declaration of Human Rights
UK	United Kingdom
UN Doc.	United Nations Document
UN	United Nations
UNCLOS	United Nations Law of the Seas Convention
UNDP	United Nations Development Programme
UNEP	United Nations Environment Programme
UNFCCC	United Nations Framework Convention on Climate Change
UNGA	United Nations General Assembly

UNHCR	United Nations High Commissioner for Refugees
UNHRC	United Nations Human Rights Council
UNSC	United Nations Security Council
UNSCR	United Nations Convention Relating to Status of Refugees
UNTS	United Nations Treaty Series
WCS	World Conservation Strategy
WIM	Warsaw International Mechanism

Chapter 1
Introduction

Abstract South Asia is a resevoir of civilizational diversity and legal pluralism that is conspicuous in its multi-layered identities of human existence in the eight nation-states consisting of Afghanistan, Bangladesh, Bhutan, India, Maldives, Nepal, Pakistan and Sri Lanka. South Asia has the distinction of having land-locked, littoral and lsland-nation-states that pose multiplicity of challeges in addressing the plight of climate change-induced human displacement and migration under the international climate change law (ICCL) and international refugee law (IRL). There is a global understanding that emerged in the political Declaration adopted on 19 September 2016 at New York on all migrants and refugees that still remains to be further negotiated in December 2018 under the UN auspices. Unfortunately, South Asian states have not acceded to the refugee law obligations despite the fact of the enormity of the climate change-driven crisis in the region. Moreover, these international legal regimes are flagrantly inadequate in addressing the overall protection requirements of the climate refugees. Further, there is neither legal and binding commitment to protect and provide safety and security to climate refugees nor there is any permanent lego-institutional framework to address the emergence of the climate change refugees or traditional humanitarian refugees in the region. In 1985, South Asian countries established a regional organization called South Asian Association for Regional Cooperation (SAARC) for developing the region into a new identity based on 'South Asian-ness' and 'regional consciousness' paradigms to galvanize international narratives for creating an equitable world order based on pluralism, cosmopolitanism and multiculturalism. It is in this conspectus, the instant book critically and cumulatively tries to examine protection and its thresholds afforded to climate refugees in South Asia under international legal standards and state practices of the SAARC countries.

South Asia is one of the cradles of civilizations in the world where people of all races and religions have come and coexisted since antiquity. This trajectory of different cultures has bestowed upon it a one-off identity that has established the region unparalleled and unique anywhere else in the world. There are eight independent countries in South Asia such as Afghanistan, Bangladesh, Bhutan, India, Maldives, Nepal,

S. Jolly and N. Ahmad, *Climate Refugees in South Asia*, International Law and the Global South, https://doi.org/10.1007/978-981-13-3137-4_1

Pakistan and Sri Lanka. In 1985, these countries established a regional organization called South Asian Association for Regional Cooperation (SAARC) for developing the region into a new identity based on 'South Asian-ness' and 'regional consciousness' paradigms to galvanize international narratives for creating an equitable world order based on pluralism, cosmopolitanism and multiculturalism. SAARC has been promoting social, economic progress and cultural development within the South Asian region and also for friendship and cooperation with other developing and developed countries. It tries to rummage the feasibility of South Asian integration premised on confluence and convergence of regional unity to voice anxieties, alarms and apprehensions to achieve international, stability and harmony.

In August 2017, the region was hit by one of the worst devastating floods in the recent history. The ravaging floods have left more than 41 million population-battling floods and displacement. Experts unanimously link the floods to the ongoing threat of climate change. The relationship between the climate and humanity is existential and rests in the excellent balance of human geography fossilized in the nethermost crust of the human existence. Moreover, this relationship is based on complementarity and reciprocity between man and nature guided to maintain the climate stability. Therefore, any attempt—inadvertent or deliberate—that is damaging and prone to the decadence in the delicate balance of human ecology would not be acceptable. The climate change, which is destructing, decimating and diluting the very roots of the original ecological balance, has now captured the popular imagination and is characterized as the monumental human development challenge for South Asia in the twenty-first century. Climate change issue came to the policy arena in the mid-1980s when UN General Assembly deliberated the impugned matter. The gravity of the situation was highlighted further by the first assessment report prepared by the Intergovernmental Panel on Climate Change (IPCC) published in 1990. The report alerted the world about the climate disaster in waiting and pointed out the different consequences of climate change. Initially, it was a contentious issue, and many refuse to believe in the science of climate change. However, the science has become increasingly particular through the multitude of studies and reports highlighting the climate change and its adverse impacts. The fourth IPCC assessment report disseminated in 2007 warned that the warming of the climate system is unequivocal and accelerating. The fifth assessment of the IPCC in clear terms pointed that the earth is warming at an exponential rate. The latest evaluation report indicates that the atmosphere and ocean have warmed, the amounts and volumes of snow and ice have petered out, sea level has soared, and the concentrations of greenhouse gases have escalated.

What are the impacts and consequences of this climate change, climate change as a nemesis will alter the coastlines, reduce the availability of food and water, expose the humanity and especially developing nations to extreme disasters and force human displacement? The Stern review highlighted the human impact of climate change and noted that the climate change would affect the essential elements of life such as access to water, food production, health and the environment for people around the world. The underlined impacts of climate change indicate a scenario where the threat of climate change is not merely environmental but social affecting the deep structure of society. However, anthropogenic greenhouse gases mostly emit from economic activities in the prosperous nation-states, and it is crucial to underline here that the impacts of climate change will be distributed unevenly and disproportionately around the world with poorer countries, vulnerable communities including women and children among the worst affected.

South Asia has a population of about 1.749 billion people in 2017 that make the region having one-fourth of the world population. The diverse geography, multiple ethnicities, varied cultural practices and climate bestow uniqueness to the region; SAARC countries do not significantly contribute to greenhouse gases in the atmosphere, but they are bound to be at an even higher disadvantage when it comes to dealing with the climate change impacts. Climate Change Vulnerability Index (CCVI) positions Afghanistan, Bangladesh, India, Pakistan and Nepal among the highest risk category of 16 nation-states that confront the 'extreme risk' due to the climate change-induced natural disasters. Climate change is hovered to make matters terrible for small farmers through an alteration in climate and agricultural regions while shifting the production patterns. The reports and studies suggest that millions of people in South Asia will be forced to move due to blistering temperatures and crop failures unless steps are taken to reduce greenhouse gas emissions. Thus, the disadvantaged groups do not have the requisite resources to cope with poverty and inequality worsened and caused by the climate change impacts. Moreover, South Asian countries have been grappling with lack of infrastructure and less technological and financial resources that have been hindering their ability to adapt and rural and urban poor are the severest hit. Women, children, elderly and marginalized people who are already at the verge of vulnerability in society would become even more prone to the adverse impact of climate change. Climate change will exacerbate and accentuate the existing inequities in the South Asian nations. Many communities will be forced to move due to rising sea levels and extreme drought and heatwaves pose a strain on resources or even excessive rainfall that becomes the norm. The intense heatwaves are predicted to make strips of South Asia uninhabitable by humans by 2100. The Small Island Developing States (SIDS) like the Maldives and Sri Lanka represent the face of climate change, are particularly vulnerable and are at the forefront of confronting the ill-effects of climate change, while Bangladesh, India and Pakistan have been facing the brunt of internal displacement of people called internally displaced persons (IDPs) living in (climate) refugees-like situations.

IPCC reports and other studies highlight the critical impacts of climate change and the forced migration to the region. The case of the Maldives, which faces the threat of total submergence with increasing sea level, symbolizes the face of climate disaster for the SAARC area. The United Nations' Global Environment Outlook (GEO-6) reports 2016 points to the alarming figure of 40 and 25 million population threatened by rising sea level in India and Bangladesh, respectively. These threats are not mere futuristic as the region has been already witnessing the extreme vagaries of climate change impact. For example, half of Bhola Island in 1995 in Bangladesh became permanently flooded, forcing half million people to relocate. Other instances comprise floods in Pakistan in 2010 and the grave floods in the Ganges basin that entirely shattered the state of Uttarakhand in India. The Cox's Bazar in Bangladesh is faced with severe coastal erosion and the ocean breaching the existing embankment.

The situation has severely impacted agricultural production, livelihood and existence. The despair of communities has already resulted in increased movement of people. For South Asia, going forward, the climate change will progressively strike at the very foundation of economic growth and human security in multiple ways and will act as a catalyst in altering the demographic processes through large-scale human displacement.

As observed by the first IPCC report that severest effects of climate change may be those on human migration as millions of people are displaced by coastal flooding, shoreline erosion, severe disasters. Fifth IPCC assessment report reiterated the above findings and discussed the issue of climate displacement in detail. The report highlighted the several complications and ramifications of climate change that masquerades risks to humanity and natural systems and that climate change has the potential to entail additional pressures for the different facets of human security, including climate change migration. The report highlighted that climate change is projected to increase the displacement of people throughout this century and South Asian countries are going to be worst affected. These reports by IPCC and other studies present multiple reasons and scenarios for climate migration/displacement in South Asia and elsewhere:

- Deprivation of housing (due to river or sea flooding or mudslides in the mountain);
- Damage of living resources and natural reserves (e.g. water, energy, food supply and/or employment adversely impacted by climate change); and
- Destruction of sociocultural resources (e.g. pillaging of cultural properties, community networks, neighbourhood demography or especially in the case of a devastating flood).

IPCC reports documented the increased vulnerability of displacement in developing nations including countries of South Asia and more impoverished communities due to high population densities and urban congestion, which increases the potential magnitude of disasters. In South Asia, the complexities associated with climate change is high as other social and economic factors operate in tandem and will be the determining force in the decision to relocate leading to displacement. The climate displacement/migration is affected by the inherent adaptive capacities, resilience level and coping abilities of the communities to environmental and climate vagaries. Hence, decision to migrate cannot be considered to be unidirectional and differs from region to region and people to people. Consequently, the certainty is missing as to what exactly climate change means for migration. The reports did not consider migration only as a negative consequence; it examined mobility and migration as a response to extreme weather events and climate change as an adaptation strategy. The UN High Commissioner for Refugees (UNHCR) identified four leading climate change-related displacement/migration scenarios.

- Weather-induced displacement, prompted by events such as hurricanes and flooding;
- Violence related to climate change-related factors;
- Increased disaster threats in high-risk zones and resultant replacement of people; and

- Gradual environmental deterioration and slow-onset disasters which could lead to coastal erosion sinking of littoral and riparian zones, regions, divisions and probable entire drowning of the low-lying island nation-states.

Different projections are made regarding climate-induced displacement, and the precision of predictions appears to be less due to its complex catena of factors. International Organization for Migration (IOM) projects the number of displaced people to be between 200 million and 1 billion from climate change alone by 2050. El-Hinnawi puts the current figure of displacement at 50 million, which is expected to reach 150 million by 2050, whereas Myers estimates the current number at 25 million which is expected to reach between 150 and 200 million by 2050. Even if we consider these estimates as an exaggeration, there is a growing realization that potential climate-induced displacement will be on a scale the world is currently ill-equipped to efficiently address and would be larger than the current number of traditional refugees managed and protected by the UNHCR worldwide.

IPCC disseminates that slow and sudden onset of disasters and the consequent impact on the lives of communities, and livelihoods will contribute to increased migration that would escalate in the years to come. The migratory patterns in South Asia are complicated and could take several forms. Firstly, the climate-induced displacement can be both internal and external. Initially, the migration will be internal. It could also be temporary or permanent. The migration patterns in South Asia could also be gendered. In the case of countries like Nepal, the cultural factors prompt men to migrate leaving women and children behind. In fact, Nepal already has villages with few working men. In countries like Bangladesh, there is documented evidence that climate-induced displacement is leading to increased female trafficking a linkage acknowledged by the Paris Agreement. In South Asia, the idea of development does not address climate concerns, and there is no state responsibility in a normative sense. South Asian countries are not parties to the Refugee Convention except Afghanistan due to its Eurocentric approach. South Asian nations have appreciated the 'Common but Differentiated Responsibilities' (CBDR) principles, but they have been diluted to a large extent by the Paris Agreement. Further, transfer of technology and lack of fiscal resources have queered pitch in South Asia in its drive of attending climate change-induced refugees.

South Asian countries are slow to recognize the role of climate change as an additional push factor of driving migration. The climate discourse in climate change has not paid much attention to climate displacement. For SAARC nations, climate change emerged as an essential agenda issue from the third SAARC summit, wherein it was backed by the adoption of the SAARC action Plan on climate change. At the national level, the issue of climate-induced migration has not been prominently debated in the norms, Plans and strategies in any of the South Asian jurisdictions, even though most of the South Asian jurisdictions have adopted national climate change policies and Plans. Afghanistan National Adaptation Programme of Action discusses 'forced migration' in the context of drought and desertification. In Pakistan, the ministry of climate change has identified climate-induced migration as a serious threat to the nation. The Climate Action Plans of India and Bangladesh do not refer to climate-induced migration in a significant way. Similarly, Bhutan, Nepal and Sri Lanka have incorporated specific concerns of climate-induced migration but fail to envisage and provide protection and assistance to climate-induced migrants moving within or across the borders. Climate change-induced human displacement and migration is a present reality and is not a simple future occurrence and natural disaster in the making for the region. There may be only a few cases of mass migration stories, which have hit the headlines; the fact is that quiet and small-scale climate migration/displacement is a reality and will accelerate in the years to come. Therefore, the existence of increased climate displacement in South Asia and across the world, prompt and adequate legal measures are needed to address and tackle the impending humanitarian crisis that has created a new class of migrants called 'climate refugees' The existing legal paradigms are inadequate, and there are many gaps in the international legal framework to address the plight of flight of climate refugees. Hence, new measures are needed for the progressive development and codification of constitutional principles thereupon. For South Asia, the recognition that climate change displacement will threaten the political boundaries has to be elevated. The governments, civil society, experts and communities should emphasis on cooperation, the establishment of flexible mechanisms and institutions, and encourage shared use of resources like water, energy. The region should take emphasis on cross-border mechanisms to assist, prepare, respond and rehabilitate people and communities facing climate change disaster that are forced to migrate/displace.

The discourse on the protection of climate refugees is scattered under international refugee law, international climate law and international human rights law. Though, international refugee law appears to be the closest mechanism to offer protective rights framework to the mass exodus of climate change-induced displacement, the effectiveness of the refugee regime is severely handicapped by the restrictive interpretation of the term 'refugee' and further lack of the 'climate change displacement ground' in the definition part under the UNCSR (1951 UN Convention relating to the Status of Refugees) (or hereinafter referred to as Refugee Convention) with its 1967 Additional Protocol that do not account for the climate displacement and consequent climate refugees. Although there are Guiding Principles on Internal Displacement adopted by the United Nations in 1998, these principles do not attend to the problems of climate refugees generated by the climate change displacement. However, African Union (AU) in 2009 has concluded and adopted the Kampala Convention (formerly the AU Convention for the Protection and Assistance to the Internally Displaced Persons (IDPs) in Africa), whereunder the problem of internal displacement triggered by the armed conflicts, natural disasters and wide-ranging development projects in Africa have also been legally acknowledged. But, unfortunately, this regional treaty is not applicable to the part of the world, i.e. South Asia.

While the level of international cooperation is necessary to address global issues, the global environmental and climate change agenda has been dominated and influenced by a set of powerful nations resulting in the marginalization of most affected countries. Additionally, the current structure of international law is based on a fragmented structure of lawmaking and institutions bringing a host of questions on coordination and cooperation. The issue of legitimacy still looms large with international institutions and lawmaking bodies, which do not operate by equality and have not incorporated the inputs and concerns of the most significant sections of jurisdictions. As a result, institution and lawmaking process have always been looked at with scepticism by developing nations.

For international law and principles to have any impact on climate change-induced migration, the international norms need to be incorporated and efficiently implemented in the domestic legal systems. The reception of international norms into a domestic legal system by the developing nations is problematic on account of multiple reasons. An international norm created out of participatory democracy embedded in the lawmaking process, the substantive provisions and institutional structures will have better receptivity. The role of developing countries in the norm-making process of international law and institutions leaves open a host of questions on legitimacy. Historically, the inputs and voices of the developing nations have been almost absent, and the majority of them have been receivers of international law.

The question of the legitimacy of participation by developing nations holds significance even today despite the numerical majority of developing south regarding global multilateral environmental negotiations. The lawmaking process under the international environmental law is based on consensus, and in theory, small island nations like the Maldives can affect the formation of international law. However, the vast difference between theory and reality has reflected in the lawmaking process of treaty negotiation, which is influenced by a group of influential nations, civil societies and international organizations. Hence, the democratic and legitimacy character of these norms are questionable.

An analysis of customary international law (CIL) will bring out the fallacy of equal state participation. Historically, the customary international law is considered as the primary source of international law. The formation of CIL is based on the generality of 'State practices' and *Opinio Juris*. The voluntary practice and acceptance of states provide an amount of legitimacy and democratic appeal to customary international law. CIL binds the existing states unless the state is a persistent objector to the practice. The current CIL binds new states. Regarding the many existing CIL practices, the participation of the vast majority of developing nations was absent on account of colonization and lack of their international personality. After the liberation and independence of these countries, the existing CIL bound these newly formed states. In the context of international environmental law, one can witness two categories of CIL and general principles. The general CIL applicable to environmental issues and the specific customary practices developed in the context of environmental problems. For instance, the principle of sovereignty was extended to the case of trans-boundary water-sharing cases and other environment disputes and relations. Established principles of international law were extended to environmental law issues rather than legal concepts being modified for ecological thinking.

Under the second category, specific customary practices about environmental protection are emerging and are concretizing. These include principles of sustainable development, precautionary principle, polluter pays and inter-generational equity. Regarding general CIL applicable to environmental contexts, the practice of developing nations has been completely missing. To that extent, one can say much of the CIL is a fiction.

Additionally, the fault line of disagreement between north and south has characterized the development trajectory of international environmental law discourse and negotiations. At the initial stage of the evolution of modern international environmental law in the 1960s, the developing Global South considered the agenda of environment protection as a distraction from their legitimate developmental trepidations. The major bone of contention was not confined to the scope and extent of environment protection. The conflict was centred on the principle of legitimacy, the legitimacy of consultation, participation, acceptance and accounting. The original anxieties of the Global South were sought to be addressed through weaving the inter-sectionalities of environment protection with an emphasis on human environment, visible from the title of Stockholm Declaration as United Nations conference on 'Human Environment'. The participation of south at the conference was achieved after series of meetings, which attempted to dissipate the fears of the south in terms of environmental protection as an obstacle to development. The negotiation at the Stockholm conference and the south involvement is reflective of the confrontation and contestation south had with the environmental agenda. The conference acknowledged the difference between 'pollution of poverty' and 'pollution of affluence'. The aftermath of Stockholm conference witnessed a gradual shift, wherein the position of south changed from confrontation to being a reluctant participant. This transformation was brought about by the multifarious political economic and social situations. The numerical majority of the south provided them a robust platform to negotiate and tilt the balance in their favour. The Rio conference, which was convened to evaluate the development of international environmental law since the Stockholm Declaration provided an apt platform to the developing nations to showcase their reinvigoration. The Rio negotiations further highlighted the different trajectory and perspectives of north and south and further cemented the structure and direction of international environmental law. The south considers north to have exploited the world without environmental concerns and claim that they owe a much higher responsibility in terms of environmental protection. They also differ in terms of priorities with north prioritizing ecological protection per se while south emphasis on the social and economic dimensions of environment protection. The Rio negotiations officially cemented the inter-linkage of environment protection with development. Right to development was declared an absolute goal, and the achievement of sustainable development was to be desired. The conflicting stand of north and south in the Rio negotiations is reflected in the north emphasizing on a legally binding document specifying the rights and obligations of states in terms of environment protection. For the south, the negotiation was not about environment protection per se. They perceived the moment as an opportunity to reflect on the economic and social conditions surrounding environment protection including poverty reduction technological and financial transfer.

The climate change with its multifarious, differential and disproportionate impact on the developing nations and marginalized sections of the society provided the developing countries the perfect platform for the enunciation of new principles reflecting the concern and call for equity and justice. Developing countries participated in the climate negotiation as an active participant. The negotiation witnessed the confrontation over environmental priorities, historical responsibility, relationship between environmental protection and development and sharing the global attempts to reduce decrease in the greenhouse gas emissions while taking into account the historical responsibility of 'Global North' countries for climate change and growing emissions in the 'Global South' countries. The north-south divide has significantly compromised the effectiveness of international law. In the context of climate change, the fault line of differences is not confined to north and south, and there exist sharp differences in the negotiating positions between countries within south and north. The environmental priorities of courtiers like India and China differed fundamentally from the vulnerable nations as small island nations reflecting on the heterogeneity of developing south. Apart from the differences persisting between the south and north regarding negotiation positions, considerable variations exist among the countries in respect of the vulnerable communities, capacities and positions. These variations among and within the south and north are significant to the problem of climate change.

These negotiations, ultimately, resulted and led to the adoption of the United Nations Framework Convention on Climate Change (UNFCCC), witnessed the crystallization of the concept of CBDR considered as a significant parameter of equity and justice. While acknowledging climate change as a common concern of humanity, the Convention incorporates the CBDR principle and espouses the contours and component of CBDR. The principle of CBDR elaborated differentiation regarding specific obligations, financial assistance and technological transfer. The CBDR was the foundational principle through the climate negotiation aimed to provide climate justice. The Kyoto Protocol, which supplemented the Convention, reflects the application of the principle of CBDR. The developing countries' commitments under the Protocol are voluntary and contingent upon the developed countries obligation to provide financial and technological assistance. The Protocol invokes the principles of inter- and intra-generational equity, the precautionary principle, CBDR, the need to promote a cooperative and transparent international economic system and the right of all parties to sustainable development. The implementation of these principles has given rise to a mosaic of complex institutional mechanisms linked to UNFCCC and Kyoto Protocol. The Paris Agreement institutionalizes a new paradigm by adopting a bottom-up nationally driven and self-differentiated approach. The success of the Paris Agreement will be assessed on how countries implement and execute their intended nationally determined contributions and aspirational goals. The climate negotiation and the established legal framework under UNFCCC and Kyoto Protocol have been touted as a victory for the south by incorporating the CBDR principle. Regardless of the formal acknowledgement of the position of the south under the climate framework, many countries are severely handicapped in their abilities by lack of resources and institutional mechanisms to adopt domestic action regarding mitigation and adaptation. Despite the enunciation of CBDR through the breadth and length of climate change regime, developing countries are challenged by resource crunch to keep up with developed countries implementing climate change measures. In this scenario,

Maldives submission to OHCR in 2008 assumes significance whether a developing country is required to initiate protective regime for the environment while attending to human rights and other socio-economic challenges. The flexible mechanisms under the Kyoto Protocol contain several challenges for developing countries. The near absence of participation of developing nations from the flexible mechanisms is viewed as an exclusion of weaker economies.

Additionally, the climate change legal framework today presents a classic case of fragmentation with different vertical and horizontal layers and clusters of rule-making and rule implementation, different multitude institutional bodies operating in different fields ranging from UNFCCC and Protocol bodies, expert bodies, UN specialized bodies, bilateral initiatives, unilateral actions and multilateral financial institutions. Operating principle and strategies of many of these institutions with their attributes, administrative challenges and distinctive political constituencies are not based on equal participation and decision powers but are dominated and determined by the voting capabilities of powerful nations. Consequently, the implementation of the climate change legal framework has always been beset with daunting challenges. At the same time, the discussion on the legitimacy and democratization of international norms is not intended to mask the legal, institutional and governance infirmities which many of the developing nations are faced with. Policy incoherence, institutional weaknesses, lack of political will and rampant corruption across institutions characterize many of the domestic systems of the developing countries.

The situation is entirely unsatisfactory and if left unregulated could erupt into a major human catastrophe in South Asia and elsewhere. The issues relating to legitimacy and implementation in the Refugee Convention, CBDR principles in climate change, lack of technology, differential treatment and heterogeneity in South Asia, domestic legal and institutional infirmities, an absence of reliable regional cooperative mechanisms have to be attended in the right earnest. The attention of international climate law has been mostly on mitigation and imposing responsibilities on nations. The attention given to the human impact of climate change was negligible. The Paris Agreement recognizes the enormous effects of climate migration and urges the international community to evolve adequate mechanisms to tackle climate displacement. A comprehensive and specific legal regime is lacking at the global level to resolve the legal issues thrown open by climate displacement. The international human rights law promises to be a significant tool in the protection of the rights of climate displacement population. However, developing and promoting human rights law require cooperation and enforcement. The developing countries have been initiating legal, policy and institutional mechanisms to mitigate and adapt to climate change. The situation presents an enormous opportunity for the developing countries to influence innovative and participatory governance structure at the domestic and international level. The solution to climate-induced migration/displacement cannot be sought in merely implementing the existing international legal rules. The issue has to be addressed in the broader paradigm of a global north-south divide among and within. Therefore, there is an indispensable requirement to address this lego-institutional hiatus in South Asia.

The book chronicles the effort to what extent people who are displaced due to the consequences of climate change and have either become refugees by crossing the international borders or climatically displaced persons (CDPs) in their homelands are protected under the current international law and the legal response of the South Asian jurisdictions to these refugee-like situations. What would happen in South Asia if profound impacts of climate impacts and consequent climate change-induced human displacement are not adequately addressed and responded promptly? What would happen to countries like Bangladesh, Maldives, Sri Lanka and India in the wake of climate change impacts in the form of poverty, human development index and disproportionate bearing on gender? How to attend the consequences and concerns flowing from climate change on technology, questions of legitimacy and heterogeneity in South Asia? The climate change-induced displacement and migration challenges have to be adequately and aggressively addressed if SAARC countries have to emerge as global leader in responding to issue of climate change, displacement and migration pandered to thereunder with a compendium of universal principle and priorities. South Asian states have to identify the trends and patterns of climate change migration and thresholds of legal protection under international climate change law and international refugee law and practices of South Asian states in addressing their lego-institutional protection commitments of the climate-induced displaced population.

The book strive to evolve norms and implementation thereof for making the optimum utilization of the available resources in South Asia while conceptualizing and contextualizing these issues in South Asia and Global South and identifying the obstructions and hiatuses in the development and implementation of South Asian norms on climate change migration and displacement. This book critically examines and assesses whether states have obligations to protect people displaced by climate change under international refugee law and international climate change law. The book discusses the issue of climate change migration and their being refugees in South Asia and analyses the legal and judicial responses initiated by South Asian nations. The instant book also explores the role of SAARC in response to climate change and its pandering to the generation of climate refugees.

1.1 Book Structure

Chapter 1—Introduction: The introductory chapter offers an overview of the book and provides a conceptual background. It brings out the issue of climate change and its relationship with climate refugees' migration/displacement with particular reference to South Asia. It explains the focus of the book, the structure and arrangement of the book.

Chapter 2—Conceptualizing the Climate Change Migration in South Asia: This chapter examines the fundamental debate of whether and how far climate change acts as a push factor in climate change migration. This chapter conceptualizes climate change-induced displacement and climate refugees thereunder in South Asia. This chapter offers an overview of the critical findings of the IPCC and other reports related to climate change impact on South Asia. It provides an overview of climate impacts, which acts as a driving factor of human mobility. The emphasis of the chapter will be on the complex interplay of factors promoting displacement, kind of movement, the current state of knowledge, the problem of recognition and protection of climate refugees and migrants of and its impact on the South Asian economy and society.

Chapter 3—Climate Change Displacement and Refugees: 'Normative Debate': The chapter maps out the various approaches and conceptions of climate refugees. The chapter critically analyses the debate of minimalists and maximalists on climate displacement. The analysis in this chapter aims to outline the theoretical foundations of the subject, before moving on to identify the core legal principles and legal regime applicable to solve the constitutional issues raised by climate migration and their evolving concept of climate refugees. The chapter also looks at the point of climate migration from a human rights dimension and discusses whether rights violated during climate migration are a human rights issue.

Chapter 4—Climate Change Migration: Legal Protection Under International Refugee Law and Climate Change Legal Regime: The chapter addresses the legal protection of the climate-induced displaced population. The primary focus of the chapter is not on whether to solve the climate change-induced displaced population as migrants or refugees. The idea is that climate change is a humanitarian crisis and whatever may be the complex political, economic and social factors influencing the extent of climate change, the situation will produce a significant number of people who could be compared to traditional refugee-like situations in terms of vulnerability and the absolute lack of control of existence and resources. Whatever may be the name assigned to climate-induced displacement, the legal response is called for. The chapter addresses the extent of protection granted to climate change-induced displaced people under international climate law and international refugee law and international human rights norms and attempts to recommend an appropriate legal framework under which the issue of climate change displacement could be addressed. Therefore, it scrutinizes explicitly the existing refugee norms and structure available to persons displaced by environmental disasters. In doing so, it highlights the gaps and limitations in the relevant regimes of international refugee law and discusses the legal avenues and options which could be explored to support the needs of persons vulnerable to displacement induced by climate change. The discussion on international climate regime will concentrate on the primary and essential features of UNFCCC, Kyoto Protocol and Paris Agreement and the COP decisions. The chapter covers the genesis of these two instruments and introduces the key provisions. It then goes on to trace the operation and elaboration of the climate change regime through consecutive meetings of Conference of the Parties (COPs) and the measures undertaken by them to deal with the issue of climate displacement. The chapter identifies and analyses the mitigation and adaptation strategies made under the Convention and the Protocol to deal with climate displacement/migration. Extensive reference is made to the principle of CBDR, which is a mainstay of the international discourse on climate change and analyses its application in climate displacement/migration scenario. Regarding international refugee law and international human rights law, the focus will be on the state responsibility of the host state and legal response of the country of origin to the CDPs within its boundaries. Moreover, it will also be examined as to what kinds of protections are available to the 'climate refugees' under the existing international refugee law framework. It will also identify the mechanism to redress the plight of climate refugees while rummaging a trajectory of innovative approaches. There are essential questions; *inter alia*, is there any possibility to have a regional or international legal regime to address the problems of climate refugees? That demands an immediate and cumulative global legitimate response whose contours are being evolved in the present book.

Chapter 5—Climate Refugees: South Asian States' Legal Protection Practices: This chapter examines the existing domestic laws and policies regarding climate change-induced displacement in South Asia. It also emphasizes developing a range of legal and policy options to deal with diverse challenges posed by climate change-induced displacement and climate refugees thereunder. Particular emphasis will be on the specific climate legislation, national refugee laws, human right laws, disaster management laws and other environmental laws operating in the sphere of climate change-induced displacement and migration. By examining the domestic laws regarding climate change displacement, the necessary rules and policies are suggested in this chapter by using comparative analysis.

Chapter 6—Climate Refugees: The Role of South Asian Judiciaries in Protecting the Climate Refugees: This chapter critically analyses the position of the judiciary in South Asian jurisdictions concerning the reception of international refugee law and international climate law in South Asian jurisdictions. The problems of climate concerns have occasioned the constitutional moments before the South Asian judiciaries which have pragmatically contributed in evolving a cornucopia of judicially created law on climate change in their respective municipal legal systems. Thus, Pakistan witnessed the judicial activism as far as the climate change scenario is concerned with judiciary recommending and overseeing the drafting of climate legislation for the country. The contents of these principles are drawn from the human rights law principles. The judiciary in India has also highlighted the need to account for climate concern in its disaster management laws and explain the protection of climate-induced displacement. The courts in the South Asian jurisdiction have suggested recommendations on the protection of internally displaced population including climate change displacement.

Chapter 7—The Role of the SAARC: Way Forward: This chapter critically evaluates the role of SAARC regarding protecting the rights of climate refugees. Energy concerns motivated the initial focus of SAARC on climate change. In 2007, SAARC initiated an action Plan for climate change and later signed the SAARC climate Declaration. In 2010, an Intergovernmental Expert Group on Climate Change was constituted to oversee the implementation of the Plan. Though the SAARC is slowly entering into the climate area, the institution is structurally and politically weak. The matters get further murkier as the position of countries in SAARC varies about climate change, and refugees with Bangladesh and Maldives have a high opinion in favour of their recognition and protection. The former UNHCR Mrs. Sadako Ogata initiated the Regional Consultation on Refugee and Migratory Movements in South

Asia in 1994; therefore, having completed a long-drawn process of regional consultations, SAARC grouping came out with a regional Declaration and National Model Law on Refugees in 2004 whose fate is still in limbo but, unfortunately, that does not address the problems arising out of the climate change and climate refugees in South Asia. Therefore, recognizing the emerging climate displacement scenario, there has been a demand to frame a SAARC charter for giving free movement rights to the refugees due to climate change within South Asia. Other suggestions include the formulation of an inclusive development Plan and its implementation among the SAARC member states. Therefore, the chapter will critically analyse these vital legal developments. Though the climate displacement is a global phenomenon, its impacts are felt locally and more at a regional level. A global treaty to tackle the climate displacement may mask the regional and local dimensions of the issue.

The book while looking at the global solution will also explore the issue from a regional agreement operating under the overarching international framework. Moreover, regionalism appears fitter because more flexibility and subsidiarity should grant a higher level of commitment from states. The book seeks to draw together the strands of the analysis and demonstrates the debates and protection gaps about climate change-induced displacement at the international and South Asian jurisdictions levels and then proceeds to make some proposals for future legislative exercise in the municipal legal systems of South Asia for formulating climate change law and policy relating to climate change-induced displacement and migration. It is suggested that these proposals would contribute to a statutory initiative on the protection of climate refugees or climate-displaced people in consonance with fairness principles while also being capable of implementation taking into account prevailing economic and political realities. The book bases its recommendations on the human rights and rule of law paradigm and highlights the dire need to seek, clarify and protect the rights of the people and communities displaced as a result of climate change.

Chapter 2
Conceptualizing the Climate Change Migration in South Asia

Abstract The climate change migration in South Asia has been silently reshaping the contours of human existence in an unprecedented fashion away from the national and global glare. The 2014 report of the IPCC on Impacts, Adaptation and Vulnerability unequivocally averred that impacts of climate change transpiring on all continents, countries and regions of the world that have presented a catch-22 situation for the humanity in making choices for its survival. Climate change has made catastrophic physical and socio-economic repercussions for the human existence universally and have emerged as the single most significant global sociopolitical and environmental concern. The climate change has been dominating the Global North scenarios in all their ecopolitical permutations, but it has not been prominently figured in the Global South socio-economic and ecocultural settings particularly in the SAARC region that has grown beyond the rubrics of a humanitarian crisis. Thus in the instant chapter, conceptualization of climate change migration has been attempted while critically exploring and evaluating the typology of climate change-induced displacement in South Asia. The emergence of a new class of displaced people called 'climate refugees' has been conceptualized with an understanding of its definitional dilemmas and dimensions from a multitude of perspectives. Therefore, it is essential to decipher the science of climate change in the context of South Asia and its susceptibility to climate change that has been causing human displacement and migration in the region.

2.1 Introduction

The climate change migration in South Asia has been silently re-shaping the contours of human existence in an unprecedented fashion away from the national and global glare.[1] The climate change migration has acquired cataclysmic proportions in the contemporary circumstances that have been expected to change the demographic

[1]Paul de Guchteneire, *Migration and Climate Change*, (Cambridge University Press, 2011) John Kingdon, *Agendas, Alternatives, and Public Policies* (2nd edn, Harper Collins College 1995) 126.

S. Jolly and N. Ahmad, *Climate Refugees in South Asia*, International Law and the Global South, https://doi.org/10.1007/978-981-13-3137-4_2

settings, stratifying the geopolitical entities into many high-risk zones[2] and impacting the sociopolitical scenarios in many parts of the world.

There existed a time when great singers like Tansen[3] could bring down rain and also light lamps with the performance of his Ragas. However, we have come far away from such times and now climate change is real and inevitable. Presently, the climate change migration has become beyond the simple understanding of a complex humanitarian crisis to include grave concerns of human and core national security narratives. Therefore, the world and humanity have been imperilled to the hilt affecting the ability of governments to react to the threatening and eroding thresholds of state legitimacy.

The IPCC report captioned *Climate Change 2014: Impacts, Adaptation, and Vulnerability*[4] unequivocally acknowledged that the impacts of climate change are occurring on all continents and across the oceans and emphasized on making choices about risks in a changing world.[5] The disastrous physical and social consequences of climate change and the universal nature of those consequences have made climate change the single most significant global political, environmental and social concern.[6] In our everyday life, we have been flooded with information about the dangerous effects of climate change. Though the debate about climate change proceeds with vigorous passion, the normative content of the debate has varied with time and space. The initial focus of the discourse was on the science and reality of climate change itself. However, the centre of the current debate is on the modalities of mitigating and adapting to climate change, rather than on the science itself. The overwhelming scientific consensus affirms the climate change and its anthropocentric nature.[7]

South Asian countries with their fragile environment and underdeveloped economic and social conditions are particularly vulnerable to the consequences of climate change.[8] The Centre for Research on Environmental Decisions (CRED) report's

---

[2]See generally Elena Correa, *Preventive Resettlement of Populations at Risk of Disaster: Experiences from Latin America* (World Bank 2011).

[3]Tansen (c. 1500–1586), also referred to as Tan Sen or Mian Tansen, was a prominent figure of North Indian (Hindustani) classical music. Among the legends about Tansen are stories of his bringing down the rains with Raga Megh Malhar and lighting lamps by performing Raga Deepak (The Tune of Lamp).

[4]IPCC, 2014: Climate Change 2014: Impacts, Adaptation, and Vulnerability, Summaries, Frequently Asked Questions, and Cross-Chapter Boxes. A Contribution of Working Group II to the Fifth Assessment Report of the Intergovernmental Panel on Climate Change [Field, C.B., V.R. Barros, D.J. Dokken, K.J. Mach, M.D. Mastrandrea, T.E. Bilir, M. Chatterjee, K.L. Ebi, Y.O. Estrada, R.C. Genova, B. Girma, E.S. Kissel, A.N. Levy, S. MacCracken, P.R. Mastrandrea, and L.L. White (eds.)]. World Meteorological Organization, Geneva, Switzerland, 3.

[5]Ibid, 4–5.

[6]Justine Thorton and Silas Beckwith, *Environmental Law* (2nd edn, Sweet and Maxwell 2004) 3.

[7]Richard Lord, *Climate Change Liability: Transnational Law and Practice* (Cambridge University Press 2012) 5.

[8]ESCAP, 'Overview of the Natural Disasters and their Impacts in Asia and the Pacific 1970–2014' (Disaster Risk Reduction Section, 2015) 7; ICHPR, 'Climate Change and Human Rights: A Rough Guide' (International Council on Human Rights Policy 2008) 1–2.

assessment of observed impacts ranks Asia as the most prominent victim of natural disasters from 1994 to 2013, accounting for nearly 41% of the global economic loss ascribed to natural disasters.[9] For the South Asian countries, climate change will result in colossal economic, social and environmental damage and threaten the very existence of the society compromising the growth potential and poverty reduction efforts.[10]

This chapter examines the fundamental debate on whether and how far climate change act as a push factor in climate change migration/climate change displacement/climate change refugee movement. Climate change *juris corpus* makes use of these nomenclatures interchangeably in the absence of universally accepted terminology. However, these terms convey different connotations that explain various facets of climate change migration. The book uses these terms interchangeably and refers to the situations where climate change act as a significant factor in the movement/mobility/displacement of the people in all parts of the world. This chapter conceptualizes climate change-induced displacement and climate refugees in South Asia. The chapter offers an overview of the critical findings of the IPCC and other international reports related to climate change impact in South Asia. It provides an overview of climate impacts, which acts as a driving factor of human mobility. The nucleus of the chapter will be on the complex interplay of factors promoting displacement, kind of movement, the current state of knowledge, the problem of recognition and protection of climate refugees and migrants of and its impact on the South Asian society.

2.2 The Science of Climate Change

Climate change is one of the most specific and focused subjects that the world is currently grappled with. The impact of the same is omnipresent, and the world as a whole is tottering under its adverse pinch. The Global South countries usually challenge that the developed Global North countries are historically responsible for the adverse changes, which the earth's environment is undergoing, because of the large and multiscale industrial activities, which they have undertaken in the past.[11] Irrespective of the blame game, which nations are engaged with, global warming and climatic change is a stark reality that requires dismantling the interconnectedness

---

[9] 'The Human Cost of Natural Disasters 2015: A Global Perspective' (Centre for Research on the Epidemiology of Disasters, 2015) 42 https://reliefweb.int/sites/reliefweb.int/files/resources/PAND_report.pdf accessed 28 April 2018.

[10] Mahfuz Ahmed and Suphachol Suphachalasai, *Assessing the Costs of Climate Change and Adaptation in South Asia* (Asian Development Bank 2014) 12.

[11] Luca Marchiori and Ingmar Schumacher, 'When Nature Rebels: International Migration, Climate Change, and Inequality', (2011) 24(2) J. Popul. Econ, 569-200, 573; Also see Duncan Clark, 'Which Nations are most Responsible for Climate Change", *The Guardian*, (London (21 April 2011)) www.theguardian.com/environment/2011/apr/21/countries-responsible-climate-change accessed 19 December 2017.

between economic growths and escalating carbon emissions at both domestic and international levels.

Scientists have advanced a multitude of reasons for the occurrence of climate change.[12] They argue that human intervention has disrupted the natural heat balance of the earth in the form of burning of fossil fuels, crop remnants and deforestation.[13] IPCC report points to an upsurge in industrial activities, since the year 1750, resulting in an upward rise in the global atmospheric concentrations of gases such as carbon dioxide, methane and nitrous oxide.[14] The existence of climate change has initially been seen with reservations on certain occasions as the global scientific community pursued a manipulative deception.[15] Over the years, climate change issue has catapulted the world attention as an unambiguous and undeniable phenomenon, and the change of this assessment was marked by the emerging consensus among the scientific community in the science of climate change.[16] The IPCC report entailed the significant enhancement in the development and dissemination of knowledge of climate change menace.[17] The first 1990 IPCC assessment report underscored the significance of climate change challenge that requires international cooperation to address and attend its repercussions.[18] The fifth Assessment report of the IPCC[19] highlighted and restated the view that the earth's climate is heating and the same is correct beyond any reconsideration.

---

[12]John T Hardy, *Climate Change: Causes, Effects, and Solutions* (John Wiley and Sons 2003) 5.

[13]Luqman Atique, Irfan Mahmood and Farman Atique, 'Disturbances in Atmospheric Radiative Balance Due to Anthropogenic Activities and its Implications for Climate Change' (2014) 1(1) American-Eurasian J Agric & Environ Sci 73–84.

[14]Timothy M Kusky and Katherine E Cullen, *Encyclopedia of Earth and Space Science* (Infobase Publishing 2010) 341; See Research Unit (Larrdis) Rajya Sabha Secretariat, *Climate Change: Challenges to Sustainable Development in India* (Occasional Paper Series 3 2008) http://rajyasabha.nic.in/rsnew/publication_electronic/climate_change_2008.pdf accessed 20 December 2018.

[15]Dean Burnett, 'Climate Change is an Obvious Myth—How Much More Evidence do you Need?' *The Guardian*, (London 25 November 2014) www.theguardian.com/science/brain-flapping/2014/nov/25/climate-change-is-an-obvious-myth-how-much-more-evidence-do-you-need accessed 20 December 2018; See Ted Goertzel, 'Conspiracy Theories in Ccience' (2010) 11 (7) Embo Reports 493.

[16]Adger, W.N., J.M. Pulhin, J. Barnett, G.D. Dabelko, G.K. Hovelsrud, M. Levy, Ú. Oswald Spring, and C.H. Vogel, 2014: Human security. In: Climate Change 2014: Impacts, Adaptation, and Vulnerability. Part A: Global and Sectoral Aspects, Contribution of Working Group II to the Fifth Assessment Report of the Intergovernmental Panel on Climate Change [Field, C.B., V.R. Barros, D.J. Dokken, K.J. Mach, M.D. Mastrandrea, T.E. Bilir, M. Chatterjee, K.L. Ebi, Y.O. Estrada, R.C. Genova, B. Girma, E.S. Kissel, A.N. Levy, S. MacCracken, P.R. Mastrandrea, and L.L. White (eds.)]. Cambridge University Press, Cambridge, United Kingdom and New York, NY, USA, 755–791, 759.

[17]Neal F Lane and Rosina Bierbaum, 'Recent Advances in the Science of Climate Change' (2001) 15 (3)Nat Resources & Env't 147–151.

[18]John T Houghton, et al. (eds) *Climate Change 1994: Radioactive Forcing of Climate Change and an Evaluation of the IPCC 1992 IS92 Emission Scenarios* (Cambridge University Press 1995).

[19]Elizabeth Carabine and Alberto Lemma, IPCC Fifth Assessment Report: What's in it for South Asia? (Over Seas Development Institute and Climate and Development Knowledge Network 2014) 2.

Scientific bodies have defined climate change in an assorted fashion, for example, Australian Academy of Science describes the 'climate change' as 'a change in the weather patterns, the rise in the sea levels, and the related change on land surfaces and glaciers that ordinarily befalls in several decades'.[20] According to IPCC, climate change has been described as 'a change in the global climate, happening over a specific period, which occurs due to natural factors as well as human activities'.[21] The manner in which the UNFCCC[22] has defined climate change is, however, different from this; where it has been referred as a 'change of climate which is solely attributable to direct or indirect human which is in addition to natural climate variations which are observed over comparable periods of time'.[23] The natural factors responsible for climate change could be internal processes or external forcings.[24] They, however, only play a minor role in the rise of the earth's temperature.[25]

The predicted increase in the global temperature by the end of the twenty-first century is between 2.8 and 5.4 °C, depending on how much the world community can control the same.[26] Therefore, it has now become essential to cite the latest IPCC report which has called global warming and climate change to be an unequivocal phenomenon and has mentioned the rise of 0.8 °C in the earth's temperature and 22 cm in the worldwide sea level, over last 100 years, as evidence for the same.[27] However, climate change is not limited to being an environmental, geographical or physical phenomenon, which the earth experiences and has tremendous economic and social repercussions too.[28] Under the Paris Agreement on Climate Change 2015,[29] countries have undertaken the commitment to keep temperatures from increasing to

[20] Australian Academy of Science, *The Science of Climate Change: Questions and Answers* (Australian Academy of Science 2015) 6.

[21] IPCC, 2014, (n 4) 4–5.

[22] United Nations Framework Convention on Climate Change (adopted on 20 January 1994) UN Doc A/Res/48/189.

[23] Ibid Art 1(2).

[24] Neal F Lane (n 17) 149.

[25] Certain Scientists, however, have claimed the contrary and contend that natural factors alone are responsible for changes which the earth's environment is undergoing.

[26] Olivier Boucher and Ulrike Lohmann, 'The sulfate-CCN-cloud albedo effect' (1995) 47(3) Tellus B: Chemical and Physical Meteorology 281–300.

[27] Mark Maslin, *Climate Change: A Very Short Introduction* (Oxford University Press 2014) 17; also see IPCC, 2014: Climate Change 2014: Synthesis Report. Contribution of Working Groups I, II and III to the Fifth Assessment Report of the Intergovernmental Panel on Climate Change, R.K. Pachauri and L.A. Meyer (eds.). IPCC, Geneva, Switzerland, 2.

[28] Martin Beckenhamp, 'The Social Dilemma of Climate Change: Socio Economic Implications', in Walter Leal Filho (ed) *The Economic, Social and Political Elements of Climate Change* (Springer 2010) 143–152; NASA, 'The Consequences of Climate Change' (*Global climate Change: Vital Signs of the Planet*) https://climate.nasa.gov/effects/ accessed 17 December 2017; Also see Ahmad El Zein and Nour Chehayeb, 'The Effect of Greenhouse Gases on earth's Temperature', (2015) 3(2) IJEMA 74–79, 74.

[29] 'Adoption of the Paris Agreement' UNFCCC Conference of the Parties Decision 1/CP.21 (30 November–13 December 2015) FCCC/CP/2015/L.9/Rev.1 https://unfccc.int/resource/docs/2015/cop21/eng/l09r01.pdf accessed 20 April 2018.

more than 2 °C by the end of this century and continuing with these commitments has become the only possible solution for avoiding the disastrous scenario which has been predicted by researchers.

2.3 South Asia and Susceptibility to Climate Change

Climate-related risks threaten the quality of life *inter alia* health, education, food security and overall well-being throughout South Asia.[30] The area is highly prone to the consequences caused due to climate change and is identified to be one of the hot spots of the most disaster-prone area in the world.[31]

Scholars have provided some hydrogeological and socio-economic factors for high vulnerability to climate change in South Asia, and some of them are:

(a) the geographical setting in South Asia;
(b) flat deltaic topography with very low altitude;
(c) the tremendous climatic inconsistency that is regulated by the monsoon and results thereof are critical water distribution over space and time;
(d) its excessive population concentration and penury and destitution occurrence; and
(e) the majority of the people is in need of crop agriculture that is significantly affected by the climate change and its other incidental variations.[32]

IPCC in its fifth Assessment report finds many observed changes in South Asia's climate.[33] It is axiomatic from the records that the statistics of cold days and nights have reduced and the figures of hot days and nights have swelled in many parts of South Asia since early 1950.[34] In South Asia, the densely populated littoral and low-lying areas including Bangladesh and India would be at the escalated threshold of risk gale proliferation placing millions of people at risk.[35] Riverine, littoral and urban floods associated with life-threatening rainfall events, sea-level risings and

[30]Carabine and Lemma, (n 19) 5.

[31]Ulka Kelkar and Suruchi Bhadwal, 'South Asian Regional Study on Climate Change Impacts and Adaptation: Implications for Human Development," Human Development Occasional Papers (1992–2007), Human Development Report Office (HDRO), United Nations Development Programme (UNDP) 2007, 1–47, 427 http://hdr.undp.org/sites/default/files/kelkar_ulka_and_bhadwal_suruchi.pdf accessed 25 December 2017.

[32]Ahsan Uddin Ahmed, '*Bangladesh Climate Change Impacts and Vulnerability*' (Climate Change Cell, Department of Environment 2006) 1.

[33]Christopher B Field, Vicente R Barros, David Jon Dokken, Katharine J Mach, Michael D Mastrandrea, T Eren Bilir, Monalisa Chatterjee, Kristie L Ebi, Yuka Otsuki Estrada, Robert C Genova, Betelhem Girma, Eric S Kissel, Andrew N Levy, Sandy MacCracken, Patricia R Mastrandrea, and Leslie L White (eds), *Climate Change 2014: Impacts, Adaptation, and Vulnerability: Contribution of Working Group II to the Fifth Assessment Report of the Intergovernmental Panel on Climate Change* (IPCC 2014); Carabine and Lemma, (n 19) 5.

[34]Carabine and Lemma, (n 19) 5.

[35]Ibid.

cyclones will pander to the general devastation of livelihoods, infrastructure and human colonies.[36] The danger of floods, sea-level rise, the damage to life and property connected to floods is the highest in India and Bangladesh.[37]

Bangladesh has been acknowledged as a country that is at the particular risk of climate change.[38] There is an unity in the scientific community regarding the impacts of climate change in Bangladesh that have been aggravating the existing natural environmental menace present in the country in the form of impulsive onset incidents like hurricanes, water-logging hazard, salinity incursion and the riverbank attrition along with gradual onset events such as coastal erosion and terrestrial loss.[39] The increase in the sea levels due to climate change has been expected to degenerate and degrade some of these phenomena and likely to submerge up to 30% of Bangladesh's seaside land by 2080.[40]

In South Asia, many other states have prognosticated that there would be serious impacts of climate change. A mean sea-level rise of 15–38 cm has been anticipated along the coastal zones of India by the mid of the present century and of 46–59 cm by the end of the current century.[41] It has been projected that a 1-m rise in the sea level would cause the displacement of 7 million people in India.[42] The studies have signified a prospect of 10–40% loss of crop production in India owing to the

---

[36]Ibid.

[37]Ibid; See also Mohammad Asaduzzaman, Mohammad Reazuddin and AU Ahmed (Eds), *Global Climate Change: Bangladesh Episode, Department of Environment* (Government of Bangladesh 1997).

[38]IPCC, 'Report on Bangladesh Launch of the Fifth Assessment Report (AR5) of the Intergovernmental Panel on Climate Change (IPCC)' (Dhaka 6–8 August 2014) https://cdkn.org/wp-content/uploads/2014/05/Report-on-IPCC-outreach-events-Bangladesh-.pdf accessed 13 April 2018; See also, Helena Wright, 'What does the IPCC say about Bangladesh?' *Dhaka Tribune* (16 February 2015) www.dhakatribune.com/uncategorized/2015/02/16/what-does-the-ipcc-report-say-about-climate-change-in-bangladesh accessed 20 January 2018.

[39]Jane McAdam, *Climate Change, Forced Migration, and International Law* (OUP 2012) 164; See AM Choudhury, Sharmind Neelormi, DA Quadir, S Mallick and AU Ahmed 'Socio-economic and Physical Perspective of Water related Vulnerability to Climate Change: Results for Study in Bangladesh' (2005) 71 Science and Culture 225–238.

[40]James Pender, 'Community-Led Adaptation in Bangladesh' (2008) 31 Forced Migration Review 54–55; See also, IPCC Working Group, *IPCC Second Assessment: Climate Change 1995: A Report of the Intergovernmental Panel on Climate Change* (IPCC 1995) 34 www.ipcc.ch/pdf/climate-changes-1995/ipcc-2nd-assessment/2nd-assessment-en.pdf accessed 20 January 2018; Cleo Paskal, *How Climate Change is Pushing the Boundaries of Security and Foreign Policy* (The Royal Institute of International Affairs 2007); See also Jane McAdam and Ben Saul, 'Displacement with Dignity: International Law and Policy Responses to Climate Change Migration and Security in Bangladesh' (2010) 53 GYIBL, 233–287, 237.

[41]*India's Initial National Communications to the United Nations Framework Convention on Climate Change*, (Ministry of Environment and Forests, New Delhi 2004) https://unfccc.int/resource/docs/natc/indnc1.pdf accessed 10 March 2018.

[42]Jyoti K Parikh and Kirit Parikh, 'Climate Change: India's Perceptions, Positions, Policies and Possibilities' (OECD 2002) 6, www.indiawaterportal.org/sites/indiawaterportal.org/files/1934784.pdf accessed 21 January 2018.

projected rise in the temperature by the year 2080–2110.[43] River basins of Sabarmati and Luni, which dominate the 60% of the areas of Rajasthan and about one-quarter of the area of Gujarat, are prone to experience severe water shortage situations in the years ahead.[44]

According to a research study conducted in 2010 where under it has been established that there are 25 possibly precarious glacial lakes in Bhutan and 20 in Nepal that on the verge of exploding floods.[45] India has also been watching the development and sweep of glacial lakes.[46] The global warming has not only triggered snowcaps, thawing and glacial lakes heading to floods that have also changed the trends of seasonal rainfall causing a colossal impact upon the agriculture and tourism-driven economy of Nepal.[47] In Nepal, the geologists have found that climate change has been causing low-intensity earthquakes in the region.[48]

The Maldives happens to be the world's a nethermost-resting country consisting of multiple islands and atolls that are only 1 m above the sea level which will make the Maldives one of the first nation-states to be drowned permanently.[49] Michael Gagain calls attention to that by the year 2100, the Maldives would face an increase of 0.47 m in the sea level. Such an increase in the sea level would seriously cause submergence of the significant areas of the archipelagic nation-state.[50] He further

---

[43] Surinder Sud, 'Agriculture May Decline due to Climate Change' *Business Standard* (New Delhi, 21 January 2008) www.business-standard.com/article/economy-policy/agriculture-may-decline-due-to-climate-change-108012101006_1.html accessed 23 January 2018.

[44] 'India's Negotiating Position in Various Climate Change Related Fora Strengthened Global Climate Change May Adversely Affect Weather, River Basins, Rainfall, Coastal Areas, Ecosystems, Forestry in India' (*Ministry of Environment, Forest and Climate Change*, 18 June 2004) http://pib.nic.in/newsite/erelcontent.aspx?relid=2023 accessed 23 January 2018.

[45] Samjwal Ratna Bajracharya, Pradeep Kumar Mool and Basanta Raj Shrestha (Eds), *Impact of Climate Change on Himalayan Glaciers and Glacial Lakes Case Studies on GLOF and Associated Hazards in Nepal and Bhutan*, (International Centre for Integrated Mountain Development 2007) 12.

[46] Jack D Ives, Rajendra B Shrestha and Pradeep Kumar Mool, *Formation of Glacial Lakes in the Hindu Kush-Himalayas and GLOF Risk Assessment* (International Centre for Integrated Mountain Development 2010) 5.

[47] Madhav Karki, Pradip Mool and Arun Shrestha, 'Climate Change and its Increasing Impacts in Nepal' (2009) 3 The Initiation 30–37; Ajaya Dixit, 'Climate Change in Nepal: Impacts and Adaptive Strategies' (*World Resource Report*, 2009) www.wri.org/our-work/project/world-resources-report/climate-change-nepal-impacts-and-adaptive-strategies accessed 21 December 2018.

[48] Some seismologists associated with the government of India say that climate change has some influence on occurrence of earthquakes. See Jayanta Basu, 'Climate Change and earthquake Link Needs More Study' (thethirdpole.net, 7 October 2016) www.thethirdpole.net/2016/10/07/climate-change-and-earthquake-link-needs-more-study/ accessed on 28 December 2017.

[49] John A Church, Neil J White and John R Hunter, 'Sea-level Rise at Tropical Pacific and Indian Ocean Islands' (2006) 53 Global and Planetary Change 155; Gemima Harvey, 'Sinking States: Climate Change and the Pacific' *The Diplomat* (22 May 2014) https://thediplomat.com/2014/05/sinking-states-climate-change-and-the-pacific/ accessed 17 December 2017.

[50] Michael Gagain, 'Climate Change, Sea Level Rise, and Artificial Islands: Saving the Maldives' Statehood and Maritime Claims through the 'Constitution of the Oceans' (2012) 23(1) Colo. J. Int'l Envtl. L. & Pol'y 79–120, 80.

states that the 15% of the capital Male would be drowned due to the increase in the level of sea by the year 2025 and further Maldives's 50% existence is also likely to be capsized by the year 2100.[51]

The economy of Sri Lanka significantly depends upon the climate-sensitive sectors like forestry, agriculture and energy generation.[52] The littoral regions are the most susceptible and exposed to climate change, and a significant amount of the population inhabits in these maritime provinces.[53] Sri Lanka is exceptionally susceptible to increasing sea levels, and climate-related adversities and catastrophes have the potential to hold up any gains accrued in the agriculture, fisheries and other services such as tourism.[54]

In South Asia, climate change has also been affecting the human health in various ways.[55] The epidemics of dengue and Japanese encephalitis have been concomitant with temperature and rainfall due to climate change.[56] Malaria incidences in Nepal and India have been associated with rainfall trends.[57] Studies have indicated a coterminous relationship between diarrhoea epidemics and a conflation of higher temperatures and heavy rainfall.[58] Cholera eruptions in the littoral populations in South Asia have been connected with escalated temperatures, El Nino effects and algal blooms.[59] In Bangladesh, people have endured heat stress, a situation that often impacts urban populations in low and middle-income nation-states where child mortality rate is higher due to an extreme temperature which has led to incremental death rate and climate change migration.[60]

---

[51] Ibid; See also Miriam Auer, 'Does the Republic of Maldives Have the Capability to Handle the Growing Tourism Demand, Whereas Facing the Difficulties of Global Warming?' (BBA Thesis, Modul University 2012) 27; T.C. Karthikheyan, 'Environmental Challenges for Maldives' (2012) 17(2) Sage Journal 343–351; Martin Parry, Osvaldo Canziani, Jean Palutikof, Paul van der Linden and Clair Hanson, *Climate Change: Impacts, Adaptation and Vulnerability: Fourth Assessment Report of the IPCC* (CUP 2007).

[52] W.A.J.M. De Costa, 'Adaptation of Agricultural Crop Production to Climate Change: A Policy Framework for Sri Lanka' (2010) 38(2) J Natn Sci Foundation Sri Lanka 79–89, 80.

[53] UNDP, 'Sri Lanka: Climate Change Adaptation' (*Climate Change Adaptation*) www.adaptation-undp.org/explore/sri-lanka accessed 16 December 2017.

[54] Editor, 'Climate Change Threat to Sri Lanka' *The Guardian* (13 November 2013) www.theguardian.com/world/2013/nov/13/climate-change-sri-lanka accessed 15 December 2017.

[55] Manpreet Singh, Mala Rao, and Colin D. Butler, 'Climate Change, health and the Future of Well Being in South Asia' in Rais Akhtar (Ed), *Climate Change and Human Health Scenario in South and Southeast Asia* (Springer 2016) 11–27.

[56] Ramesh C Dhiman, Scenario of Malaria and Dengue in India: A Way Forward in Rais Akhtar (Ed), *Climate Change and Human Health Scenario in South and Southeast Asia* (Springer 2016) 91–98.

[57] Aditya Dash, 'Rainfall and Malaria Transmission in North-Eastern India' (2017) 101(5) Pathogens and Global Health 457–459; See also Koko Warner, Mohamed Hamza, Anthony Oliver-Smith, Fabrice Renaud, and Alex Julca, 'Climate Change, Environmental Degradation and Migration' (2010) 55(3) Natural Hazards 689–715.

[58] Carabine and Lemma (n 19)5.

[59] Ibid.

[60] Ibid.

South Asia is also home to some of the fastest growing cities and urban centers in the world. By the year 2020, in India, Mumbai will be the second largest city in the world, closely followed by Delhi and Dhaka, respectively.[61] With Karachi and Kolkata—five of the world's 11 megacities will then be in South Asia.[62] In Mumbai, more than 50% of the population is packed into about 2,000 densely populated slums. These slums are at greater risk of flooding particularly where human habitation and settlements lack basic protective infrastructure. Therefore, there are particular challenges in making cities and urban human concentrations climate resilient.[63] These impacts are bound to contribute to climate change migration vulnerability and disturb the geodemographic balance of human existence that will further result in human displacement. Thus, climate change cannot be limited to a single reason rather it being caused by a vortex of factors that have been adumbrated hereinabove.

2.4 Climate Change and Global Human Displacement

The anticipated effects of climate change include an acceleration in the relentlessness of natural disasters, desertification, land degradation, salinization, riverbank and littoral land erosion, sea-level rise, the frequency and intensity of floods, and increase in precipitation, affecting the environment, health social and national security.[64] However, the specific concern that has the potential to cause climate change migration and human displacement in the region must categorically be evaluated and arrested.[65] The conservative projections point out that by the year 2050, globally between 250 million and 1 billion people will be displaced for environmental/climatic reasons.[66]

The mounting climate-related disasters and their occurrence and severity propelled by the climate change along with a sizable people living in climate-change-prone areas, and their triggering vulnerability are causes of the increasing human displacement.[67] It is globally estimated that people are more likely to be displaced

---

[61] World Bank, 'Why is South Asia vulnerable to Climate Change?' (*Relief Web,* 1 December 2009) 1–3 https://reliefweb.int/report/india/why-south-asia-vulnerable-climate-change accessed 19 December 2017.

[62] Ibid.

[63] Ibid.

[64] Donald A Brown, 'Climate Change Refugees: Law, Human Rights and Ethics', in Laura Westra and Satvinder Juss (Eds), *Towards a Refugee Oriented Right of Asylum* (Routledge 2015) 43–66 see also, Colleen Murphy, Paolo Gardoni, Robert McKim, *Climate Change and Its Impacts: Risks and Inequalities*, (Springer 2018).

[65] See UNHCR, 'Summary of Deliberations on Climate Change and Displacement (Bellagio 22–25 February 2011)' www.unhcr.org/4da2b5e19.pdf accessed 18 April 2018; Susin Park, 'Climate Change and the Risk of Statelessness: The Situation of "Sinking Island States"' (Division of international Protection UNHCR 2011) PPLA/2011/04.

[66] Cecilia Tacoli, 'Crisis or Adaptation? Migration and Climate Change in a Context of High Mobility' (2009) 21(2) Environ Urban 513–525,513.

[67] Ibid.

twice by environmental disasters now than they were in the 1970s.[68] The Green peace study highlights that an average of 25.4 million annual displacements every year as a result of natural disasters.[69] Thus, out of the 24.2 million people, there are newly IDPs who were displaced by sudden-onset disasters in 2016, and 23.5 million people who were displaced by weather-related disasters, including storms and floods.[70] This is more than three times the number of people newly displaced by conflict and violence. This figure is astonishing because war refugees receive a great deal of media concentration and legal safeguard compared to CDPs.[71]

Though the increased intensity of disasters may be the primary push factor in driving the migration or displacement, climate displacement/migration is not exclusively an ecological problem. It is a multicausal problem[72] where environmental and socio-economic vulnerability and coping capacity, which changes with time and space, work jointly to displace vulnerable people.[73] Professor of Geography Steve Lonergan noted in 1998: '*there is too frequently gullible acceptance of a direct causal link between environmental dilapidation and degradation and population dislocation*'.[74] It must be accepted that the destruction of the climate is socially and spatially structured; only through a fundamental appreciation of the atmosphere in the comprehensive politico-cultural context of a continent or country can one start to appreciate the role it performs as a reason in human migration.[75] As per the IOM, the interconnection of climate change migration and human displacement is extremely complicated and multifaceted[76] that is fraught with methodological uncertainties

[68] Julie-Anne Richards and Simon Bradshaw, 'Uprooted by Climate Change Responding to the Growing risk of Climate Displacement' (Oxfam Briefing Paper 2017) 12 www.oxfam.org/sites/www.oxfam.org/files/file_attachments/bp-uprooted-climate-change-displacement-021117-en.pdf accessed 23 April 2018.

[69] Joanne Runkel and Alexis Conklin, *Climate Change, Migration, and Displacement: The Underestimated Disaster* (University of Hamburg 2017) 6.

[70] Richards and Bradshaw(68) 12.

[71] Ibid 12.

[72] Catherine Locke, W Neil Adger and P Mick Kelly, 'Changing Places: Migration's Social and Environmental Consequences', (2000) 42(7) Environ Sci Policy for Sust Dev 24–35.

[73] Shweta Jayawardhan, 'Climate Change, Vulnerability, and Displacement' (2017) 17(1) JSD 103–142, 104; See also Nathan James Bennett, Jessica Blythe, Stephen Tyler and Natalie C Ban, 'Communities and Change in the Anthropocene: Understanding Social-Ecological Vulnerability and Planning Adaptations to Multiple Interacting Exposures' (2016) 16(4) Reg. Environ. Change, 907–926, 907; Deepti Samant Raja and Nirmita Narasimhan, 'Inclusive Disaster and Emergency Management for Persons with Disabilities' (*The Center for Internet and Society*, 2013) 3 https://cis-india.org/accessibility/blog/emergency-services-report.pdf accessed 23 April 2018.

[74] Steve Lonergan, 'The Role of Environmental Degradation in Population Displacement' (1998) 4 Environmental Change and Security Project Report, 5–15, 8.

[75] Ibid, see also Cecilia Tacoli (66) 516.

[76] Camillo Boano, Roger Zetter and Tim Morris 'Environmentally Displaced People Understanding the Linkages between Environmental Change, Livelihoods and Forced Migration' (2008) 1 FMR, Briefing 1, 1–44, 9.

and difficulties[77] and dependent on factors like human security, strategies for earning livelihood, conflicts and socio-economic development, poor governance, lack of access to resources and social tension.[78]

Robert McLeman of the University of Ottawa takes out the triggers of forced migration into two different groups comprising of climate processes and climate events.[79] Climate events are abrupt and dramatic risks such as storms, typhoons, hurricanes, monsoon floods and glacial lake outburst floods.[80] Climate processes, on the other hand, have the sluggish commencement of changes like sea-level upsurge, desertification, salinization of arable land, food insecurity and increasing water shortage that make the human existence and survival most difficult.[81] These triggers of forced migration result in two kinds of displacement patterns: slow and rapid climate displacement.[82] IPCC report on adaptation to climate change also notes fast and slow types of climate-induced migration as a reaction to severe climate incidents and migration caused due to longer-term climate erraticism, capriciousness and changeability.[83] Rapid onset climate displacement is when people are displaced due to natural disasters such as hurricanes, cyclones, storms, typhoons, temperatures, flooding and landslides, which render places uninhabitable.[84] Slow-onset displacement involves steadily worsen people's livelihoods to the position where they cannot endure an adequate standard of living in their lands of origin due to factors such as drought, soil erosion and sea-level rise.[85] Migration following sudden-onset events often seems to be large scale, but temporary while slow-onset events cause

---

[77] Jane McAdam (ed), *Climate Change and Displacement: Multidisciplinary Perspectives* (Bloomsbury 2010) 2.

[78] UNHCR, 'Summary of Deliberations on Climate Change and Displacement' (Bellagio 22–25 February 2011) 2 www.unhcr.org/4da2b5e19.pdf accessed 18 April 2018.

[79] Robert A Mcleman, *Climate and Human Migration: Past Experiences, Future Challenges* (Cambridge University Press 2013); See also Oli Brown, 'Climate Change and Forced Migration: Observations, Projections and Implications' Human Development Report 2007/11, 10 http://hdr.undp.org/sites/default/files/brown_oli.pdf accessed 25 April 2018.

[80] Ibid.

[81] Ibid.

[82] Runkel and Conklin (n 69) 6 (Also called migration due to sudden natural disasters and migration linked to gradual changes in the environment); See also Jayawardhan (n 73) 105.

[83] Philip Dane Warren, 'Forced Migration After Paris Cop 21: Evaluating the Climate Change Displacement Coordination Facility' (2018) 116 (8) Colum. L. Rev, 2103- https://columbialawreview.org/content/forced-migration-after-paris-cop21-evaluating-the-climate-change-displacement-coordination-facility/ accessed 21 April 2018.

[84] Sarah Opitz Stapleton, Rebecca Nadin, Charlene Watson and Jan Kellett, *Climate Change, Migration and Displacement: The Need for a Risk-Informed and Coherent Approach* (UNDP 2017) 9.

[85] Jayawardhan (n 73) 103; See also Susana Beatriz Adamo, 'Vulnerable People in Fragile Lands: Migration and Desertification in the Drylands of Argentina: The Case of the Department of Jáchal' (Ph.D. thesis, University of Texas at Austin 2003) http://www.file://Users/sau/Downloads/adamosb036.pdf accessed 25 April 2018.

more gradual and permanent movement.[86] Climate-induced displacement/migration which takes place due to sudden disasters is easy to track, and data gathering is also more efficient in comparison with displacement which takes place due to a gradual degradation of the environment.[87] In the case of the latter, displaced migrants usually move to their new habitat in small groups and the cause of their displacement is also difficult to be tracked.[88] The process of climate-induced displacement is also categorized as 'voluntary' and 'forced' based on the time frame available for preparation and intent.[89] However, the question remains in the context of rapid environmental degradation, and increased disasters whether an informed choice operates making the entire distinction blurred.[90]

The UN General Assembly Resolution 64/162 of December 2009 acknowledged the natural disasters as a factor of human displacement and precipitated concerns that climate change could aggravate the adverse effects of both sudden- and slow-onset disasters.[91] While the climate change is not tending to be the sole or the primary cause, it can be a significant push factor in activating migration and human displacement. While the specific scale, location and timing of such migrations are volatile, there is increasing evidence that they will be substantial and will grow in years to come. Climate change can be recognized as an impact booster and escalator.[92] For instance, besides its negative effects, climate change may exacerbate the risk of conflict, which can, in turn, trigger further human displacement.[93] Subsequently, such migration may lead to an enhanced struggle over limited natural resources and the deprivation of livelihoods which may further proliferate the risk of conflict and violence, producing additional human displacement.[94] The Security Council (SC) debated on July 20, 2011, the adverse impact of climate change for international peace and security by

---

[86]Graeme Hugo, 'Climate Change-Induced Mobility and the Existing Migration Regime in Asia and the Pacific' in Jane McAdam (ed), *Climate Change and Displacement: Multidisciplinary Perspectives* (Bloomsbury Publishing 2010) 9–37.

[87]Runkel and Conklin, (n 69) 15.

[88]Ibid 9.

[89]Ibid.

[90]Jon Barnett and Michael Webber, 'Accommodating Migration to Promote Adaptation to Climate Change' (2009) World Bank Policy Research Working Paper 5270, 6–7 www.researchgate.net/publication/46443903_Accommodating_Migration_to_Promote_Adaptation_to_Climate_Change accessed 23 April 2018; See UNHCR, 'Forced Displacement in the Context of Climate Change: Challenges for States Under International Law' (Bonn 1–12 June 2009) (19 May 2009).

[91]Protection of and Assistance to Internally Displaced Persons, UNGA Res 64/162 (18 December 2009) UN Doc A/RES/64/162; Chaloka Beyani 'Climate Change and Internal Displacement' (Brookings Institution 2014) 1 www.brookings.edu/wp-content/uploads/2016/06/Climate-Change-and-Internal-Displacement-October-10-2014.pdf accessed 19 April 2018.

[92]Government Office for Science London, 'Migration and Global Environmental Change' (20 October 2011) 32–33 www.gov.uk/government/uploads/system/uploads/attachment_data/file/287717/11-1116-migration-and-global-environmental-change.pdf accessed 29 December 2017; Chaloka Beyani (n 91) 4.

[93]Chaloka Beyani (n 91) 1.

[94]Ibid, See also Rafael Reuven, 'Climate Change-Induced Migration and Violent Conflict' (2007) 26(6) Political Geography 656–673.

underscoring the security ramifications of climate change, including how it could obfuscate or exacerbate new and present security apprehensions in precarious and defenceless nation-states, possibly jeopardizing the global economic viability.[95]

There is overwhelming humanitarian and protection apprehension that requires to be addressed, and the international community shoulders the responsibility to scrutinize and impact policy changes, and at the same time also introduce legal and humanitarian responses.

2.5 Climate Change and Human Displacement in South Asia

It is obvious from the ongoing deliberations that the South Asian region can become the most suffered sub-continent in the world due to climate change.[96] The Internal Displacement Monitoring Centre (IDMC) has signified India to be one of the foremost nation-states that are bound to confront the human displacement due to natural calamities.[97] Of the CDPs in 2015, 41% were displaced in South Asia alone[98] that has been witnessing an unparalleled increase in the number of climatic disasters including floods, heatwaves and storms, which are affecting thousands of peoples' property and environment. The sub-continent has also commenced experiencing steady changes in the climate such as the rise in sea levels, which are resulting in submerging of coastal areas of the South Asian countries.

It becomes essential to study the linkage between climate change and displacement of the affected population because environmental factors often get ignored while analysing patterns of human displacement/climate change migration.[99] The relationship between human displacement and climate change migration is partly so impulsive that the climate change matrix is complicated and theoretical in its appreciation of impact on societies wedded with a variety of resources and adaptability to external astonishments. Partially, the decisions of individual migrants to leave their country of origin differ widely about determining the causality between economic

---

[95] Ibid; Chaloka Beyani (91) 4.

[96] The Sub-Continent has a nation like Bangladesh which is often stated to be a country which is likely to be hugely affected by climate change displacement, because it is a classic example of a delta nation which is low-lying and is also densely populated. A big proportion of this population lives in coastal areas or areas which are prone to floods. See Warren (n 83).

[97] GRID 2017: The Global Report on Internal Displacement' (*iMDC*, 2017) www.internal-displacement.org/global-report/grid2017/ 10 accessed 20 April 2018.

[98] Michelle Yonetani, 'Global Estimates 2014: People Displaced by Disasters' (iMDC, September 2014) 40 https://reliefweb.int/sites/reliefweb.int/files/resources/201409-global-estimates.pdf accessed 20 April 2018.

[99] Gunvor Jónsson, 'The Environmental Factor in Migration Dynamics—A Review of African Case Studies' International Migration Institute, University of Oxford Working Paper 21/2010, 3–4 www.oxfordmartin.ox.ac.uk/downloads/WP21%20The%20Environmental%20Factor%20in%20Migration%20Dynamics.pdf accessed 21 April 2018.

pull and climate push that often lacks objectivity. Ultimately, the segregation of the role of climate change from other socio-economic and enviro-cultural factors necessitates analytical and determined measures to dispel the sordid state of affairs. Hence, the portrayal of causative and linear demarcation between climate change migration and forced displacement is a challenging and arduous proposition.[100]

South Asia has been experiencing migration that ranges from temporary to permanent and from voluntary to forced migration since historical times.[101] However, the impacts of climate change have been aggravating further the existing adverse effects of natural calamities in South Asia. Although there is no data available to map the role of the climate change in overall climate change migration and human displacement in South Asia, climate change is understood to affect the human mobility patterns in at least four ways:

(i) The strength of natural calamities both abrupt and sluggish commencement through substantial rainfall pandering to flash or river flooding that leads to enhanced human displacement and climate change migration;
(ii) The repercussions of the incremental global warming, climate variations, and other impacts of the climate change on public health, livelihood opportunities, food security, undermining of ecosystem services including water availability and soil fertility[102];
(iii) Escalating sea levels that make littoral areas inhospitable for human survival; and
(iv) The contest and struggle for sparse and limited natural resources possibly leading to growing tensions and even and in turn triggering the human displacement and climate change migration.[103]

In South Asia, there are multifaceted drivers of human displacement[104] and climate change as a driver may primarily be invisible in the migration/displacement discourse and for a region seriously influenced by the natural calamities, climate change as a ground can be difficult to be construed.[105] The matters are also getting complicated since climate change has the effect of exasperating the current sensibilities, inclinations and vulnerabilities that make it contested to segregate climate change as a cause of migration and human displacement.[106] For example, politi-

[100]Oli Brown (n 79) 6.

[101]Partha S Ghosh, *Unwanted and Uprooted: A Political Study of Migrants, Refugees, Stateless and Displaced of South Asia* (Sanskriti Publishers 2004).

[102]Andrew Morton, Philippe Boncour and Frank Laczko, 'Human Security Policy Challenges' (2008) 31 FMR 5.

[103]Frank Laczko and Christine Aghazarm (eds), *Migration, Environment and Climate Change: Assessing the Evidence* (IOM 2009) 15.

[104]Catherine Locke, W Neil Adger and P Mick Kelly 'Changing Places: Migration's Social and Environmental Consequences' (2000) 42(7) Environment 24–35.

[105]McAdam and Saul(n 40) 237.

[106]António Guterres, 'Migration, Displacement and Planned Relocation' (*UNHCR*, 31 December 2012) www.unhcr.org/news/editorial/2012/12/55535d6a9/migration-displacement-planned-relocation.html accessed 22 April 2018.

cal, organizational, legal environment and peculiar factors such as ethnicity, family size,[107] economic stratification, resources available to adapt to the environment and climate consequences act in both facilitating and preventing the intensity and magnitude of displacement.[108] A study by the Action Aid reveals that in areas affected by Cyclone Aila, delay/lethargy and failure of governance resulted in socio-economic crisis forcing the people to migrate.[109] The same study also found that increasing salinity in Khulna is leading to migration for economic but also social reasons[110] that require a panoramic perspective to address the issue.

In South Asia, most of the studies on climate displacement have been conducted in Bangladesh. It has been projected that seasonal floods displace and uproot 500,000–1 million people annually in Bangladesh and erosion of land of river banks also affects a large chunk of people every year.[111] Human displacement in Bangladesh as a result of the adverse effects of climate change cooperating and combining with bad decision-making regarding development and is challenging to disentangle natural disasters and calamities from climate change processes. For example, mercantile activities like shrimp cultivation had deteriorated flood ridges and defences and emasculated the local agricultural production and livelihood opportunities.[112] Many CSOs and NGOs articulated that greater responsibility for such human displacement rests with the government and its inimical policies to the climate.[113] In Bangladesh, human displacement, therefore, is different and crucial in the wake of the abrupt inception of events like cyclones, floods and riverbanks' erosion and crumple.[114] The people intend to migrate short distances and try to return to their homes at the earliest. However, returning is very difficult in such areas in case of repeated inundations.[115]

---

[107]Douglas S Massey, William G Axinn, and Dirgha J Ghimire, 'Environmental Change and out-Migration: Evidence from Nepal' (2016) 32(2) Pop. & Env't 109–136.

[108]James Morrissey, 'Rural–Urban Migration in Ethiopia' (2008) 31 Forced Migration Review 28, 29; See also Elizabeth Fussell, 'The Long Term Recovery of New Orleans' Population after Hurricane Katrina' (2015) 59(10) Sage Journal 1231–1245 (The review revealed a need for a comprehensive social, spatial and temporal framework for explaining inequality in population recovery and displacement); Binu Jacob, 'Disaster Mythology and Fact: Hurricane Katrina and Social Attachment' (2008) 123(5) Public Health Reports 555–565; Kaushal Roy, and Umme Tania Sultana, 'Climate Change Disasters and Rural poverty: Case of Coastal Bangladesh' (Third International conference on Bangladesh Environment, Dhaka, January 2010) 13.

[109]Md Shamsuddoha, SM Munjurul Hannan Khan, Sajid Raihan and Tanjir Hossain, 'Displacement and Migration from Climate Hot-spots: Causes and Consequences' (Center for Participatory Research and Development and Action Aid Bangladesh, Dhaka 2012) 21.

[110]Ibid, 20; Richards and Bradshaw (n 68) 10–11 Kaushal Roy(n 108) 13.

[111]McAdam and Saul (n 40) 238.

[112]Ibid, 240.

[113]Ibid, 240.

[114]Sanjoy Biswas and Md. Akterul Alam Chowdhury, 'Climate Change Induced Displacement and Migration in Bangladesh: The Need for Rights Based Solution' (2012) 157–180, 159 www.mcrg.ac.in/rw%20files/RW39_40/13.pdf accessed 20 January 2018.

[115]Ibid.

Therefore, in some situations, communities and individuals end up in a state of displacement with perennial and perpetual manner.[116]

The climate change does not recognize the geopolitical settings and international borders and affects all parts of the globe without any set course. Therefore, in other parts of South Asia also, climate change displacement is a grave issue.[117] 'Climate change is very genuine in the Maldives, and its effects the survival of the Maldives and on a compendium of rights such as the right to housing, right to safe water, and right to livelihoods are being felt and advocated on many islands', said Chaloka Beyani, the Special Rapporteurs on the human rights of IDPs.[118] The SAARC countries are slow in recognizing the role of climate change in migration and displacement. Climate change is still conspicuous by its absence in the human displacement discourse in South Asia and most of the studies on climate change come from the Maldives and Bangladesh, which do not adequately document empirical evidence and address the climate factors in human displacement in entire South Asia. The crisscrossing challenges of climate change migration, human displacement, national and global volatility offer a one-off problem for South Asia in the years ahead. These factors undermine the conventional understandings of the national security narratives and climate refugee protection narratives in South Asia.

The 2004 Indian Ocean earthquake and tsunami sent enormous waves to South Asia that has affected Sri Lanka, the Maldives and India in multiple ways.[119] In Sri Lanka, in the space of 20 min of the tsunami, around 35,000 were dead which was almost equivalent to the death toll from 20 years of civil war which decimated over 100,000 homes[120] and displaced approximately 500,000 people. In the Maldives, some 200 islands were devastated by the waves between 15,000 and 30,000 people (5–10% of the population) were initially displaced.[121] Later, the government's IDPs unit estimated that there were at least 12,000 long-term IDPs in the country.[122] Over 10,000 Indians died in the disaster, with at least 647,556 people getting displaced.[123]

---

[116] Abrar R. Chowdhury and Syed Nurullah Azad, *Coping with Displacement: Riverbank Erosion in North-West Bangladesh* (North Bengal Institute for Alternative Research and Advocacy 2004) 33, 34.

[117] Justin Ginnetti and Chris Lavell, 'The Risk of Disaster-Induced Displacement in South Asia' (IDMC, 2015) 18).

[118] Editor, 'UN expert urges Maldives to tackle displacement caused by climate change' *UN News* (21 July 2011) https://news.un.org/en/story/2011/07/382322-un-expert-urges-maldives-tackle-displacement-caused-climate-change accessed 28 December 2017.

[119] Cottle, *Global Crisis Reporting*, (McGraw-Hill Education (UK), 2008) 51.

[120] Jayadeva Uyangoda, 'Ethnic Conflict, the Sri Lankan State and the Tsunami Marin' in Marion Couldrey and Tim Morris (eds), *Tsunami: Learning from the Humanitarian Response* (Refugee Center Studies 2005) 30–31.

[121] Heidi Brown, 'Lost Innocence: The tsunami in the Maldives' in Marion Couldrey and Tim Morris (eds), *Tsunami: learning from the humanitarian response* (Refugee Center Studies 2005) 48.

[122] Ibid 48.

[123] Human Rights Watch, 'After the Deluge: India's Reconstruction Following the 2004 tsunami' (2005) 17(3) HRW, 1 www.hrw.org/reports/2005/india0505/index.htm accessed 28 December 2017.

In Pakistan, the October 2005 earthquake had a devastating impact on its northern part that also included the Pakistan-administered Kashmir (PAK). An estimated 2.8 million people who were affected by the earthquake became IDPs.[124] Again a two-month heavy rainfall in 2010 with its unprecedented floodwaters affected some 20 million people, submerging almost one-fifth of Pakistan under water.[125] The disaster thereby claimed 2,000 lives and destroyed 1.6 million homes across the provinces of Baluchistan, Khyber Pakhtunkhwa, Punjab and Sindh.[126] The IDMC report stated that approximately 11 million people were IDPs, thereby Pakistan witnessed one of the most significant human displacement events within 2008–2012.[127] In Nepal earthquake of 2015, also called *Gorkha earthquake,* about 9,000 people were killed and many thousands more were injured.[128] IDMC estimates as of April 2016 that there are 2.6 million people who are still displaced by the earthquakes.[129]

Regarding South Asia, countries are increasingly exposed to the gradual and slow onset of disasters including heatwaves, droughts and floods which will lead to massive human displacement. In the summer of 2015, Pakistan and India endured one of the most severe heatwaves in the history that exterminated more than 3,700 people in both the countries collectively.[130] Further, the inundating of the Sundarban region and the Maldivian islands[131] and another part of South Asia has become almost inescapable due to the pace at which the global temperature is increasing.

By the end of the present century, the densely populated subcontinent's temperature would cross the habitable thresholds and the region will become susceptible to deadly heatwaves.[132] Around 15% of the SAARC population will be confronting the extreme and unhealthy weather conditions.[133] An increase in the temperature of 2 °C above pre-industrial levels would merely mean that the cities like Karachi in

[124] US Agency for International Development, 'South Asia—earthquake Fact Sheet #25, Fiscal Year (FY) 2006' (*Relief Web,* 17 November 2005) https://reliefweb.int/report/india/south-asia-earthquake-fact-sheet-25-fiscal-year-fy-2006 accessed 27 December 2017.

[125] Editor, 'A Year after Pakistan's Devastating Floods, UN Continues Effort to Rebuild Lives' *UN News* (28 July 2011) https://news.un.org/en/story/2011/07/382952-year-after-pakistans-devastating-floods-un-continues-effort-rebuild-lives accessed 27 December 2017.

[126] Ibid.

[127] Michelle Yonetani, 'Global Estimate 2012: People Displaced by Disasters,' (iMDC, 2013) 15.

[128] See also John P Rafferty, 'Nepal earthquake of 2015' (*Encyclopedia Britannica*, 18 April 2018) www.britannica.com/topic/Nepal-earthquake-of-2015 accessed 25 April 2018.

[129] Marita Swain, 'Nepal: One Year after the earthquakes: An End to Displacement is Still Years Away' (Relief Web, 25 April 2016) https://reliefweb.int/report/nepal/nepal-one-year-after-earthquakes-end-displacement-still-years-away accessed 27 December 2017.

[130] Richards and Bradshaw (n 68) 22.

[131] Brian Black (ed), *Climate Change: An Encyclopedia of Science and History* (ABC-CLIO 2013) 290.

[132] Eun-Soon Im, Jeremy S Pal and Elfatih AB Eltahir, 'Deadly Heat Waves Projected in the Densely Populated Agricultural Regions of South Asia' (2017) 3(8) Science Advances http://advances.sciencemag.org/content/3/8/e1603322.full accessed 20 December 2017.

[133] Justin Worland, 'Climate Change Will Make Parts of South Asia Unlivable by 2100, Study Says' *Time* (2 August 2017) http://time.com/4884648/climate-change-india-temperatures/ accessed 21 April 2018.

Pakistan and Kolkata in India will have to experience almost every year circumstances similar to the 2015 heatwave.[134] Studies show that heat-based stress is more responsible for climate-induced displacement than flooding.[135] Moreover, the temperature levels by the end of the present century would be such that 4% human population of this region would find it difficult to survive without air-conditioning machines.[136] In Afghanistan too, heatwaves are becoming more extreme and more frequent. Climate change-driven problems like drought, water scarcity and heat and soil degradation greatly endanger the agriculture and food security situation in Afghanistan that leads to substantial intensification in human displacement.[137]

Droughts have severely hit the SAARC South Asian region in recent years. The worst-affected countries are India, Pakistan and Afghanistan. 15% of the Indian population i.e., 130 million people have been affected by the worst drought that India witnessed in a century.[138] As far as the Indian Scenario is concerned, the pattern of climate change migration does not represent consistency with only 2% of drought-affected households' migration in 1983 and 1994 was reported.[139] Significant numbers of people are seen to participate in rural-to-rural movement in response to environmental change or disaster.[140] The condition of the Pakistani and Afghani population is more critical than this because the resources there are scarcer than in India. Afghanistan is also under threat from ongoing conflicts, making life even more terrible. In Afghanistan, 10% population is suffering from the worst drought since the year 1971. During 1980–2015, this country witnessed a maximum number of deaths due to natural disaster after Haiti.[141] At the same time, 3 million Pakistanis are at risk of starvation and most of this population is rural. Hundreds of thousands of people have vacated the regions lying in the Southern Thar Desert because of the devastation of crops and also livestock. This area was formerly home to more than a million people.[142]

The SAARC region also keeps experiencing floods, as it homes some of the most essential and mighty rivers, which have also been adversely affected by the damaging climatic scenarios. In 2017, in India and Bangladesh alone, over 1,200 have lost their

[134] Tom KR Matthews, Robert L Wilby and Conor Murphy, 'Communicating the Deadly Consequences of Global Warming for Human Heat Stress' (2017) 114 (15) Proceedings of the National Academy of Sciences of the United States of America 3861–3866.

[135] Richards and Bradshaw (n 68) 23.

[136] Justin Worland (n 133).

[137] Runkel and Conklin (n 69) 23.

[138] 'Asia' (*Drought Disasters*) www.unicef.org/drought/asia.htm accessed 21 April 2018.

[139] Clionadh Raleigh, Lisa Jordan and Idean Salehyan 'Assessing the Impact of Climate Change on Migration and Conflict' (*Social Development: The World Bank*, 2008) 344.

[140] Aromar Revi, 'Climate Change Risk: An Adaptation and Mitigation Agenda for Indian Cities' (2008) 20(1) Environ. Urban 207–229, 209.

[141] Runkel and Conklin (n 69) 23.

[142] UNICEF, 'Drought Disasters' www.unicef.org/drought/asia.htm accessed 22 December, 2017.

lives to flooding, and 950,000 houses have been destroyed, affecting more than 41 million people.[143]

On an average, 1/5 of Bangladesh confronts and braces the floods annually.[144] The central part of the Bangladeshi coastline, which is dominated by rivers like Ganga, Brahmaputra and Meghna, has also seen an increase in disasters like Typhoons and storms in recent years.[145] A cyclone called Aila displaced 2.3 million people in India and around 850,000 in Bangladesh.[146] The Sundarban region and the Mangrove forests are also witnessing a threat from the rising sea levels, and approximately 15 million people are susceptible to losing their homes because of this.[147] It has been asserted that 70,000 people out of the 4.1 million living in the Indian part of the Sundarbans islands would be rendered homeless by the year 2020.[148] The Northeastern Indian state of Assam, in September 2012, witnessed a displacement of 1.5 million people, because of massive flooding. In Afghanistan, the climate crisis has not received adequate attention.[149] As pointed by Runkel, the climate change pattern shows that when snow-falling happens in the Northern parts of Afghanistan, later in the season the temperatures are not cold enough, therefore, it does not settle and melts swiftly in spring that accelerates flooding and protracts the period of water shortage in the summer.[150]

Additionally, migration in South Asia is highly gendered with more out-migration of males from Nepal and females from Sri Lanka. The impacts of climate-induced migration on women and girls are not being monitored by government agencies in South Asia, resulting in increased vulnerability and exploitation.[151] Studies reveal that in Bangladesh migrated male family members were unable or unwilling to send money back to their households leaving the women exposed to vulnerability and abuse.[152] Keeping in mind, the factors discussed above it can be said that there is no doubt about the fact that the South Asian region is extremely vulnerable when

---

[143]Chloe Farand, 'Floods in India Bangldesh, and Nepal Kill 1200 and Leave Millions Homeless' *The Independent* (29 August 2017) www.independent.co.uk/news/world/asia/india-floods-bangladesh-nepal-deaths-millions-homeless-latest-news-updates-a7919006.html accessed 17 April 2018.

[144]Runkel and Conklin (n 69) 19.

[145]Ibid, 19.

[146]Arpita Bhattacharyya and Michael Werz, *Climate Change, Migration, and Conflict in South Asia Rising Tensions and Policy Options across the Subcontinent* (Centre for American Progress and Heinrich Boll Stiftung 2012) 2; See Katha Kartiki, 'Climate Change and Migration: A Case Study from Rural Bangladesh', (2011) 19(1) Gender & Development 23–37, 24.

[147]Runkel and Conklin (n 68) 19.

[148]Architesh Panda, 'Climate Refugees: Implications for India' (2010) 45(20) EPW 76.

[149]Daud S Saba, 'Afghanistan: Environmental Degradation in a Fragile Ecological Setting' (2001) 8(4) Int. J. Sust. Dev. World. Eco. 279–289.

[150]Runkel and Conklin (n 69) 23.

[151]Scott Lockie, 'Climate Related Disasters and Displacement: Homes for Lost Homes, Lands for Lost Lands', in Jose Miguel Guzman, George Martine, Gordon McGranahan, Daniel Schensul and Cecilia Tacoli, (eds) Population Dynamics and Climate Change (IIED, UNFPA 2009) 119–132.

[152]Teresa Anderson, Md.Shamsuddoha and Ajaya Dixit, 'Climate Change Knows No Borders: An Analysis of Climate Induced Migration, Protection Gaps and Need for Solidarity in

it comes to environmental disasters and climate change.[153] It merely implies that it is predestined for the SAARC region in coming years that will witness an increase in both internal and international migrations. Further, specific South Asian areas, which are already undergoing conflicts because of other factors that will have to endure even more severe consequences and also security concerns.[154]

Among the researchers and policymakers, forced migration/human displacement is being seen more and more as a precautionary adaptation strategy to severe climate change threats.[155] While climate change migration can be a manifestation of helplessness, it can also be a method to accomplish human rights with a sense of security, particularly, if it is well-formulated. Having a potentially positive adaptive strategy for individuals, households and communities that may bring significant development and socio-economic benefits for migrants with their families and their spaces and places of mobility.[156] Regarding South Asia, displacement and migration are not used as a positive adaptation strategy significantly.[157]

2.6 Typology of Climate Change-Induced Displacement in South Asia

The climate change[158] in SAARC nations has brought an upsurge in both the frequency and intensity of chronic environmental disasters including floods, droughts and landslides, aggravating the human displacement.[159] South Asia, in 2015 and 2016, witnessed the displacement of 7.9 million and 3.6 million people respectively, because of flooding and landslides, which occurred due to an increase in the intensity

South Asia' (Action Aid 2016) 17 www.unescogym.org/wp-content/uploads/2017/01/ActionAid_Climate-change-knows-no-borders.pdf accessed 20 April 2018.

153 World Bank, 'Why is South Asia vulnerable to Climate Change?' (*Relief Web,* 1 December 2009) https://reliefweb.int/report/india/why-south-asia-vulnerable-climate-change accessed 19 December 2017.

154 Bhattacharyya and Werz (n 146) 6.

155 Cecilia Tacoli, (n 66) 514.

156 Jane McAdam, *Climate Change, Forced Migration, and International Law* (OUP 2012) 164.

157 Robert Stojanov, Ilan Kelman, A.K.M. Ahsan Ullah, Barbora Duží, David Procházka and Klára Kavanová Blautová, 'Local Expert Perceptions of Migration as a Climate Change Adaptation in Bangladesh' (2016) 8(12) Sustainability 1223–1237, 1235.

158 Luca Marchiori and Ingmar Schumacher, 'When Nature Rebels: International Migration, Climate Change, and Inequality' (2011) 24(2) J. PoP. Eco, 569–600, 573; See also Justin T. Locke, 'Climate Change-Induced Migration in the Pacific Region: Sudden Crisis and Long-Term Developments' (2009) 175(3) Geogr. J., 171–180, 172.

159 Maria Waldinger and Sam Fankhauser, 'Climate Change and Migration in Developing Countries: Evidence and Implications for PRISE Countries' (2015) Centre for Climate Change Economics and Policy and Grantham Research Institute on Climate Change and the Environment, 7 http://eprints.lse.ac.uk/64526/1/Climate-change-and-migration-in-developing-countries_final.pdf accessed 18 April 2018.

of monsoon and storms.[160] The August 2017 has seen floods due to the extreme monsoon that affected around 43% million people in different parts of Bangladesh, Nepal and India.[161] Following types of climate-induced migrant/displaced populations have observed to be present in the sub-continent.

2.7 Internal Displacement/Migration

In case of climate-induced displacement alters, the cross-border movement has been observed to be limited, and the populations usually prefer to relocate themselves within their own country.[162] Apart from relocating to new habitats, moving back to the areas where the population getting affected by climate change was residing before the occurrence of climate hazards can be stated to be another type of internal migration.[163]

Initial reports of IPCC cited migration as a possible outcome of climate change. The latest reports (fourth assessment report, 2007) have emphasized that adaptation to the existing circumstances has become the preference of most of the populations who have been getting affected by climate change.[164] People residing in disaster-prone areas or areas which have become degraded due to climate change usually become habitual of the risks associated with living in a particular area and incorporate them into their lifestyles. Such absorption of climate change patterns happens through individual- or community-level coping mechanisms. Though, such mechanisms are dependent on several factors like financial assets, which they possess their position in the social hierarchy, political relationships and policies of the government.[165] Internal migration dependent on the strength of the above-mentioned coping mechanisms takes place in situations involving both economic and environmental difficulties faced by people, both of which are by-products of climate-induced hazards.[166]

Regarding South Asia, most of the data available on internal displacements occurring after a disaster are from Bangladesh. Migration has become a coping mechanism for climate change in this country.[167] It is a disaster-prone nation,[168] and occurrence

---

160 Richards and Bradshaw (n 68)17.

161 Ibid, 4.

162 Bhattacharyya and Werz (n 146) 4.

163 Richard Black, 'Environmental Refugees: Myth or Reality?' (2001) UNHCR, New Issues in Refugee Research Working Paper No 34 www.unhcr.org/research/RESEARCH/3ae6a0d00.pdf accessed 18 April 2018.

164 Clionadh Raleigh, Lisa Jordan and Idean Salehyan 'Assessing the Impact of Climate Change on Migration and Conflict' (Social Development: The World Bank 2008) 3.

165 Ibid,17.

166 Ibid.

167 Carabine and Lemma (n 19) 18.

168 Clark L Gray and Valerie Mueller, 'Natural Disasters and Population Mobility in Bangladesh' (2012) 109(16) Proceedings of the National Academy of Sciences of the United States of America 6000–6005.

of a large number of disasters has in a way aided researchers to study community patterns of climate-induced displacement/migration. The people of Bangladesh were found to have moved to a distance of two miles on an average, from their previous residence, after the disaster, especially from rural areas to urban areas.[169] It can be said to be a persistent belief among the Bangladeshi population is that their land will become habitable again after the disaster and that it can be reclaimed when the impacts of disasters have reduced. The populations do not hesitate from displacements, after a reoccurrence of a disaster, and this has been observed to be a common characteristic among Bangladeshi settlements.[170] In contrast to the majority, which decided to move back to their original homes, a significant number of the migrants were, found to have become permanent squatter settlers in urban areas, with a growth rate of 4% per year by 2007.[171] In *India,* the extreme Cyclone Phailin, that lashed the coastal state of Orissa frightening mass destruction, has prompted large-scale migration and dislocation of fishing communities.[172] Consequently, the submergence of the Lohachara Island in India's Sundarban has compelled the people to move to the adjoining Sagar Island.[173]

Among the category of internal migration/displacement, rural outward migration has become inevitable because of the gradual deterioration, which has resulted in environmental conditions like droughts, soil degradation and deforestation. Bangladesh, in recent years, has noted an increase in the depth of standing water, which has prevented crop cultivation during Kharif season.[174] Surveys conducted in households of the Indian state of Odisha have indicated a fall of 67% in the production levels of Kharif crops.[175] All this, apart from affecting jobs and livelihood opportunities, has also resulted in migration to rural areas. In a 2007 study done on the Asian region, which was based on the data collected in 1974 from Nepali households of Chitwan Valley, it was observed that outward migration has close connections with

---

169 Dominic Kniveton, Christopher Smith, Richard Black and Kerstin Schmidt-Verkerk, 'Challenges and Approaches to Measuring the Migration– Environment Nexus' in Frank Laczko and Christine Aghazarm (eds), *Migration, Environment and Climate Change: Assessing the Evidence* (IOM 2009) 41,112, 73.

170 Chowdhury and Azad (n 116) 33–34; See also Biswas and Chowdhury (n 114) 157–180.

171 Saudia Anwer, 'Climate Refugees in Bangladesh: Understanding the Migration Process at the Local Level' (Diakonisches Werk der EKD eV for Brot für die Welt, 2012) 23.

172 Neeta Lal, 'A Precarious Fate for Climate Migrants in India' *Inter Press Service* (New Delhi, 19 May 2016) www.ipsnews.net/2016/05/a-precarious-fate-for-climate-migrants-in-india/ accessed 24 April 2018.

173 Architesh Panda, 'Climate Induced Internal Migration in India: Emerging Challenges' (*Migration Narratives: The SHRAM Blog*, 3 June 2016) www.shram.org/blogs/?p=1417 accessed 26 April 2018.

174 Ministry of Environment and Forest Bangladesh, 'NAPA Project Profile: Promoting Adaptation to Coastal Crop Agriculture to Combat Salinization' (2005) Project No 11, 35 https://unfccc.int/resource/docs/napa/ban01.pdf accessed 23 April 2018, see generally Udaya Sekhar Nagothu, Attila Nemes, Jatish Chandra Biswas and Motaleb H. Sarker (eds), *Climate Change Impacts, Vulnerability and Adaptation: Sustaining Rice Production in Bangladesh* (Bioforsk 2014).

175 Kelkar and Bhadwal (n 31) 18.

the inability to gather firewood.[176] It was interpreted as a type of climate insecurity, which is a massive factor behind the human movement. The study, however, has been criticized, as being incomplete because the reason cited for the migration appears to be insignificant.[177]

Rural out-migration, which becomes rural in-migration in other agricultural areas, has not been much in focus and often results in environmental degradation of the area, which receives such population. India is no exception to the trend of rural out-migration and the drought the country faced in the year 2012, which resulted in lower crop yields, along with excessive rains which destroyed other agricultural produce, can be stated to be factors behind this kind of trends. Serious crop failures as a result of unpredictable weather conditions have resulted in the suicide of at least 240,000 farmers between 1995 and 2009.[178] On the other side, those who survived chose to migrate. Along with rural outward migration, countries in the SAARC region are also susceptible to increase rural-to-urban migration. In Bangladesh, Dhaka and Chittagong are the critical destinations for such movement. In India, climate change and the consequently enlarged drought in Western India and semi-arid peninsular India heading to further submergence of the landless people and small and marginal farmers, who are in no small extent forced to relocate more often to cities.[179] Such migration to the growing megacities will exasperate the already recognized peril of 'overcrowding, pollution, poverty, impossible demands for energy and water'.[180]

2.8 Cross-Border Migration

Climate-induced migration taking place from one border to another is not a concept alien to the South Asian sub-continent, primarily because of the porous borders and the historical, familial and cultural ties which exist here between countries.[181] An excellent example of the same is increasing the number of Bangladeshi and Nepali migrants in India particularly to West Bengal and Assam, most of which have left their original home because of the increased number of natural disasters and crop

---

[176] Pratikshya Bohra and Douglas S Massey, 'Processes of Internal and International Migration from Chitwan, Nepal' (2009) 43(3) Int Migr Rev 621–651.

[177] Frank Laczko and Christine Aghazarm (eds), *Migration, Environment and Climate Change: Assessing the Evidence* (IOM 2009) 15; Golam Rabbani, Fathimath Shafeeqa, Sanjay Sharma, *Assessing the Climate Change Environmental Degradation and Migration Nexus in South Asia* (IOM 2017) 172.

[178] Eberhard Weber, 'Economic Reform, Social Development and Conflict in India', (2012)4(3) Regional Sciecne Policy and Practice, 207–230 see generally Shahin Razi, 'Rural Distress and Rural Migration' (2014) 62(11) Kurukshetra 3–6.

[179] Aromar Revi, 'Climate Change Risk: An Adaptation and Mitigation Agenda for Indian Cities' (2008) 20(1) Environment and Urbanization 207–229, 213.

[180] Bhattacharyya and Werz (162) 28.

[181] D Suba Chandran, 'Border Management in South Asia: Volatile, Violent and Porus' (*IPCS*, 22 August 2013) www.ipcs.org/comm_select.php?articleNo=4096 accessed 23 April 2017.

failures which their country faces.[182] The fences on the India–Bangladesh borders, which were built in the year 1980s and other border arrangements have proved to be a hindrance in the way of illegal migration. Apart from stopping migration, the fences have also made cross-border socio-economic change difficult resulting in fostering of bribery and human trafficking.[183] The intended results have, not been achieved, and the fencing has also caused permanent settlement of migrants, once they manage to cross the borders, leading to security implications, loss of social capital and crucial support systems because they entrap the population which had once migrated.[184] Further, the rising intolerance towards Bangladeshi migrants in India's Assam and West Bengal is an example of cross-border migration/displacement that has been stoking tensions in the region.[185] Cross-border migration involves a large number of hardships and impact migrants, their area of origin and destination.[186] Migrants and their families have to face lousy working conditions, and no legal protection is offered to them. The only resolution to the instant impugned problem is to provide improved sociopolitical and economic conditions to the migrants in their new area of residential settlement. Therefore, the SAARC region has the vast population that is responsible for sending migrants from Pakistan, Nepal, Sri Lanka and India to other parts of the world mainly to Asia and the Pacific region[187]; but migrants from these countries do not have the space to get the protection of their rights in the host countries. In 2011, SAARC Charter of Democracy[188] reiterates 'faith in fundamental human rights and the human dignity as ordained in the Universal Declaration of Human Rights (UDHR) and the SAARC Constitutions. Moreover, the SAARC Social Charter advocates a comprehensive commitment to advance the social and economic welfare of the people of the region. Thus, the concept of climate refugee and its concept and definition dimensions become germane in the current context.

---

[182] Chandan Kumar Sharma, 'The Immigration Issue in Assam and Conflicts Around it' (2012) 13(3) Asian Ethnicity 287–309.

[183] Bhattacharyya and Werz (162) 29.

[184] Ibid.

[185] Chirantan Kumar, 'Migration and Refugee Issue between India and Bangladesh' (2009) 1(1) Scholar's Voice: A New Way of Thinking 64–82, 76.

[186] Rajeev Sharma and Ankya Bhushan, 'Illegal Immigration of Bangladeshis into India: National Security Challenges' (23rd World Congress of IPSA, Montreal, 19–24 July 2014) 1–24, 12 http://paperroom.ipsa.org/papers/paper_42931.pdf accessed 20 April 2018.

[187] Asian Development Bank, *Addressing Climate Change and Migration in Asia and the Pacific* (Asian Development Bank Manila 2012).

[188] SAARC Charter of Democracy (adopted February 2011) http://globalsummitryproject.com.s197331.gridserver.com/archive/saarc/saarc-sec.org/SAARC-Charter-of-Democracy/88/index.html# accessed 23 April 2018.

2.9 Climate Refugee: Concepts, Definition and Its Dimensions

Currently, there are many perspectives on what constitutes climate migration/displacement and how it is defined; the terminology which is found scattered throughout the literature that ranges from 'climate-induced migrant', 'climate displaced' to 'climate refugees'.[189] Each definition carries with it a specific connotation, and there is no universally established categorization of such people who are forced to relocate and flee because of human-induced environmental/climate degradation.[190]

The term 'environmental refugee' was first conceived in the 1970s by Lester Brown of the World Watch Institute, the term, gained popularity in the 1990s.[191] The popularity of the terminology owes a great deal to the works of prominent legal experts who have explored the potential human tragedy and attempted to provide normative contours to the concept of environmental/climate displacement.

Essamel Hinnawi in the year 1985, defined the 'environmental refugees' as:

> ...those people who have been forced to move or leave their traditional habitat, temporarily or permanently, because of a marked or conspicuous environmental disruption (natural and/or triggered by people) that jeopardized and imperiled their existence and/or seriously affected the quality of their life.[192]

Three main categories of climate-induced migrants have been identified by him:

1. Temporarily displaced migrants who get an opportunity to move back to their original home after the *status quo* has been restored;
2. Permanently displaced individuals who resettle somewhere else;
3. And migrants' leaving their original home because it gets degraded to such a level that it no longer remains habitable.[193]

Norman Myers has defined the term 'environmental refugees' as:

> 'people who can no longer get or gain a secure livelihood (or living)in their homelands because of soil erosion, drought, deforestation, desertification and other environmental problems, together with associated and incidental problems of population pressures and profound poverty. In their desperation or despair, these people feel they have no alternative or option but to seek sanctuary elsewhere, however, hazardous the attempt. Not all of them have fled

---

189Frank Biermann and Ingrid Boas, 'Preparing for a Warmer World: Towards a Global Governance System to Protect Climate Refugees' (2010) 10(1) Glob. Environ. Politics, 60–88.

190Angela Williams, 'Turning the Tide: Recognizing Climate Change Refugees in International Law' (2008) 30 Law & Poli'y 502–529, 516.

191James Morrissey, "Rethinking the 'Debate on Environmental Refugees': From 'Maximilists and Minimalists' to 'Proponents and Critics.'" (2012) 19(1) JPE 36–49; See also Camillo Boano (n 76) 7.

192Essam El-Hinnawi, *Environmental Refugees* (United Nations Environment Programme 1985) 4; See Boano (n 76) 7.

193Ibid, Hinawi.

or left their countries, many being internally displaced. But all have abandoned their homelands on a semi-permanent if not permanent basis, with no or little hope of a foreseeable return.'[194]

Following these expositions, usage of this term is on the rise despite there being no agreement regarding its use under international law.[195] The term has been increasingly questioned on the basis that it leads to a reductionist view of the complexity of real-life situations.[196] The people who got affected by the climate change do not want to be recognized as refugees due to the inherent vulnerability and despondency in such terminology.[197] These definitions do not make a distinction between migration/displacement taking place internally and cross-border. Any attempt to define by including group migration of climate refugees could be construed as a circumvention of the municipal law. For instance, "the United Nations Guiding Principles on Internal Displacement, 1998 indicates that the "'national authorities' have the fundamental duty and responsibility to extend protection and humanitarian assistance to the IDPs within their jurisdiction."[198] By avoiding encroachment on internal displacement, some commenters have defined the group barely to concentrate only on cross-border displacement. For instance, the nonbinding Nansen Initiative concentrates absolutely on cross-border climate change migration.[199] Focusing exclusively on cross-border displacement limits the scope that is inherent within the ambit of the developed world with their incidental security obsessions including xenophobic narratives which might be entailed in the form of reactions.[200]

The UNHCR has refrained from using the term and has cautiously defined 'environmentally displaced persons', instead, in the following words:

'who are displaced from or who feel obliged to leave their usual place of residence, because their lives, livelihoods and welfare have been placed at serious risk as a result of adverse environmental, ecological or climatic processes and events'[201] The UNHCR intended to avert ambiguities and confusions without making any mention

---

194 Norman Myers, 'Environmental Refugees: a Growing Phenomenon of the 21st Century' (2002) 357 Philos Trans R Soc Lond B Biol Sci. 609–611; See also Boano (n 76) 7.

195 Koko Warner, 'Global Environmental Change and Migration: Governance Challenges, Global Environmental Change' (2010) 20(3) Sci. Direct 402–413.

196 Cecilia Tacoli (n 66) 516–517.

197 Colette Mortreux, Jon Barnett, 'Climate Change, Migration and Adaptation in Funafuti, Tuvalu' (2009) 19(1) Global. Environ. Chang. 105–112.

198 See UNHCR Guiding Principles on Internal Displacement (11 February 1998) UN Doc E/CN.4/1998/53/Add.2.

199 The Nansen Initiative, 'Agenda for the Protection of Cross-Border Displaced Persons in the Context of Disasters and Climate Change' (The Nansen Initiative 2015 (1) https://nanseninitiative.org/wp-content/uploads/2015/02/PROTECTION-AGENDA-VOLUME-1.pdf accessed 16 April 2018.

200 David Hodgkinson, Tess Burton, Heather Anderson and Lucy Young, 'The Hour When the Ship Comes in: A Convention for Persons Displaced by Climate Change' (2010) 36 Monash Univ L Rev 69–120.

201 Jerry I.-H. Hsiao, 'Climate Refugee and Disappearing states: In Need for a New Legal Regime?' (2017) 5(5) Cul. Rel. Stud., 268–276, 270.

of cross-border migration and displacement owing to persecution based on armed conflict and human rights transgressions.[202]

About climate change, though the term has frequently been employed to convey a situation of urgency, the term is not strictly accurate under the current international law relating to refugee protection and climate change. Under international law, no specific definition of a climate refugee exists, and United Nations has also not endorsed its usage. Further, recognition has also not been granted to climate refugees under any other category of refugees under international legal framework about refugee protection of climate change.[203]

With an increase in the number of such people, it has, become important to understand the legal connotation and grant legal protection to them. Attempts have been made at domestic level in certain countries to acknowledge the rights of such people, but the same has not been much of a success. In 2014, legal status was granted to a family of displaced people by the New Zealand Immigration and Protection Tribunal.[204] The family had migrated from a Small Island Nation called Tuvalu, which is under a threat of receding into the ocean because of rising sea levels, and made legal claims, calling themselves protected people, claiming humanitarian grounds. The court kept into consideration the harm which the young members of the family could have suffered because of climate change while giving the decision.[205] The decision, however, have not been consistent with the other rulings and the High Court of New Zealand which gave a verdict against recognizing a family from Kiribati in the year 2015.[206] This decision has been delivered despite the fact that the circumstances existing in this island nation are so dire that the government is making Plans to evacuate the entire country for the survival of its people. The family, could not prove persecution under the law and hence was deported back.[207]

It becomes important to note here that the UNCSR and the 1967 Additional Protocol on the UNCSR do not have explicit provisions for granting protection to populations affected by environmental/climate factors despite it being a well-established international treaty. Even IDPs have not been recognized under the UNCSR, but they have been protected under the 1998 United Nations Guiding Principles on Internal Displacement. Article 1 of the UNCSR defines the term 'Refugee', and mentions the need of 'well-founded fear of persecution without which refuge cannot be granted under the convention. This prima facie appears to be the sentiment of a single individual. At the same time, the impugned expression can be interpreted to apply it

---

[202]Boano (n 76) 10.

[203]Ibid, 10–11.

[204]*AD (Tuvalu)* [2014] NZIPT 501370-371 (New Zealand).

[205]Carol Farbotko. 'Representation and Misrepresentation of Climate Migrants', in Benoit Mayer and Francois Crepeau (eds), *Research Handbook on Climate Change, Migration and the Law* (Edward Elgar Publishing 2017) 79.

[206]*Teitiota v Chief Executive of the Ministry of Business, Innovation and Employment* SC7/2015 [2015] NZSC 107 (Supreme Court of New Zealand).

[207]Kimberley Curtis, 'Climate Refugees: Explained' (*UN Dispatch*, 24 April 2017) www.undispatch.com/climate-refugees-explained/ accessed 27 December 2017, see Carol Farbotko (n 205) 79–80.

to a whole group of individuals and history has seen such occasions in the form of persecution of Jews in Nazi Germany. This means that the article 1, when interpreted in this manner,[208] would mean that the definition is not capable of covering environmentally displaced individuals.[209] Nation-states, are not very keen on giving refuge to individuals who get displaced due to environmental factors because of the large numbers, which get affected due to such disasters and gradual climate changes. Therefore, such individuals prefer to shift to areas within their own country, which are habitable. It is pertinent to note that the Guiding Principles on Internal Displacement also have not addressed all the environmental/climate factors, which have an impact on displacement.[210]

Under the climate change legal framework, climate displacement has been catalogued as part of adaptation strategy. Paris Agreement has not used the term 'climate refugees' and instead employed the term 'human mobility'. Therefore, Paris Agreement deliberated the migration due to environmental factors, and it has incorporated the three crucial elements for environmental displacement and human mobility issues:

1. In the Agreement's preamble, climate change has been recognized as a common concern of humanity, and it contains a reference to migrants. Parties have been asked to respect, promote and consider their respective obligations towards migrants while addressing climate change-related issues.[211]
2. The main text of the Agreement has incorporated many references to the protection of climate migrants and has also highlighted the importance, which their livelihoods deserve.[212]
3. Under the Agreement, a request has been made to the Executive Committee of the Warsaw International Mechanism (WIM) on Loss and Damage for the establishment of a task force on displacement. The task force has been authorized to execute the responsibility of 'developing recommendations for integrated approaches to obviate, minimalize and address human displacement relating to the adverse impacts of climate change'.[213]

---

[208]Under Article 1 (2) of the UNCSR, the term "refugee" shall apply to any person who 'As a result of events occurring before 1 January 1951 and owing to well-founded fear of being persecuted for reasons of race, religion, nationality, membership of a particular social group or political opinion, is outside the country of his nationality and is unable or, owing to such fear, is unwilling to avail himself of the protection of that country; or who, not having a nationality and being outside the country of his former habitual residence as a result of such events, is unable or, owing to such fear, is unwilling to return to it'.

[209]Laura Westra, *Environmental Justice and The Rights Of Ecological Refugees* (Earthscan 2009) 5; Also see Jayawardhan (n 73) 106.

[210]Koko Warner (n 195) 404.

[211]Paris Agreement (n 29) Preamble.

[212]Ibid; UNHCR 'COP 22 Technical Inputs: Climate & Disaster Displacement' (Marrakesh 7–18 November 2016) www.unhcr.org/583455267.pdf accessed 27 April 2018.

[213]UNHCR, 'Frequently Asked Questions on Climate Change and Disaster Displacement' (*UNHCR*, 6 November 2016) www.unhcr.org/news/latest/2016/11/581f52dc4/frequently-asked-questions-climate-change-disaster-displacement.html accessed 27 December 2017.

It can aptly be put forward that the climate change migration and displacement have severely affected the SAARC region in the last few decades. In South Asia, the climate change migration, displacement and policy paradigms are on the lowest pedestal of national priority due to the absence of public awareness regarding the interconnectedness between climate change, migration and displacement. But, unfortunately, there is no definition of the term 'climate refugees' under the refugee law and climate law framework. The contemporary refugee law and climate law scholarship, experts and policymakers have elaborated the idea of climate refugees in different manners that are deviant to the international human rights protection framework. The most prominent concern in the refugee protection arena is the protection of climate refugees in the existing international legal order. Therefore, the SAARC as a regional organization must conceive, evolve and develop a legally binding and ethically attributable understanding on climate change migration and human displacement and must address the implicit and explicit challenges and the same will be attended to in the chapter *infra*.

2.10 Conclusion

Climate change has been pervading and impacting all continents, countries and regions of the world including South Asia internationally. It has the potential of exacerbating migration and displacement at an unprecedented magnitude. South Asia is a climate change-prone region, and its vulnerability has become more conspicuous than ever before due to countries like the Maldives, Sri Lanka, Bangladesh, Nepal and India. The people of SAARC region has experienced cataclysmic climate disasters such as tsunami in 2005 and the scorching heat-waves in 2015 that have made them the victim and vulnerable both at the same time. The Maldivian islands and Sundarban Mangrove forests are on the verge of submergence, and some Himalayan glaciers are entirely at the melting point. Afghanistan has been undergoing a major humanitarian crisis, and above all, natural forces have also been equally relentless and merciless, and climate change jeopardized its existential survival. 10% Afghan population has been experiencing the worst drought since 1971, and during 1980–2015, it has witnessed a record number of human casualties owing to natural calamities after Haiti. In Pakistan, there is a lot of climate mayhems that keep 3 million people of Pakistan on the verge of starvation, and the majority of them belong to rural areas of Pakistan. Thus, hundreds of thousands of people have abandoned the regions situated in the Southern Thar Desert that was once home to more than 1 million people due to the annihilation and destruction of livestock and crops. These developments have the potential to disturb the survival and existence of the people in the region significantly and do not leave any option for the people but to migrate to safer places.

There is an emerging understanding that migration and displacement have been construed as one of the coping and adaptation strategies of climate change. Even the fourth IPCC report assessed that growing number of people have now preferring to adapt the current climate conditions by migrating to safer areas. Therefore, in South Asia, a vast majority of the people may prefer to relocate in the wake of climate change, which can turn into a humanitarian crisis. The topic has attracted global scholarship, CSOs and international organizations like the IPCC, IOM and the UNHCR to study the climate change. The climate change displacement process is complex. The scenario ranges from internal to external and temporary to permanent. Many among the climate refugees urge to return to their roots after the effects of the disaster have diminished. It is evident from the Bangladeshi climate change patterns where the majority of the people preferred to return to their roots, in contrast, to merely 10–25% population that permanently stayed back in their new settlements in urban areas. Therefore, climate change migration patterns are propelled by the governmental policies that also cast repercussions on the economy and climate conditions of the regions that host such climate refugees.

The South Asian regions have been enduring the ill-effects migration in the form of socio-economic degeneration, violence and conflict. India, the largest country in South Asia, has been confronting the problem of illegal migrants in the state of Assam from Bangladesh who have purportedly been reported to have been creating problems and have become a security threat to the unity, integrity and internal security of India. This situation can become grave in the wake of climate change. The UNHCR correctly stated, 'climate change germinates the seeds of conflict, but it also causes human displacement much worse when it occurs'. In South Asia, climate change migration aggravated by factors have been appreciated by the IOM while taking into consideration the various causes of violence and security. But, there is a need to adopt a holistic approach that addresses the multitude of issues altering the lives of the people in the region due to climate change, such as human security narratives, state concerns, conflict paradigms, livelihood strategies and socio-economic development models. In the SAARC region, CDPs usually prefer to relocate themselves within the country of residence instead of crossing the international borders.

The concept and understanding of 'climate refugees' have not been appreciated at the international level within the current refugee and climate law architecture. There is no explicit or implicit incorporation, recognition or understanding in the UNCSR regarding the protection of climate refugees, CDPs or climate migrants despite their worldwide mounting emergence. Some efforts have been made in the Paris Agreement where under protection to avert, and minimise climate refugees has been contemplated, but that does not adequately cater to their needs. Nevertheless, it must also be kept in mind that the SAARC region has an inherent strength and potential of having a wealth of natural resources, indigenous knowledge and native expertise for the management of climate adaptability and mutability that requires being ensconced and embraced to attend the issue of climate change migration and displacement and other incidental problems thereto or connected therewith in the region. Therefore, it has become imperative to critically analyse the emerging climate change challenges in South Asia for averting a crisis like situations that might ascend in the future. There are improbabilities and reservations regarding the climate change impact on migration trends, but the climate change is still invisible and obscure in the contemporary South Asian climate change discourse.

Chapter 3
Climate Change Displacement and Refugees: 'Normative Debate'

Abstract The mass human displacement caused by climate change and its ramifications have generated and gestated the normative debate about climate change-induced displacement. The climate change is destined to emplace new humanity of displaced populations in future, and it has been exacerbating the current human displacements as well. The stability of humanity is bound to be adversely affected and socioeconomic, politico-cultural and lego-institutional drivers, all at the same time. Therefore, climate change has been producing refugees in multiple manners, compelling people to relocate or migrate from their homes, contributing to the emergence of conflict situations and jeopardizing human security dimensions. Such scenarios are bound to shrink the natural resources, make reserves scarce and inaccessible for the present and posterity. Hence, there are questions regarding the climate change-induced displacement, identification of contours of the normative debate regarding human mobility and recognizing the climate change rights as human rights. Therefore, the instant chapter dwells upon the different approaches, concepts and debates in underscoring the climate refugees in the absence of international, regional and national laws and policies on them. It critically analyses the interfaces of minimalists and maximalists on climate change-induced displacement and its theoretical constructs while rummaging the core legal norms and issues of constitutional obligations within the applicable legal regime to climate refugees. The present chapter visits climate migration from human rights dimension and articulates the place of rights in transition while determining their nature as human rights-based normative debate.

3.1 Introduction

It has become an annual phenomenon where reports highlight millions of people who are compelled to flee from their roots, homelands and countries owing to droughts, earthquakes, floods, windstorms and other multiple ravages of climate change. In 2016 alone, there were an estimated 23.5 million people forced to flee due to extreme

S. Jolly and N. Ahmad, *Climate Refugees in South Asia*, International Law and the Global South, https://doi.org/10.1007/978-981-13-3137-4_3

weather-related catastrophes.[1] The concepts of climate change-induced displacement, migration or climate refugees are sporadically scattered in the writings of scholars.[2] In the contemporary circumstances, climate change migration has presented a cluster of causes that have made the task of identifying climate refugees and locating the legal place for them extremely difficult. Multiple narratives are prevalent in the discourse on climate-induced displacement. At one extreme, climate-induced displacement is considered as a manifestation of 'climate injustice' wherein the vulnerable population from the Global South are victimised by climate change founded on the socio-political discrimination. The injustice narrative emphasis on the responsibility of Global North for the protection and assistance to climate-displaced population. The other side of the discourse project climate migration as a positive strategy of adaptation needs to be promoted with earnest measures. The studies and reports also call for greater focus on climate change as a driver of security-relevant migratory flows. The discourses range from projecting the issue as serious 'crisis of nature' to rely on the complex inclusive set of concepts needed to holistically and affectively approach the characterisation of climate displacement.

The chapter maps out the various approaches and conceptions of climate refugees. The chapter critically analyses the debate of minimalists and maximalists on climate displacement. The analysis in this chapter aims to outline the theoretical foundations of the subject, before moving on to identify the core legal principles and legal regime applicable to solve the constitutional issues raised by climate migration and their evolving concept of climate refugees. The chapter looks at the point of climate migration from a human rights dimension and discusses whether rights violated during climate migration are a human rights issue. The basic question is without a definition and empirical consensus, is there a specific problem that needs attention?

3.2 Climate Change-Induced Displacement: Myth or Reality?

The contemporary debate on refugees has been environmentally silent, climatically casual and regionally cynical in South Asia and elsewhere in the absence of credible empirical evidence. If the first climate change-induced displacements and forced migration of people are witnessed in the Lateu village in Vanuatu, the Shishmaref village in the Sarichef Island in Alaska, the Carteret Islands in Papua New Guinea

[1]Julie-Anne Richards and Simon Bradshaw, 'Uprooted by Climate Change: Responding to the Growing Risk of Displacement' (2017) Oxfam International Briefing Paper, 10 https://www.oxfam.org/sites/www.oxfam.org/files/file_attachments/bp-uprooted-climate-change-displacement-021117-en.pdf accessed 25 June 2018.

[2]Olivia Dun and François Gemenne, 'Defining "Environmental Migration"', (2008) FMR 31 10–11 http://www.fmreview.org/climatechange/dun-gemenne.html accessed 25 June 2018.

(PNG) and Lohachara Island in the Hooghly River in India,[3] the conflict in Darfur has displaced more than 4 million people, and 200,000 refugees in Chad have posed an ecological disaster due to the partial climate change.[4]

The report of the IPCC has highlighted colossal impact of climate change on human displacement and migration that would trigger the displacement of millions of population.[5] Ever since the IPCC report underscored the impact of climate change migration, the impacts of other drivers of climate change-induced human displacement and migration have become ever more eclipsed.[6] The IPCC report projected that millions of people could be on the track of mobility that would provide the people an escape from the most pernicious and horrendous impacts of climate change.[7] However, climate change migration would also provide people with an opportunity to stabilize their stay in the new homes and destinations and prevent a cycle of migration.[8] Therefore, such an expansion of migration would reduce the vulnerability of the people on the move and change trends in the migration. This could be responses to both worst climate events and elongated climate variability and change that might make migration an efficacious adaptation strategy[9] based on what we call 'prognostic migration'.[10] The current IPCC report emphasizes the complication in predicting the

---

[3]Oli Brown, 'Climate Change and Forced Migration: Observations, Projections and Implications' (2007) UNDP's Background Paper for the 2007 Human Development Report 2007/17, 16–17 http://hdr.undp.org/sites/default/files/brown_oli.pdf accessed 25 June 2018.

[4]Ban Ki Moon, 'A Climate Culprit in Darfur' *Washington Post* (16 June 2007) http://www.washingtonpost.com/wp-dyn/content/article/2007/06/15/AR2007061501857.html?noredirect=on accessed 25 June 2018.

[5]Brown (n 3) 13–14.

[6]Jon Barnett and Michael Webber, 'Accommodating Migration to Promote Adaptation to Climate Change' (2010) World Bank's Policy Research Working Paper 5270, 22–26 http://documents.worldbank.org/curated/en/765111468326385012/pdf/WPS5270.pdf accessed 25 June 2018.

[7]Adger, W. Neil et al. Human security. In: Field, Christopher B. et al. (Eds.) Climate Change 2014: Impacts Adaptation, and Vulnerability. Part A: Global and Sectoral Aspects. Contribution of Working Group II to the Fifth Assessment Report of the Intergovernmental Panel on Climate Change, (Cambridge University Press: 755–791). Helen Davidson and Adam Vaughan, 'Climate Change Report: 'The Worst is Yet to Come'—As it Happened' *The Guardian* (31 March 2014) https://www.theguardian.com/environment/2014/mar/31/climate-change-report-ipcc-governments-unprepared-live-coverage accessed 28 May 2018.

[8]Robert A. McLeman and Lori M. Hunter, 'Migration and Adaptation to Climate Change' (2009) Wiley Interdisciplinary Reviews-Climate Change, Population Program POP2009-14, 1–3 http://citeseerx.ist.psu.edu/viewdoc/download?doi=10.1.1.576.8279&rep=rep1&type=pdf accessed 25 June 2018.

[9]Working Group II, 'Climate Change 2014: Impacts, Adaptation, and Vulnerability, Part A: Global and Sectoral Aspects' Fifth Assessment Report of the Intergovernmental Panel on Climate Change, http://www.ipcc.ch/pdf/assessment-report/ar5/wg2/WGIIAR5-PartAFINAL.pdf accessed 25 June 2018.

[10]The idea of 'prognostic migration' denotes that human migration is taking place in advance on the basis of an assessment of extreme climate change impacts that are preferred, projected and presented as adaptation and mitigation strategies instead of waiting for sudden forced displacement due to the subsequent climate change impacts.

real number of people who might migrate due to climate change. These complications are evident from the differing predictions on the extent of climate displacement.

The predictions on climate change displacement[11] rendered by Professor Norman Myers of Oxford University contend that global warming would cause displacement of 200 million people by the year 2050 due to the disturbances in monsoon systems, sea-level rise, coastal floods and droughts.[12] The same has been supported by the estimates made by Biermann and Boas,[13] and Faist and Schade.[14] The estimates ranging from the complex interrelationship between climate change and migration have triggered an alarmist approach to conservative calculations. For instance, the IOM's acknowledgement of insufficient current knowledge[15] about the multicausal nature of climate change migration represents a challenge in identifying environmental triggers of human mobility that substantially differs from the Hugo's proposition of 'from totally voluntary migration to totally forced migration'[16] where reality is in limbo if it is measured against the Black's conceptualization[17] of climate change migration factors resulting into implications for the legal frameworks. Through these predictions, which substantially differ from each other, the international community gradually recognizes that extreme environmental degradation and climate change might result in large-scale human displacement and migration for which global population is not fully prepared.[18] Therefore, present evidence on record alludes to climate change-induced displacement and migration in a scenario where the debate regarding climate-induced migration has been led by its future dynamics and repercussions. It has posed the question whether the prognostications about climate-induced migration and displacement are true or false?

---

[11]Walter Kalin, 'Conceptualising Climate-Induced Displacement' in Jane McAdam (ed), *Climate Change and Displacement: Multidisciplinary Perspectives* (Hart Publishing, Oxford, 2010) 81–103.

[12]Norman Myers, 'Environmental Refugees: An Emergent Security Issue' (13th Economic Forum, Prague, 23–27 May 2005) www.osce.org/documents/eea/2005/05/14488_en.pdf accessed 31 May 2018.

[13]Frank Biermann and Ingrid Boas, 'Preparing for a Warmer World: Towards a Global Governance System to Protect Climate Refugees' (2010) 10(1) Glob. Environ. Politics 60–88 https://www.researchgate.net/profile/FrankBiermann/publication/227627225 PreparingforaWarmerWorldTowardsaGlobalGovernanceSystemtoProtectClimateRefugees/links/5422bc260cf238c6ea6b866f/Preparing-for-a-Warmer-World-Towards-a-Global-Governance-System-to-Protect-Climate-Refugees.pdf accessed on 25 May 2018.

[14]Thomas Faist and Jeanette Schade, 'The Climate-Migration Nexus: A Reorientation' in *Disentangling Migration and Climate Change* (Springer Netherlands, 2013) 3–25.

[15]Frank Laczko and Christine Aghazarm, *Migration, Environment and Climate Change: Assessing the Evidence* (International Organisation for Migration, 2009) 17.

[16]Susan F. Martin, Environmental Change and Human Mobility: Trends, Law and Policy (2017) 42 Comparative Population Studies 187–218, Graeme Hugo, 'Environmental Concerns and International Migration' 1996 30(1) Int'L Migration Rev 105–131.

[17]Richard Black, 'Environmental Refugees: Myth or Reality?' (2001) UNHCR Working Paper 34 http://www.unhcr.org/research/working/3ae6a0d00/environmental-refugees-myth-reality-richard-black.html accessed on 25 June 2018.

[18]Andrew Morton, Philippe Boncour and Frank Laczko, 'Human Security Policy Challenges', (2008) 31 FMR 5–6 http://www.fmreview.org/sites/fmr/files/FMRdownloads/en/FMRpdfs/FMR31/05-07.pdf accessed 25 June 2018.

3.3 Identifying the Contours of Normative Debate on Climate Refugees

The climate change human migration is a formidable challenge for the humanity. The projections for mass migrations owing to the impacts of climate change would alter the roots of 200 million[19] people by 2050 in an astounding transformation that has crafted a new category of displaced people called 'climate refugees' that does not have a well-settled definition. Even almost three decades ago, IPCC put to record the growing humanitarian apprehensions[20] regarding the ever-expanding lexicon of climate change nomenclature 'climate refugees' within the global humanitarian community.

The concept of 'climate refugee' is problematical and debatable, as the concept does not have legal status under the international refugee law or 1951 Refugee Convention (UNCSR) with its 1967 Additional Protocol.[21] It is also debatable because there is a scant understanding among the scholars as to how to attend and address the problems and difficulties it poses. There are many perspectives on what a climate migrant is how it is defined; this terminology ranges from 'environmentally induced migrant', to 'environmental refugee' to the more recent 'climate change refugee'.[22]

Historically speaking, the environment and migration have been premised within the architecture of the absence and presence and the notion of hiding and seek.[23] Initially, scholars and experts of environmental change perceived environment as a triggering factor of human mobility. These factors included natural disasters, industrial accidents and development projects.[24] A Unied Nations Environment Programme (UNEP) study led by Hinnawi substantiated these dimensions and gave the world the definition of environmental refugees.

"Environmental refugees are those people who have been forced to leave their traditional habitat, temporarily or permanently, because of a marked environmental

---

[19]Norman Myers 'Environmental Refugees: A Growing Phenomenon of the 21st Century' (2002) Philosophical Transactions Royal Society: Biological Sciences 609–13 https://www.ncbi.nlm.nih.gov/pmc/articles/PMC1692964/pdf/12028796.pdf accessed on 25 May 2018.

[20]Giovanni Bettini, Sarah Louise Nash and Giovanna Gioli, 'One Step Forward, Two Steps Back? The Fading Contours of (in)justice in Competing Discourses on Climate Migration' (2016) Geogr. J, 1–11 http://lib.icimod.org/record/32247/files/HiAwareBettini_et_al16.pdf accessed on May 21, 2018.

[21]Bruce Burson, 'Protecting the Rights of People Displaced by Climate Change: Global Issues and Regional Perspectives' in Bruce Burson (ed), *Climate Change and Migration: South Pacific Perspectives* (Institute of Policy Studies, 2010) 150–179.

[22]Biermann and Boas (n 13) 61.

[23]Ettiene Piguet, 'From "Primitive Migration" to "Climate Refugees": The Curious Fate of the Natural Environment in Migration Studies' (2013) 103(1) Annals of the Association of American Geographers 148–162.

[24]Graeme Hugo, 'Climate Change-Induced Mobility and the Existing Migration Regime in Asia and the Pacific' in Jane McAdam (ed.), *Climate Change and Displacement: Multidisciplinary Perspectives* (Bloomsbury Publishing, 2010) 9–37.

disruption (natural and/or triggered by people) that jeopardized their existence and/or seriously affected the quality of their life."[25]

The climate change as a trigger of human mobility shot into prominence in the 1980s,[26] when the global security dimensions were in a significant redefinition.[27] In the USA, initial discourse on environmental change was recognized as a significant driver of climate change that recorded in the work of Thomas Homer-Dixon[28] who established a relationship between environmental change, migration and conflict and cautioned about the 'waves of environmental refugees'[29] and subsequently Robert Kaplan propagated these notions,[30] supported by Norman Myers and Jennifer Kent.[31]

Myers and Kent, in their definition of environmental refugees, climate change was added as a triggering factor. They defined 'environmental refugee' as 'people who can no longer gain a secure livelihood in the homelands because of drought, soil erosion, desertification and other environmental problems, together with associated problems of population pressures and profound poverty'.[32]

Bierman and Boas who categorized a separate class of climate refugees advanced the previous definitions. Climate refugees were defined as 'people who have to leave their habitats, immediately or in the near future, because of sudden or gradual alterations in their natural environment related to at least one of three impacts of climate change: sea-level rise, extreme weather events, and drought and water scarcity'.[33]

Asian Development Bank (ADB) report on Climate Change and Migration in Asia and the Pacific[34]does not provide exact definition of the terms 'climate migrant' or 'climate refugee'; rather, it provides an understanding couched in general words of expressions like 'displaced persons', 'economic displacement' or 'physical dis-

[25]Essam El-Hinnawi 'UNEP: Two Decades of Achievement and Challenge' (UNEP Nairobi, 1992) 4 https://files.eric.ed.gov/fulltext/ED386388.pdf accessed on 25 June 2018.

[26]Andrew Baldwin, Chris Methmann and Delf Rothe, 'Securitizing 'Climate Refugees': The Futurology of Climate-Induced Migration' 2(2) Critical Studies on Security 121–130.

[27]Richard Ullman, 'Redefining Security', (1983) 8(1) Int'l Security 129–153 https://is.muni.cz/el/1423/jaro2017/BSS177/um/Ullman_Redefining_Security.pdf accessed on 25 June 2018.

[28]Thomas Homer-Dixon, 'On the Threshold: Environmental Changes as Causes of Acute Conflict' (1991) 16(2) International Security 76–116 https://www.jstor.org/stable/2539061?seq=1#page_scan_tab_contents accessed on 25 May 2018.

[29]Ibid.

[30]Robert D. Kaplan, 'The Coming Anarchy' (1994) 273(2) The Atlantic Monthly 44–76 https://www.theatlantic.com/magazine/archive/1994/02/the-coming-anarchy/304670/ accessed on 28 May 2018.

[31]Norman Myers and Jennifer Kent, 'Environmental Exodus: An Emergent Crisis in the Global Arena' (Climate Institute, Washington DC, 1995) http://climate.org/archive/PDF/Environmental%20Exodus.pdf accessed 15 May 2018.

[32]Myers, 'Environmental Refugees' (n 19) 609.

[33]Biermann and Boas (n 13) 62.

[34]Asian Development Bank, 'Addressing Climate Change and Migration in Asia and the Pacific' (ADB, 2012) https://www.adb.org/sites/default/files/publication/29662/addressing-climate-change-migration.pdf accessed 18 April 2018.

placement'. Among these formulations,[35] the 'displaced persons' terminology is more akin to a situation where people got displaced due to the hostile environmental circumstances.[36] As per the ADB report, approximately 37 million people from India, 22 million from China and 21 million from Indonesia will be at risk from rising sea levels by the year 2050.[37] It establishes a comprehensive account of the potential impacts of climate change-induced displacement and migration in Asia and the Pacific while providing an informed analysis of a growing trend that demands the immediate attention of the national governments and global community.[38]

The IOM attempted to formulate a working definition of 'environmental migrants'[39]; IOM's working definition runs as under:

> Environmental migrants are groups or persons of persons who, predominantly for reasons of progressive or sudden change in the environment that adversely affects their lives or living conditions, are obliged to leave their habitual homes, or choose to do so, either temporarily or permanently, and who move either within their country or abroad.[40]

It may aptly be gleaned from the working definition that IOM definition captures the paradoxes and complexities of the climate change and natural disasters that are the major factors of human displacement and migration. However, climate change prognosis and predictions for the current century indicate that majority of the people expected to be on the move due to climate change-induced displacement[41] resulting in the forms of extreme precipitations and temperatures that will also impact the human livelihoods in a multiple order. Therefore, the IOM attempts to discourage the use of expressions 'environmental or climate change refugee' that has been there in the media for quite some time.[42] The IOM contends that environmentally induced migration is rampant in the developing countries[43] making climate change move-

---

[35] Karen Elizabeth McNamara, 'Conceptualizing Discourses on Environmental Refugees at the United Nations' (2007) 29(1) Population and Environment 12–24.

[36] Asian Development Bank, 'Environment Operational Directions 2013–2020: Promoting Transitions to Green Growth in Asia and the Pacific' (ADB, 2013) https://www.adb.org/sites/default/files/institutional-document/33869/environment-operational-directions-2013-2020.pdf accessed on 23 May 2018.

[37] Ibid.

[38] John Campbell, 'Climate Change and Population Movement in Pacific Island Countries' in Bruce Burson (ed), *Climate Change and Migration: South Pacific Perspectives* (Institute of Policy Studies, 2010) 29–50.

[39] International Organisation for Migration, 'Moving to Safety: Migration Consequences of Complex Crises' (2012) IOM's International Dialogue on Migration 20, 10–12.

[40] International Organisation for Migration, *Glossary on Migration* (IOM, 2011) 33.

[41] Ben Saul and Jane McAdam, 'An Insecure Climate for Human Security? Climate-Induced Displacement and International Law' in Alice Edwards and Carla Ferstman (eds), *Human Security and Non-Citizens* (Cambridge University Press, Cambridge, 2009) 357–403.

[42] Roger Zetter, 'Legal and Normative Frameworks', (2008) 31 FMR 62–63 http://www.fmreview.org/sites/fmr/files/FMRdownloads/en/climatechange/zetter.pdf accessed 25 June 2018.

[43] International Organization for Migration, 'Migration, Climate Change and the Environment' (2009) IOM Policy Brief 4 https://www.iom.int/sites/default/files/our_work/ICP/IDM/iom_policybrief_may09_en.pdf accessed on 25 June 2018.

ments permanent[44] in the form of sudden-onset disasters and slow-onset impacts of environmental degradation,[45] climate change variability, and vulnerability.[46] The IOM further argues that it is problematic to portray the cause of human mobility or migration as distinctively climatic or environmental grounds.[47] There are many other factors such as sociopolitical, ecocultural and demographic dynamics that interacts with climatic or environmental factors. Human mobility could be voluntary or involuntary or forced. The IOM thus makes a distinction between 'displacement' and 'relocation' because the notion of displacement is perceived as a migration where people flee or leave their homes involuntarily[48] also known as forced migration owing to political turmoil or sudden-onset of environmental disasters. Thus, the IOM delineates 'relocation' as 'planned migration' that is executed by the administrative authorities at the national and local stages within the municipal hierarchy.

Primarily, the definitional dilemma is directly linked to the conceptualization and typologies of environmental migration, its estimates and forecasts, and the policy responses aimed at addressing it.[49] In the absence of a clear definition of 'climate refugees', it is not possible to identify which populations are of concern and in need of aid and assistance nor can precise estimations be made of the number of people who got displaced or occasioned to relocate due to environmental causes.[50] Therefore, the terms and typologies also matter for the populations themselves because of the messages and meaning they carry within the architecture of climate change-induced displacement and migration with legal protection options.

The issue of definition formulation may have different nomenclatures, but the point is that it must serve the physical protection and humanitarian objectives of 'climate refugees' or 'environmental migrants' as a separate category under the

[44]Catherine Phuong, *The International Protection of Internally Displaced Persons* (Cambridge University Press, Cambridge, 2005); Nils Geissler, 'The International Protection of Internally Displaced Persons' (1999) 11 IJRL 451–478.

[45]Steve Lonergan, 'The Role of Environmental Degradation in Population Displacement' (Spring 1998) 4 Environmental Change Security Project Report 5–15 https://www.oceanfdn.org/sites/default/files/The%20Role%20of%20Environmental%20Degradation%20in%20Population%20Displacement.pdf accessed 15 May 2018.

[46]Kyung-wha Kang, 'Climate Change, Migration and Human Rights' (conference on Climate Change and Migration: Addressing Vulnerabilities and Harnessing Opportunities, Geneva, 19 February 2008) https://newsarchive.ohchr.org/EN/NewsEvents/Pages/DisplayNews.aspx?NewsID=9162&LangID=E accessed 25 June 2018.

[47]Sarah Opitz Stapleton, Rebecca Nadin, Charlene Watson and Jan Kellett, 'Climate change, migration and displacement: The Need for a Risk-Informed and Coherent Approach' (United Nations Development Programme, Overseas Development Institute, 2017) 12–14 https://www.odi.org/sites/odi.org.uk/files/resource-documents/11874.pdf accessed 14 June 2018.

[48]Marta Bivand Erdal and Ceri Oeppen, 'Forced to Leave? The Discursive and Analytical Significance of Describing Migration as Forced and Voluntary' (2017) 44(6) J. Ethnic & Migr. Stud. 981–998, 982.

[49]Mostafa Mahmud Naser, 'Climate Change, Environmental Degradation, and Migration: A Complex Nexus' (2012) 36(713) Wm. & Mary Envtl. L. & Pol'y Rev 742–744.

[50]Ana Iglesias, Cynthia Rosenzweig, Gunther Fischer, Martin Parry, 'Millions at Risk: Defining Critical Climate Change Threats and Targets' (2001) 11 Global. Env'l Change 181–183.

existing mechanisms or new instruments. The contours of the 'normative debate' on 'climate refugees'[51] have to be identified by appreciating and understanding the theoretical framework of normative ethics[52] within which the climate refugees have been located regarding their human rights protection and humanitarian treatment in the host state. The normativity depicts the key characteristics of the theoretical framework that serves as the foundation of the expectations regarding the conduct of the SAARC countries in the international community. There are normative contours such as democracy, equality, human rights and social solidarity[53] that must form the foundation of 'South Asian regional consciousness' as an inherent part of state practices of the SAARC jurisdictions both at intra-South Asian (regional) and extra-South Asian (global) levels. The norms of reception of climate change refugees in the tradition of democratic pluralism and liberal cosmopolitanism[54] must serve the normative framework of protecting the climate refugees.

Therefore, the SAARC has to initiate policy reforms which are varied and proactive in providing humanitarian aid and assistance to refugees who flee from climate change-induced human displacement.[55] South Asian policy trajectory must prioritize the integration of the climate change refugees into local host populations[56] while adhering to the principles of diversity and multiculturalism in the host societies and must take into consideration geopolitical and geostrategic concerns and national economic agenda of the reception countries. These norms create an ethical dichotomy between the normative structures of the traditional nation-states who subscribe to the right to have national self-determination and advocacy of applicable global norms that might conflict with conventional norms of the nation-states, which address only the domestic rights.[57]

Therefore, the definition of 'climate refugee' in the present book is proposed as 'a climate refugee is a person forced to flee, leave or move his or her habitual residence or home and migrate or relocate himself or herself permanently or temporarily to a new country owing to the well-founded reasons of anthropogenic climate change, or sudden or gradual climate disturbances or uninhabitable conditions in their coun-

---

[51] Jessica Cooper, 'Environmental Refugees: Meeting the Requirements of the Refugee Definition' (1998) (6) N.Y.U. Envtl. L. J. 480–529.

[52] Ian Manners 'The Normative Ethics of the European Union' (2008) 84(1) International Affairs 45–60.

[53] Smruti Pattanaik, *South Asia: Envisioning a Regional Future* (Institute for Defence Studies and Analyses, New Delhi, 2011) 76–206.

[54] Daniel Bray, 'Pragmatist Ethics and the Will to Believe in Cosmopolitanism' (2013) 5(3) International Theory 446–476.

[55] The Nansen Conference, Climate Change and Displacement in the 21st Century, Oslo, Norway, 5–7 June 2011, 17 http://www.unhcr.org/4ea969729.pdf accessed 25 June 2018.

[56] George Martine, 'Population Dynamics and Policies in the Context of Global Climate Change' in José Miguel Guzmán George Martine, Gordon McGranahan, Daniel Schensul and Cecilia Tacoli (eds), *Population Dynamics and Climate Change* (International Institute for Environment and Development, 2009) 9–30, 26–27.

[57] Brian Opeskin, 'The Influence of International Law on the International Movement of Persons' (2009) UNDP's Human Development Reports Research Paper 2009/18 http://hdr.undp.org/sites/default/files/hdrp_2009_18.pdf accessed 22 May 2018.

try of origin'.[58] The proposed definition does not differentiate between voluntary or forced, internal (within the home country) or external (outside the home country) migration. The proposed definition is devoid of ethical commitments, moral imperatives, human rights framework,[59] and fundamental freedoms narrative of the people as these reasons are also the grounds of migration.[60] However, protection to 'climate refugees' must be contemplated on the premise of 2005 UN Responsibility to Protect Framework (R2P)[61] adopted at the UN World summit and agreed upon by all the nation-states to protect their populations from ethnic cleansing, genocide, war crimes and crimes against humanity as a collective responsibility while upholding the principle of non-intervention[62] in the municipal affairs of the state and the same is applicable to humanitarian and political refugees. Therefore, similar idea of protecting the 'climate refugees' must also be debated in any definition of 'climate refugee' while taking into account the negative impacts of climate change. Therefore, it may aptly be attributed that definition exercise from many quarters has created confusion and volatile state of understanding of who exactly is a 'climate refugee' that remains a distant goal and these definitions are not able to distinguish between environmental displacement and climate displacement.

3.4 Rethinking the 'Debate on Environmental Refugees': From 'Maximalists and Minimalists Perspectives'

A voluminous amount of the literature around the definition and contours of climate change refugees has been developed. However, most of the semantics used in the context are abstract and vague ranging from the alarmist predictions propounded by the media and politics to completely ignoring the milieu of climate change displacement. Venturing into eliciting a consensus on the topic is not only adventurous but also blurred between multiple stakeholders who bring heterogeneous perspectives, ideas, paradigms, methodologies and interests to the discourse.[63]

---

[58] It is the understanding of the authors to evolve and conceive a definition of 'climate refugee' while placing the same in the public domain to extract more clarity and certainty thereupon for a universal application.

[59] Brooke Havard, 'Seeking Protection: Recognition of Environmentally Displaced Persons under International Human Rights Law' (2007) 18 Vill. Envt'l L. J. 65–82.

[60] Benoît Mayer, 'Governing "Climate Migration"' in Walter Leal Filho (ed), *Handbook of Climate Change Adaptation* (Springer, Berlin, Heidelberg, 2014) 1–14.

[61] Sheri P. Rosenberg, 'Responsibility to Protect: A Framework for Prevention', Global Responsibility to (2009) 1 Protect 442–444.

[62] Charter of the United Nations (adopted 24 October 1945) 1 UNTS XVI (UN Charter) art 2(7).

[63] Silja Klepp, 'Climate Change and Migration: Policy, Politics, and Governance, Future Climate Change Scenarios', *Oxford Research Encyclopedia of Climate Change* (2017) 1–35 http://climatescience.oxfordre.com/view/10.1093/acrefore/9780190228620.001.0001/acrefore-9780190228620-e-42?print=pdf accessed 23 March 2018.

At this point, two dominant narratives characterize the entire debate on climate refugees and are divided into minimalists[64] and maximalists.[65] Minimalists take a position that has been attaining momentum in international policy paradigms as their critique resonates with human security[66] narrative emerging in the late 1990s, which gyrates around an individual away from nation-states. Minimalists consider the term 'environmental refugees' to be legally non-binding[67] and consider climate change only as one among the variable in the complex setting and points to the problematic aspects of directly attributing climate change to displacement and migration.[68] Their main ground of argument lies in simplifying the complex scenario of migration by overemphasizing on climate change.[69] For Castles, the term climate refugees are too 'simplistic, one-sided and misleading'.[70] Their opposition is also based on the fear of climate reductionism and determinism, which only highlights that individuals and communities are bound to their environmental settings, especially climate, and this dependency determines the trajectories of human and social development.[71] This determinist approach completely ignores the inherent resilience and adaptive capacity of societies to change in the wake of climate change.[72] They drew enormous support from the various empirical evidence and research studies from countries like Bangladesh wherein primary pattern of migration has observed to be of short distance and duration.[73] Media and public opinion may feel elated with the concept of refugees. However, minimalists object to the usage of the term climate refugees, as it brings a picture of vulnerability and helplessness; the communities who are forced to flee from their territory may not be wanted to be identified as refugees

[64]The minimalists perceive that environmental change is a causal mutable but emphasizes that it is impossible to analytically and empirically segregate environmental change as the cause of migration.

[65]The maximalists perceive that environmental change is the direct cause of migration and advocates a holistic approach in addressing them without any classification.

[66]Andrew Baldwin, Chris Mathmann and Delf Rothe, 'Securitizing Climate Refugees: The Futurology of Climate-induced Migration', (2014) 2(2) Critical Studies on Security 121–130.

[67]Astri Suhrke, 'Environmental Degradation and Population Flows' (1994) 47(2) J. Int'l Affairs 473–496; Hugo (n 16) 105–131.

[68]Koko Warner, 'Global Environmental Change and Migration: Governance Challenges' (2010) 20(3) Global Environmental Change 402–413.

[69]Black (n 17) 12.

[70]Stephen Castles, 'Environmental change and forced migration: making sense of the debate: New Issues in Refugee Research' (2002) Oxford Refugee Studies Centre Working Paper 70, 8 http://tinyurl.com/cjrr4cy accessed 11 May 2018.

[71]Brian Black, *Climate Change: An Encyclopedia of Science and History* (ABC-CLIO, California, 2013).

[72]Cecilia Tacoli, 'Crisis or Adaptation? Migration and Climate Change in a Context of High Mobility' (2009) 21(2) Environment Urbanization 513–525, 516 http://journals.sagepub.com/doi/abs/10.1177/0956247809342182 accessed 14 April 2018.

[73]Castles (n 70) 6; JoAnn McGregor, 'Refugees and the Environment' in Richard Black & Vaughan Robinson (eds), *Geography and Refugees: Patterns and Process of Change* (Wiley, 1993) 158.

themselves.[74] Applying the doctrine of personal self-determination, it is important to account for the concerns and perspectives of these displaced populations. However, with all these antagonisms against the application of climate refugees, some of the minimalists attempted to contextualize the use of climate refugees. Bilsborrow is not inclined to invoke the refugee glossary in all climate change displacement scenarios. However, he advocates the perception that the term 'environmental refugee' should be exclusively ascribed to only those migrants who were uprooted or compelled to flee the extreme hostile living conditions and widespread destruction and collapse of the physical environment.[75] Minimalists also argue that blanket recognition of climate refugees will have the effect of deviating from the obligation of developed countries to provide assistance to Global South nation-states in acclimating, alleviating the adverse effects of the climate change as a durable solution provided by the developed countries and rely on the defence of providing protection to the refugees.[76] Nevertheless, the recognition of refugees can also facilitate the promotion of climate justice by imposing differentiating responsibility on the developed nations to provide protection to the most vulnerable and disproportionately affected climate refugees.

Among the maximalist category, El-Hinnawi occupies the prominent position. He is credited with coining the term 'environmental refugees' to the political lexicon. He adopts a broader definition, which includes a situation where people are forced to leave their habitat due to identified environmental degradation. His definition covers both internal and cross-border migration as well as temporary and permanent movement.[77] Maximalists saw a direct and strong link between environmental degradation, large-scale massive migration and conceivable conflicts and argue that all migrants are refugees without realizing the fact that there are distinctions[78] of a problematic[79] dimension in the sense of making environment discernible on social, political and economic contexts. Many of the contentions of the maximalists were aimed at ensuring climate justice and also to provide opposition to anti-asylum loy-

---

[74]Colette Mortreux and Jon Barnett, 'Climate Change, Migration, and Adaptation in Funafuti, Tuvalu' (2009) 19 Global. Envt. Change 105–112; António Guterres, 'Climate Change, Natural Disasters and Human Displacement: A UNHCR Perspective' (2008) UNHCR's Policy Paper http://www.unhcr.org/4901e81a4.pdf accessed 05 May 2018; Carol Farbotko and Heather Lazrus, 'The First Climate Refugees? Contesting Global Narratives of Climate Change in Tuvalu' (2012) 22(2) Global Environmental Change 382–390; Karen Elizabeth McNamara and Cristopher Gibson, 'We do not Want to Leave Our Land: Pacific Ambassadors at the United Nations Resist the Category of "Climate Refugees"' (2009) 40(3) Geoforum 475–483.

[75]Richard Bilsborrow, 'Rural Poverty, Migration, and the Environment in Developing Countries: Three Case Studies' (1992) World Bank's Background Paper for World Development Report, 55–56 http://documents.worldbank.org/curated/en/777691468767386516/pdf/multi0page.pdf accessed 15 February 2018.

[76]McNamara, Gibson (n 74) 475–483.

[77]El-Hinnawi (n 25).

[78]James Morrissey, 'Rethinking the Debate on Environmental Refugees' (2012) 19 Journal of Political Ecology 39.

[79]Andrew Geddes, *The Politics of Migration and Immigration in Europe* (Sage Publications, London, 2003).

alists who relied on increased border control.[80] Hence, maximalists 'take up and support the proposition that all climate-induced migration whether it falls under the category of forced or voluntary should receive international legal protection. Similar to minimalists, maximalists too exhibit a wide deviation in their approach to climate displacement. While some of them attributed a simple, direct correlation between climate change and displacement, many of them emphasize certain parameters to qualify as refugees. Meyer, the prominent among the maximalists, approached the climate displacement as the inability to secure a livelihood in their homeland due to environmental degradation brought by the climate change.[81] Docherty and Giannini draw attention to trans-border movement and excluded any kind of internal movement from the ambit of climate displacement.[82] El-Hinnawi and Jacobson focused on the typology and patterns of displacement as temporary and permanent.[83]

Further, the differences in the narratives also extend to the purpose for which a particular terminology is adopted. Maximalists adopt an alarmist and broader definition for climate refugees to force the world political community to respond. For them, the possibility of a new and expanded category of the term brings excitement to investigate.[84] The war of climate narrative between maximalists who projected an alarmist figure of future climate refugee crisis and the minimalists who disputed the oversimplified approach of attributing a direct causality of climate change in migration got tilted in favour of minimalists who got a tight grip on the discourse with the release of Foresight report on Migration and Global Environmental Change that unambiguously underscored the range and complexity of the communications between these factors of migration in which climate change acts only as one of the robust variable responsible for migration.[85] The report evidently stated that 'the range and complexity of the interfaces between these drivers means that it will rarely be possible to differentiate individuals for whom environmental causes are the sole driver'[86] In spite of the recent popularity and support received by the minimalist, there are authors who take up the position that the debate is still relevant due to the

---

[80] James Morrissey, 'Environmental change and forced migration' (Workshop on Environmental Change and Forced Migration, Oxford Refugee Studies Centre, University of Oxford, London, 8–9 January 2009) https://www.rsc.ox.ac.uk/publications/environmental-change-and-forced-migration-a-state-of-the-art-review accessed 04 April 2018.

[81] Norman Myers 'Environmental refugees in a globally warmed world' (1993) 43(11) Bio Sciences 752–761.

[82] Bonnie Docherty, Tyler Giannini, 'Confronting a Rising Tide: A Proposal for a Convention on Climate Change Refugees' (2009) 33 Harv. Envt'l L. Rev 349–403.

[83] Black, 'Environmental Refugees' (n 17) 1.

[84] Diane Bates, 'Environmental Refugees? Classifying human migrations caused by environmental change', (2002) 23(5) Population and Environment 465–477.

[85] Andrew Baldwin, Chris Methmann and Delf Rothe, 'Securitizing 'Climate Refugees': The futurology of Climate-Induced Migration', (2014) 2(2) Critical Studies on Security 121–130.

[86] Ibid; Government Office for Science, 'Migration and Global Environmental Change: Future Challenges and Opportunities' (2011) Foresight's Final Project Report https://assets.publishing.service.gov.uk/government/uploads/system/uploads/attachment_data/file/287717/11-1116-migration-and-global-environmental-change.pdf accessed 15 March 2018; Yvonne Su, 'The Concept of Climate Migration' (2017) 30(4) J. Refugee. Stud. 624–662.

vast discrepancy between the doomsday prediction of maximalists and the sceptics approach followed by the minimalists.[87] Warner highlights that currently, available evidence may find climate as an ambiguous driver of migration. The future potential of climate change on migration could be profound and fundamental to open up new blocks and corridors of large-scale mass migration. It is to this future potentiality the book concentrates.

What surfaces from these narratives is the paucity of unanimity and precision afflicting the expression 'climate refugees'. Both the minimalists and maximalists are blameworthy while observing the issue from a perspective that is preventive and obstructive without appropriately understanding the framework of whole circumstances and situations. The lack of clarity and consensus in appreciating the term 'climate refugees' is the inconceivable and implausible complexity and intricacy to exhaustively and straightforwardly ascribing the preference for migration to climate change as it is forever situation reliant and locally identifiable.[88] There is a crucial requirement to study and research the context critically to appreciate the local and regional variations of climate change migration; otherwise, global community will be struggling with the issue of climate displacement in a fragmentary fashion. However, the lack of precise definition should not be construed as an excuse to shroud the lego-institutional responses. It is a reality that climate change will cause mammoth human displacement whose situations could easily be equated with refugees in terms of susceptibility, defencelessness and liability, and total lack of management of resources makes it practical to invoke the term 'refugees' to attribute it to climate-induced displaced persons provided they are able to meet the criteria advocated by the maximalists. A parallel could be drawn regarding the expression 'refugee' as defined in the UNCSR where under direct correlation between people fleeing from their homelands owing to the 'well-founded fear being persecuted' and other socioeconomic conditions. The former set of conditions is more specific as one cannot entirely bypass them, and later set of conditions is conclusive in their decision to migrate. The reason that must be understood is that human transactions and engagements would forever be propelled by an assortment of dynamics that contemplates the solution in the assignment of distinction to the corollaries and consequences of a condition.

3.5 Approaches to Climate Change Human Mobility: An Evolving Concept

The nation-states, communities and societies across the globe have been experiencing the negative effects of climate change and the same time have been expanding a com-

[87] Cathrine Brun, Piers Blaikie, *Alternative Development: Unravelling Marginalization, Voicing Change* (Routledge, Abingdon, 2016) 208.

[88] Black, 'Environmental Refugees' (n 17) 6; Gaim Kibreab, 'Environmental Causes and Impact of Refugee Movements: A Critique of the Current Debate' (1997) 21(1) Disasters 20–38.

pendium of solutions and conceptualizing the approaches to cope with the climate change-induced migration.[89] These solutions and methods can produce pragmatic and tangible results provided these are engrafted into a binding regional and global legal framework. There is an inescapable necessity of integrating the rule of law into the substantive and processual body of climate change at the global, regional and municipal stages for the protection of rights, mitigation of risk, construction of resilience, empowerment of the populations and facilitation of affirmative migration.[90]

The evolved approaches of climate displacement pertain to migration, securitization, adaptation and mitigation.[91] A security narrative on climate change portrays climate migration as a matter of security. Multiple narratives are woven around the discourse. First approach highlights climate change as an existential threat to showcase and force the governments to take appropriate climate mitigation strategies. The second narrative perceives climate change migration from the state-centric security dogma and adversely observes climate change migration founded upon often ill-conceived threat assessment. These assessments span from the intimidation to migrants' political infidelity, cultural superiority and fierce conflicts over religious aspects and socioeconomic resources stemming from the plural nature of migrants. Thus, this narrative inclines to depict migrants as a possible terrorist and flourishes on fear and threat perception. The final approach examines climate change migration from the semantics of human security narrative that emphasis on potential cooperation and collaboration among the countries. Therefore, the semantics of security narrative highlights the urgency of the climate activism. Two views are prevailing about climate change displacement and adaptation discourse: first view perceives displacement as an outcome of incapability to adapt and second treats displacement as an adaptive strategy. It is deemed that increasing and fostering possibilities for human mobility can decrease the vulnerability of populations and their area. In few instances, migration may be the only alternative for people. However, whether migration could act as a constructive strategy to adapt depend on multiple factors. The situation is context specific and will have a divergent explanation for both countries of nationality and reception. For example, the migration of distressed communities enduring climate change to more developed countries in their vicinity could be an adaptive strategy for the country of nationality. However, the similar arguments cannot be advanced for the country of reception in case of migration squeezes the resources and generates conflicts in the destination area. Although migration might be one facet of adaptation strategy, entire issue of climate change displacement cannot be seen from adaptation angle alone. The discourse further involves various legal and policy strands and approaches that have been formulated to protect the

[89]Cosmin Corendea, 'Protection under Climate Law: A Hybrid Approach' (*Refugee Law Initiative*, 29 May 2018) https://rli.blogs.sas.ac.uk/2018/05/29/protection-under-climate-law-a-hybrid-approach/ accessed on 31 May 2018.

[90]Cosmin Corendea, 'Hybrid Legal Approaches Towards Climate Change: Concepts, Mechanisms and Implementation' (2016) 21(1) Ann. Surv. Int'l & Comp. L. 5, 29–42, 35.

[91]François Gemenne and Julia Blocher, 'How can Migration Serve Adaptation to Climate Change? Challenges to Feshing out a Policy Ideal' (2017) 183(4) Geogr. J. 336–347.

rights of climate migrants or climate refugees. The environment, climate change and migration are interwoven in which the issues of adaptation, securitization and human rights got inextricably enmeshed.[92] It is a multifarious interconnection that necessitates to be pursued in a holistic manner by assessing other attributing causes *inter alia* human security, human rights, livelihood avenues, human development, and economic growth, and conflict resolution.[93] Generally, migration is perceived and construed as an adaptation strategy.[94] Migration could be a coping system and survival method for climate migrants and climate refugees.[95] But migration and massive movement of people could have significant environmental ramifications for countries of origin and destination and the migratory routes in between pandering to accelerated environmental degradation.[96]

Besides the dominant adaptation approach, Tine Tolstrup maps the current approaches and discourses to climate change-induced displacement and migration propounded by Faist and Schade.[97] This include *inter alia*:

(a) *Ecosystem Discourse*: This is a central discourse and provides foundation to all other debates on climate change that deteriorate the ecosystems pandering to human mobility and thus contends strongly for climate change mitigation to avert the deterioration of the ecosystems.[98]
(b) *Refugee Discourse*: It emphases on human security and governance narrative by revisiting the international law framework and demanding recompense.[99]
(c) *Conflict Discourse*: It is a subject matter of peace, geostrategic and security studies and institutions, respectively, that consider climate change migration a trigger with regard to territoriality and intra-state conflict posing threat to global security.[100]

---

[92]Elisabeth Abiri, 'Migration and Security from a North-South Perspective: Sweden and Malawi' in David Graham and Nana Poku (eds), *Migration, Globalisation and Human Security* (Routledge, London and New York, 2000) 70–74.

[93]David T. Graham, 'The People Paradox: Human Movements and Human Security in a Globalizing World, in David Graham and Nana Poku (eds), *Migration, Globalisation and Human Security* (Routledge, London and New York, 2000) 185–190, 186.

[94]Matthew Lister, 'Climate Change Refugees' (2014) 17(5), CRISPPP 618–634.

[95]Clionadh Raleigh, Lisa Jordan and Idean Salehyan, 'Assessing the Impact of Climate Change on Migration and Conflict' (World Bank, Washington DC) 16–18.

[96]Roger Zetter, 'Protecting Forced Migrants: A State of the Art Report of Concepts, Challenges and Ways Forward, Migration Reports' (2014) Federal Commission on Migration, 18–27 https://www.ekm.admin.ch/content/dam/data/ekm/dokumentation/materialien/mat_schutz_e.pdf accessed 29 January 2018.

[97]Quoted in Tine Tolstrup, 'Understanding climate migration: An analysis of political discourses and scientific disputes on linkages between climate change and migration in the UNFCCC negotiations' (Master's Thesis, Faculty of Science, University of Copenhagen, 20–22, 6 June 2016).

[98]Ibid.

[99]Ibid.

[100]Ibid.

(d) *Relocation Discourse*: It stresses on adaptation and mitigation[101] of human distress by taking preventive measures through relocation strategies as a pragmatic resolution to climate change.

It is, indeed, understandable from such multiple categorizations that the political orientation of the discourse on climate change migration has posed significant implications.[102] Therefore, the above categorization brings to fore fundamental differences in the approach that the policy-makers use desultory strategies to formulate the issue as per their perceptions.[103] However, these approaches have taken a unidimensional position ignoring the multicasual nature of climate change, which requires multi-pronged strategy recognising climate change both as a problem as well as solution.

In the wake of the intrinsic limitations inherent in these approaches to tackle and address the heightened concerns of climate displacement holistically, discourses in the recent times have concentrated on the model which outlays the paradigm of climate change as a human rights concern. Therefore, the theoretical concept of climate change migration should not be limited to an 'imaginative resource'[104] rather, the power relations pertained to climate change migration discourses, practices and politics must be impugned.[105] Such a premise demands the political networking for climate change debate that encounters and confronts the questions relating to the frequent discrimination, sloppy picture of the plunging climate change refugee or migrant, and derides the maximalists' perceptions that can prompt ostensible security apprehensions of the Global North world. At the same time, one should bear in mind the fact that climate change at its core is embedded in inequalities and injustices in the social stratification in the 'Global South' societies who have least contributed to climate change. Thus, any solution to climate impact must accommodate the CBDR as enunciated in the UNFCCC's burden-sharing principle[106] applied to both mitigation and adaptation and to avoid climate change adaptation apartheid[107] by preserving the human rights and to explore global environmental concerns and opportunities for succeeding generations must be taken by the policy-makers and stakeholders.

---

[101] Ibid.

[102] John Vogler, *Climate Change in World Politics* (Palgrave Macmillan, UK, 2016) 22–36.

[103] Chloé Anne Vlassopoulos, 'Defining Environmental Migration in the Climate Change Era: Problem, Consequence or Solution?' in Thomas Faist and Jeanetter Schade (eds), *Disentangling Migration and Climate Change* (Springer Netherlands, 2013) 145–63.

[104] Klepp (n 63)12.

[105] Alex Arnall, Uma Kothari and Ilan Kelman, 'Introduction to Politics of Climate Change: Discourses of Policy and Practice in Developing Countries' (2013) 18(43) Geogr. J., 98–101.

[106] Robyn Eckersley, 'The Common But Differentiated Responsibilities of States to Assist and Receive "Climate Refugees"' (2015) 14(4) Eur. J. Political Theory 1–21 https://www.researchgate.net/publication/277938867_The_common_but_differentiated_responsibilities_of_states_to_assist_and_receive_'climate_refugees' accessed 4 June 2018.

[107] Margaux J. Hall and David C. Weiss, 'Avoiding Adaptation Apartheid: Climate Change Adaptation and Human Rights Law' (2012) 37(2) Yale J. Int'l L. 319–325, 345–346.

3.6 Climate Change Rights as Human Rights

Academic and policy discourse on the linkages between climate change and human rights is relatively a recent phenomenon. However, the foundations of the coexistence have been entwined ever since the global community realized the need for environmental protection.[108] The interdependency is visible in the wordings of Principle 1 of the Stockholm Declaration,[109] which correlated fundamental rights of man to the quality of the environment.[110] However, a shift in understanding can be perceived by studying the Rio Declaration, which reduced humans to being the 'central concern of sustainable development' and 'entitled to a healthy and productive life in harmony with nature'; thus, separating human rights aspect from environmental protection.[111] The environmental protection frameworks envisaged in the post-Rio scenario were formulated in a similar framework where human beings were treated as users and beneficiaries of nature and natural resources, rather than active right holders under international law. The contemporary debate of human rights commitments in the perspective of climate change can be traced to the well-established body of the literature connecting both ideas[112] and efforts to establish a right to a clean and healthy environment.[113]

Multiple factors have generated numerous grounds that have produced much of the recent interest in the human rights aspects of climate change. The fundamental failure of the negotiation-based international policy developments to address the climate change in an accountable manner and its inability to focus on the vulnerability of individuals and victims have forced a paradigm shift to consider the issue from

[108]See Francesco Francioni, 'International Human Rights in an Environmental Horizon' (2010) 21(1) Eur. J. Int'l L 41–55; Louis Sohn, 'The Stockholm Declaration on the Human Environment' (1973) 14 Harv. Int'l L. J. 413–451.

[109]United Nations Conference on the Human Environment (adopted 15 December 1972) UNGA A/RES/2994 (Stockholm Declaration) Principle 1.

[110]"Man has the fundamental right to freedom, equality and adequate conditions of life, in an environment of a quality that permits a life of dignity and well-being".

[111]Francioni (n 108) 45; Basil Ugochukwu, 'Climate Change and Human Rights: How? Where? When?' (2015) 82 CIGI Papers 1–10 https://www.researchgate.net/publication/303564554_Climate_Change_and_Human_Rights_How_Where_When, 3 accessed 2 May 2018.

[112]See Alan E. Boyle & Michael R. Anderson (eds), *Human Rights Approaches to Environmental Protection,* (Oxford University Press, Oxford, 1996); Romina Picolotti and Jorge Daniel Taillant (eds), *Linking Human Rights and the Environment* (University of Arizona Press, 2003); Dominic McGoldrick, 'Sustainable Development and Human Rights: An Integrated Conception', (1996) 45 Int'l & Comp.L.O., 796–818.

[113]Convention on the Rights of the Child (adopted 20 November 1989) 1577 UNTS 3 (CRC) art 24; Indigenous and Tribal Peoples Convention (adopted 27 June 1989) 28 ILM 1382 art 4, 7, 15; African Charter on Human and Peoples' Rights (adopted 27 June 1981) (1982) 21 ILM 58 (African Charter) art 24; Additional Protocol to the American Convention on Human Rights in the Area of Economic, Social and Cultural Rights (adopted 17 November 1988) 28 ILM 156 (Protocol of San Salvador) art 11; see generally John Lee, 'The Underlying Legal Theory to Support a Well-Defined Human Right to a Healthy Environment as a Principle of Customary International Law' (2000) 25 Colum. J. Envtl. L. 283.

a human rights perspective. The focus on the human rights was discernible from the plight of the Inuit[114] who petitioned before the Inter-American Commission on Human Rights claiming that the negative impacts of climate change violated their fundamental human rights.[115] Although the petition was declined, it drew global[116] as well as academic[117] attention to the links between climate change and human rights. In the light of the controversy, The Permanent Forum on Indigenous Issues under the Economic and Social Council (ECOSOC) recommended a human rights-based approach (HRBA) to guide the formulation and implementation of global climate policies.[118] Another notable incident which shook the world community was the campaign by Small Island states (particularly Maldives), for linking climate change and human rights.[119] The Male Declaration resulted in a debate in the UN Human Rights Council[120] and subsequently to a report released by the Office of the United Nations High Commissioner for Human Rights (OHCHR) focusing on the interconnectedness between climate change and human rights in a systematic way.[121] This move inspired regional organizations to adopt similar resolutions, seeking to study the link between climate change and human rights.[122] Despite efforts from

---

[114] Sheila Watt Cloutier, 'Climate Change and Human Rights' (*Carnegie Council for Ethics in International Affairs*, 22 April 2004) https://www.carnegiecouncil.org/publications/archive/dialogue/2_11/section_1/4445 accessed 20 June 2018.

[115] Sheila Watt-Cloutier, 'Petition Seeking Relief from Violations Resulting from Global Warming Caused by Acts and Omissions of the United States' (Organisation of American States, 2005) http://www.ciel.org/Publications/ICC_Petition_7Dec05.pdf accessed 13 December 2017.

[116] Joanna Harrington, 'Climate Change, Human Rights and the Right to be Cold' (2007) 18 Fordham Envtl. L. Rev 513–535, 532.

[117] Hari M. Osofsky, 'The Inuit Petition as a Bridge? Beyond the Dialectics of Climate Change and Indigenous Peoples' Rights' (2007) 31(2) Am. Indian L. Rev. 675–699; see generally Randall S. Abate, 'Climate Change, the United States and the Impacts of Arctic Melting: A Case Study in the Need for Enforceable International Environmental Human Rights' (2007) 26 Stan. Envt. L. J. 4–76; Timo Koivurova, 'International Legal Avenues to Address the Plight of Victims of Climate Change: Problems and Prospects' (2007) 22 J. Envtl. L. & Litig. 267–299; Sara C. Aminzadeh, 'A Moral Imperative: The Human Rights Implications of Climate Change' (2007) 30(2) Hastings Int'l & Comp. L. Rev. 231–265.

[118] United Nations Permanent Forum on Indigenous Issues, 'report on the seventh session' (2008) Economic and Social Council E/2008/43, e/c.19/2008/13, 3–4 https://www.un.org/development/desa/indigenouspeoples/unpfii-sessions-2/seventh-session-of-the-unpfii.html accessed 19 March 2018.

[119] Malé Declaration on the Human Dimension of Global Climate Change (adopted 14 November 2007) (Male Declaration) http://www.ciel.org/Publications/Male_Declaration_Nov07.pdf accessed 15 December 2017.

[120] United Nations Office of the High Commissioner for Human Rights, 'Human Rights and Climate Change' (adopted 28 March 2018) Resolution 7/23 http://ap.ohchr.org/documents/E/HRC/resolutions/A_HRC_RES_7_23.pdf accessed 15 December 2017.

[121] United Nations Office of the High Commissioner for Human Rights, 'Report on the Relationship Between Climate Change and Human Rights' (adopted 15 January 2009) UN Doc. A/HRC/10/61 http://www.refworld.org/docid/498811532.html accessed 20 May 2018.

[122] Organization of American States, 'Human Rights and Climate Change in the Americas' (adopted 3 June 2008) AG/RES. 2429 (XXXVIII-O/08) https://www.oas.org/dil/AGRES_2429.doc accessed on 16 May 2018.

the international community and support shown by nations,[123] no concrete binding legal documents could be formulated about the relation of climate change and human rights.[124]

Based on the study of actions taken by the UN Human Rights Council (UNHRC) and other international organizations, it can be concluded that a soft law has developed to support and critically appreciate the issue of climate change on the anvil of human rights. The UNHRC confirmed that human rights obligations, norms and principles have the possibility to not only highlight the seriousness of the scenario but also to propel, steer and strengthen policy-making in the field of climate change.[125] It also stressed upon the need for all states to boost international dialogue and cooperation to address the adverse bearings of climate change on the enjoyment of human rights including the right to development[126] and the right to health,[127] among others.[128] The outcome document of the 2012 United Nations Conference on Sustainable Development reaffirmed the importance of human rights for achieving sustainable development.[129] The Conference of the Parties to the UNFCCC has recognized that 'the effects of the adverse impact of climate change have a range of direct and indirect implications for the effective enjoyment of human rights'.[130] Conference of Parties 22 (COP 22) acknowledged the role of human rights principles to guide the implementation of Paris Agreement and in promoting coherence, effectiveness,

---

[123]United Nations Human Rights Council, 'Human Rights and Climate Change' (adopted 25 March 2009) Resolution 10/4 http://ap.ohchr.org/documents/E/HRC/resolutions/A_HRC_RES_10_4.pdf accessed 04 November 2017.

[124]Siobhan McInerney-Lankford, Mac Darrow and Lavanya Rajamani, 'Human Rights and Climate Change: A Review of the International Legal Dimensions' (2011) 8 World Bank Study https://elibrary.worldbank.org/doi/pdf/10.1596/978-0-8213-8720-7 accessed 05 May 2018; see generally Tiffany T.V. Duong, 'When Islands Drown: The Plight of "Climate Change Refugees" and Recourse to International Human Rights Law' (2010) 31(4) U/Pa. J. Int'l L. 1239–1266.

[125]United Nations Human Rights Council, 'Human Rights and Climate Change' (adopted 17 October 2011) Resolution 18/22, UN Doc. A/HRC/RES/18/22 https://www.ohchr.org/Documents/Issues/ClimateChange/A.HRC.RES.18.22.pdf accessed 23 May 2018.

[126]United Nations Human Rights Council, 'Human Rights and Climate Change' (adopted 25 June 2014) Resolution 26/L33, UN Doc. A/HRC/26/L.33/Rev.1 https://documents-dds-ny.un.org/doc/UNDOC/LTD/G14/065/81/PDF/G1406581.pdf?OpenElement accessed 25 April 2018.

[127]United Nations Human Rights Council, 'Human Rights and Climate Change' (adopted 22 July 2015) Resolution 29/15, UN Doc. A/HRC/29/L. 21 https://documents-dds-ny.un.org/doc/UNDOC/GEN/G15/163/60/PDF/G1516360.pdf?OpenElement accessed 23 April 2018.

[128]United Nations Human Rights Council, 'Human Rights and the Environment' (adopted 12 April 2011) Resolution 16/11, UN Doc. A/HRC/RES/16/11 http://www2.ohchr.org/english/bodies/hrcouncil/docs/16session/A.HRC.RES.16.11_en.pdf accessed 21 April 2018; United Nations Human Rights Council, 'Human Rights and the Environment' (adopted 19 April 2012) Resolution 19/10, UN Doc. A/HRC/RES/19/10 https://documents-dds-ny.un.org/doc/RESOLUTION/GEN/G12/131/59/PDF/G1213159.pdf?OpenElement accessed 21 April 2018.

[129]United Nations Development General Assembly, 'The Future We Want: Biodiversity and Ecosystems—Driving Sustainable Development: United Nations Development Programme Biodiversity and Ecosystems Global Framework 2012–2020' (adopted 27 July 2012) Resolution 66/288, UN Doc. A RES/66/288.

[130]United Nations Human Rights Council, 'Resolution 29/15' (n 127).

equality and climate justice.[131] Also, the UNFCCC has also proclaimed its support to Geneva Pledge for Human Rights in Climate Action, which focuses on developing the inter-linkage between climate change and human rights. Though legally non-binding, the voluntary initiative commits to promoting human rights principle guiding the international and domestic climate policies.[132]

In the recent years, there has been growing attention paid by international organizations and policy mechanisms in highlighting the complex interrelation between climate change displacement and human rights. A study published by the IOM opined that implementing a human rights perspective is vital in searching for an effective solution to protect and reconstruct the lives of displaced individuals and communities.[133] Similarly, the Cancun Adaptation Framework emphasized coordination and cooperation, the cardinal principle of international human rights law[134] to address climate change-induced displacement, migration and planned relocation, at all levels of the policy framework.[135] Additionally, the Special Rapporteur on the human rights of migrants noted the role of climate change in determining the type and pattern of migration at local and global level.[136] Human rights strategies have been sought in the case of climate change displacement/refugee movement owing to a host of factors. A refugee, in the classical political sense, is understood to be a person who flees from the country of his origin or residence owing to the fear of persecution and seeks asylum in a different nation.[137] Owing to the restrictive definition under the Refugee Convention, people fleeing due to climate change is not recognized as refugees, and their protection is demonstrably unfulfilled or violated.[138] A study conducted by the UNHCR in 2009 disclosed that an average of 22.5 million people had been displaced globally due to the climate change.[139] Despite such intimidat-

---

[131] United Nations Framework Convention on Climate Change, Conference of Parties 22 (Marrakech, Morocco, 7–18 November 2016).

[132] Promoting the Geneva Pledge for Human Rights in Climate Action (Human Rights and Climate Change Working Group) http://climaterights.org/our-work/unfccc/geneva-pledge/ accessed 19 April 2018.

[133] Laczko and Aghazarm (n 15) 387–391.

[134] UN Charter (n 62) art 1.

[135] United Nations Framework Convention on Climate Change, 'Decision 1/CP.16 The Cancun Agreements: Outcome of the work of the Ad Hoc Working Group on Long-term Cooperative Action under the Convention' (adopted 15 March 2011) FCCC/CP/2010/7/Add.1 https://unfccc.int/resource/docs/2010/cop16/eng/07a01.pdf accessed 20 May 2018.

[136] United Nations General Assembly, 'Report by the Special Rapporteur on the human rights of migrants, Francois Crepeau: Climate Change and Migration' (adopted 13 August 2012) A/67/299 https://documents-dds-ny.un.org/doc/UNDOC/GEN/N12/460/71/PDF/N1246071.pdf?OpenElement accessed 12 April 2018.

[137] United Nations Convention Relating to the Status of Refugees (adopted 28 July 1951) 189 UNTS 137 art 1.

[138] James C Hathaway, 'Food Deprivation: A Basis for Refugee Status?', (2014) 81(2) Social Research 327–339, 328 https://repository.law.umich.edu/cgi/viewcontent.cgi?article=2075&context=articles accessed 12 March 2018.

[139] UNHCR, *Climate Change and Disasters*, (14 December 2009) http://www.unhcr.org/climate-changeand-disasters.html accessed 26 December 2017.

ing numbers, states remained unconcerned with the problem of climate refugees[140]; even though UNHCR recognized the problem.[141] A fundamental issue that follows the concept of climate refugee and their global acknowledgement is the reluctance of nation-states to admit them in their territories. Countries such as New Zealand and Australia[142] have refused entry to climate refugees when faced with the question of their admission and protection. Most common arguments by countries against environmental refugees are the floodgate reasoning,[143] the self-defence reasoning, and the sovereignty argument.[144] However, considering the vast number of displacements due to climate change, it is axiomatic that such people be given due protection and human rights law provides the platform under international law in filling up the existing protection gaps to promote climate justice.

Climate justice as an idea is based on the understanding that climate change has a disproportionate effect on the marginalized and vulnerable population.[145] The idea also brings along the historical responsibility of the developed Global North for the climate change and concentrates on differential responsibility, distribution of resources and protection of victims.[146] In the context of climate displacement, the affected communities and populations are among the most vulnerable who have contributed the least to the climate problem.[147] However, the existing dominant approaches to climate migration deeply aligned to adaptation consider climate dis-

---

[140] Stefan Lovgren, 'Climate Change Creating Millions of "Eco Refugees" UN Warns' (*National Geographic News*, 18 November 2005) http://news.nationalgeographic.com/news/2005/11/1118_051118_disaster_refugee_2.html accessed 23 December 2017.

[141] United Nations High Commission for Refugees, 'Statement by Sadako Ogata, United Nations High Commissioner for Refugees' (United Nations Conference on Environment and Development (UNCED), Rio de Janeiro, 10 June 1992).

[142] *Mohammed Matahir Ali v Minister of Immigration*, [1994] FCA 887 (Aus.).

[143] The nation-states contend that acknowledging one person as an 'environmental refugee', would open a Pandora box for millions who are placed in similar situations of deprivation and recognizing whom would not be financially viable for the State. See, Jessica Rodger, 'Defining the Parameters of the Non-Refoulement Principle' LLM Research Paper, Victoria University Wellington, 2001.

[144] The nation-states have reasoned that the sole discretion to decide whether a foreigner should be given entry to their territories rests upon them. See also: *Nishimura Ekiu v US*, 142 US 651; Catherine Phuong, 'Identifying States' Responsibilities towards Refugees and Asylum Seekers' (2005) ESIL Research Forum 1–12 http://www.esil-sedi.eu/sites/default/files/Phuong.PDF accessed 21 January 2018.

[145] Olivier Godard, *Global Climate Justice: Proposals, Arguments and Justification* (Edward Elgar Publishing, 2017) 12; Damilola S. Olawuyi, 'Advancing Climate Justice in International Law: An Evaluation of the United Nations Human Rights-Based Approach' (2015) 11(1) FAMU 103–125, 103 http://commons.law.famu.edu/famulawreview/vol11/iss1/3 accessed 11 May 2018.

[146] Simon Caney, 'Two kinds of Climate Justice: Avoiding Harm and Sharing Burdens' (2014) 22(2) Journal of Political Philosophy 125–149, 147; Julia Häggström, 'Climate Justice as a Foundation for Climate Migration Policy' (Lund University, 2015) 15 http://lup.lub.lu.se/luur/download?func=downloadFile&recordOId=8244061&fileOId=8244071 accessed 12 January 2018.

[147] Ibid; Biermann and Boas (n 13); Eike Albrecht and Malte Paul Plewa, 'International Recognition of Environmental Refugees' (2014) 2 Envt'l Pol'y & L. 78–85.

placement as a development challenge[148] such an approach can mask the inequalities and disproportionate burden experienced by the communities in the most vulnerable nations.[149] A human rights strategy can account for this vulnerability in envisaging the kind of short-term or long-term climate displacement prevention and recovery strategies based on the CBDR, which the current approaches under adaptation, migration studies and security narratives are unable to incorporate.[150] Human rights approach will also strengthen interpretation and explanation for climate justice by advancing the transparent, participatory and accountable machineries while attributing sufficient consideration to the particular needs for the sense of protection among the girls, women, indigenous peoples and other persons who might be disproportionately adversely influenced by climate change.[151]

Once it is argued that human rights strategies are significant in effectively addressing the global scenario of climate change displacement, it becomes crucial to understand the manner in which climate change displacement affects human rights. The first question that arises is which rights are violated due to alteration in climatic conditions and the consequent displacement? Secondly, how can a human rights approach be helpful in addressing the ongoing and anticipated climate displacement catastrophe?

The inter-linkage between climate change displacement and human rights operates at multiple levels. Firstly, climate displacement itself is created by a scenario where the region has been reduced to the extent of inhabitability, which violates the basic human rights including the right to life, access to necessities like food, water shelter. However, this is rather a unidimensional approach, and climate displacement could be a preventive strategy wherein climate change is the dominant push factor in adaptation to climate change. Secondly, a climate change-induced displacement can have a detrimental and disastrous effect on population and communities both at the source of displacement and at the host state affecting the most vulnerable and marginalized people.[152]

---

[148] Giovanni Bettini, 'Climate Migration as an Adaption Strategy: De-Securitizing Climate-Induced Migration or Making the Unruly Governable?' (2014) 2(2) Critical Studies on Security 180–195.

[149] Friederike Gesing, Johannes Herbeck, SiljaKlepp, 'Denaturalizing Climate Change: Migration, Mobilities And Space' (2014) University of Bremen's Artec Paper No. 200, 7 https://www.uni-bremen.de/fileadmin/user_upload/sites/artec/Publikationen/artec_Paper/200_paper.pdf accessed 11 December 2018.

[150] Eike, Plewa (n 147) 79; see generally Tanja Dreher and Michelle Voyer, 'Climate Refugees or Migrants? Contesting Media Frames on Climate Change Justice in the Pacific' (2015) 9(1) Environmental Communication: A Journal of Nature and Culture 58–76.

[151]

Susannah Fisher, 'The Emerging Geographies of Climate Justice' (2012) Centre for Climate Change Economics and Policy Working Paper No. 94 http://www.lse.ac.uk/GranthamInstitute/wp-content/uploads/2014/02/WP83-emerging-geographies-climate-justice.pdf accessed 11 January 2018; Mary Robinson Foundation Climate Justice, 'Human Rights, Migration, and Displacement related to the Adverse Impacts of Climate Change' (2016) Office of the High Commissioner for Human Rights' Discussion Paper, 2 https://www.ohchr.org/Documents/Issues/ClimateChange/EM2016/HumanRightsMigrationDisplacement.pdf accessed 05 January 2018.

[152] Burson, 'Protecting the Rights of People Displaced by Climate Change' (n 21) 170–171.

The rights affected by climate displacement includes

- Right to life[153];
- Right to cultural heritage[154];
- Right to livelihood[155];
- Right to food, clothing, and housing[156];
- Right to self-determination.[157]

The most fundamental right that is affected due to climate change displacement would perhaps be right to life. Under international law, the UDHR,[158] International Covenant on Civil and Political Rights (ICCPR),[159] the Convention on Rights of Child (CRC)[160] and three regional instruments: the European[161] and American Conventions[162] and the African Charter[163] grant right to life to every individual. The Human Rights Committee has stated that 'inherent right to life' provided under ICCPR cannot be construed in a restrictive fashion, and states should take positive measures for its protection.[164] Further, the OHCHR has stated that climate change poses direct as well as an indirect threat to human lives, particularly their right to life.[165]

---

[153]International Covenant on Civil and Political Rights (adopted 16 December 1966) 999 UNTS 171 (ICCPR) art 6; Mary Robinson Foundation Climate Justice, 'Protecting the Rights of Climate Displaced People' (2016) Position Paper https://www.mrfcj.org/wp-content/uploads/2016/07/Protecting-the-Rights-of-Climate-Displaced-People-Position-Paper.pdf accessed 12 April 2018; Siobhan McInerney-Lankford, 'Climate Change and Human Rights: An Introduction to Legal Issues' (2009) 33 Harv. Envtl. L. Rev. 431-436.

[154]International Covenant on Economic, Social and Cultural Rights (adopted 16 December 1966) 993 UNTS 3 (ICESCR) art 15(2).

[155]Ibid art 7(a)(ii).

[156]ICESCR (n 154) art 11.

[157]ICCPR (n 153) art 1.

[158]Universal Declaration of Human Rights (adopted 10 December 1948) UNGA Res. 217 A(III) (UDHR) art 3.

[159]ICCPR (n 153) art 6.

[160]CRC (n 113).

[161]Convention for the Protection of Human Rights and Fundamental Freedoms (adopted 4 November 1950) ETS 5 (ECHR) art 2.

[162]American Convention on Human Rights (adopted 22 November 1969) 1144 UNTS 123 (ACHR) art 4.

[163]African Charter (n 113) art 4.

[164]United Nations Human Rights Committee (UNHRC), 'Nature of the General Legal Obligation on States Parties to the Covenant' (General Comment No. 31, 2004) UN Doc. CCPR/C/21/Rev.1/Add.13 http://www.refworld.org/docid/478b26ae2.html accessed 10 April 2018; UNCHRC, 'Equality of Rights Between Men and Women: article 3' (General Comment No. 28, 2000) UN Doc. CCPR/C/21/ Rev.1/Add.10 http://www.refworld.org/docid/45139c9b4.html accessed 10 April 2018; Manfred Nowak, *UN Covenant on Civil And Political Rights: CCPR Commentary* (NP Engel, 2005).

[165]United Nations Office of the High Commissioner for Human Rights (OHCHR), 'Report on the Relationship Between Climate Change and Human Rights' (n 121)8.

Various scholars have argued that climate change has the potential to aggravate insecurity and violent conflict on a large scale.[166] The Inuit case[167] and the male resolution[168] also showcase the potential threat to the right to life due to adverse impacts of climate change. In its 2008 submission to the OHCHR, Maldives stated that the island nation faces flood threats due to rising sea level and that even partial flooding is likely to result in injury, drowning and loss of life.[169] Human rights organizations have also argued that environmental degradation and consequent climate change-induced displacement can hinder socioeconomic and cultural rights, including the rights to the highest achievable benchmarks of health[170] and components of a sufficient and reasonable standard of life that also includes the right to water[171] and food.[172] OHCHR in its report has stated that the right to health falls among the universal, indivisible rights guaranteed to all people.[173] The UDHR, CRC, the Convention on the Elimination of All Forms of Discrimination against Women and the Declaration on the Rights of Indigenous Peoples oblige states to respect, promote, protect and fulfil the right to health.[174]

International communities are slowly acknowledging the linkage between climate change and human rights to health. The UNFCCC outlines the negative influences

---

[166]Oli Brown and Alec Crawford, 'Rising Temperatures, Rising Tensions: Climate Change and the Risk of Violent Conflict in the Middle East' (International Institute for Sustainable Development, 2009) 1 https://www.iisd.org/pdf/2009/rising_temps_middle_east.pdf accessed 17 December 2017; United Nations Environment Programme, 'From Conflict to Peace-Building: The Role of Natural Resources and the Environment' (UNEP, 2009) 5 https://postconflict.unep.ch/publications/pcdmb_policy_01.pdf accessed 17 December 2017; Brahma Chellaney, 'Climate Change and Security in Southern Asia: Understanding the National Security Implications' (2007) 152(2) The RUSI Journal 62–69, 63.

[167]Watt-Cloutier, 'Petition Seeking Relief' (n 115).

[168]Male Declaration (n 119).

[169]United Nations Framework Convention on Climate Change, 'Views Regarding the Work Programme of the Ad Hoc Working Group on Long-term Cooperative Action under the Convention' (Bangkok, 31 March–4 April 2008) UNFCCC/AWGLCA/2008/MISC.1 https://unfccc.int/resource/docs/2008/awglca1/eng/misc01.pdf accessed 19 December 2018.

[170]CESCR, 'The Right to the Highest Attainable Standard of Health' (General Comment No. 14, 2000) E/C.12/2000/4 http://www.refworld.org/pdfid/4538838d0.pdf accessed 17 December 2018.

[171]El Hadji Guisse, 'report of the Special Rapporteur of the Sub-Commission on the Right to Drinking Water Supply and Sanitation on the relationship between the enjoyment of economic, social and cultural rights and the promotion of the realization of the right to drinking water supply and sanitation' (CESCR's General Comment No. 15, 2002) UN Doc E/CN.4/ Sub.2/2002/10 http://www.refworld.org/docid/4538838d11.html accessed 16 February 2018.

[172]*Social and Economic Rights Action Center/Center for Economic and Social Rights v Nigeria*, 96 AJIL 937 (Decision Regarding Communication 155/96 in Ogoniland case, African Commission on Human and People's Rights, 2001).

[173]United Nations General Assembly, 'Analytical Study on the Relationship between Climate Change and the Human Right of Everyone to the Enjoyment of the Highest Attainable Standard of Physical and Mental Health' (UNHRC's 32nd Session, 2016) A/HRC/32/23 https://documents-dds-ny.un.org/doc/UNDOC/GEN/G16/092/02/PDF/G1609202.pdf?OpenElement accessed 24 April 2018.

[174]Ibid.

of climate change as transformation to the physical atmosphere with detrimental impacts on human health and welfare.[175] Therefore, it exhorts the high contracting parties to initiate actions to minimize the factors and adverse impacts of climate change, including on health,[176] commissioning the appropriate and applicable tools and devices such as impact assessments.[177] Further, the Paris Agreement calls for all states to 'respect, promote and consider their respective obligations on human rights', including the right to health when addressing climate change.[178]

Another important concern regarding the effects of climate displacement pertains to the right of self-determination, which has been acknowledged as a fundamental principle of international law.[179] Also, the emphasis lay on this right in various international instruments shows that it is an indispensable condition for the effective assurance and adherence of individual human rights'[180] It has been recognized as an erga omnes obligation[181] simultaneously negative and positive duties such as to respect 'the right to self-determination' and to advocate the realisation of the 'right to self-determination', respectively.[182] About climate change, it is increasingly argued that the impact of environmental degradation leads to violation of the right to self-determination particularly that of Small Island Developing States (SIDS) leading to the disappearance of states itself.[183]

As previously highlighted, climate change leads to numerous human rights violation, including loss of territory. The territory is one of the fundamental criteria of

---

[175]United Nations Framework Convention on Climate Change (adopted 29 May 1992) 1771 UNTS 107 (UNFCCC) art 1.

[176]Ibid art 3.

[177]UNFCCC (n 175) art 4.

[178]UNFCCC, 'Adoption of the Paris Agreement' (21st Conference of Parties, Paris, 30 November–11 December 2013) Report No. FCCC/CP/2015/L.9/Rev.1 http://unfccc.int/resource/docs/2015/cop21/eng/l09r01.pdf accessed 17 December 2017.

[179]United Nations General Assembly, 'The Right of Peoples and Nations to Self-Determination' (1952) Resolution 637A (VII) http://www.refworld.org/docid/3b00f0791c.html accessed 14 November 2018; *Legal Consequences for States of the Continued Presence of South Africa in Namibia (South West Africa), Notwithstanding Security Council Resolution 276* (Advisory Opinion) 1971 http://www.icj-cij.org/files/case-related/53/053-19710621-ADV-01-00-EN.pdf accessed 21 June 2018 [313]–[312]; *Western Sahara* (Advisory Opinion) 1975 http://www.icj-cij.org/files/case-related/61/061-19751016-ADV-01-00-EN.pdf accessed 21 June 2018 [12]–[74].

[180]United Nations Human Rights Committee (UNHRC), 'The right to self-determination of peoples (Art. 1)' (General Comment No. 12, 1984) UN Doc. HRI/GEN/1/Rev.6, http://www.uio.no/studier/emner/jus/humanrights/HUMR5508/v12/undervisningsmateriale/General%20Comment%20No12-self-determination.pdf accessed 10 April 2018.

[181]*East Timor (Portugal v Australia)* (Merits) [1995] ICJ Rep 25.

[182]UNHRC, 'General Comment No.12' (n 180) 6; ICCPR (n 153) art 1(3); ICESR (n 154) art 1(3); Advisory Opinion, 'Legal Consequences of the Construction of a Wall' (n 179) 88.

[183]United Nations Office of the High Commissioner for Human Rights (OHCHR), 'report on the Relationship Between Climate Change and Human Rights' (n 121) 14; Alliance of Small Island States (AOSIS), 'Declaration on Climate Change' (New York, 21 September 2009) https://sustainabledevelopment.un.org/content/documents/1566AOSISSummitDeclarationSept21FINAL.pdf accessed 21 May 2018; see generally Jane McAdam, 'Environmental Migration' in Alexander Betts (ed), *Global Migration Governance* (Oxford University Press, Oxford, 2011) 153–188.

state recognition[184] and consequently the right to self-determination. Without land, the rights of people (both individual and collective) are no longer protected by the State, and hence, such group becomes vulnerable to further violation of human rights. It is true that international instruments do not mainly refer to responsibility of states in reference to climate change impacts on self-determination (or any other human rights)[185]; however, the OHCHR has maintained that despite lack of clear legal framework, states owe a duty to avert the threat posed by climate change on the right to self-determination.[186] In the present scenario, where the world faces an imminent threat from climate change, discussions around the right to self-determination are the need of the hour. It is essential to comprehend how this right can be upheld in cases of climate change migration/displacement and also consider the cultural and spiritual dimensions of the right.[187]

Moving further, one arrives at the question of the vulnerability of different social groups regarding climate change displacement. It is widely believed that specific groups are more threatened by the deteriorating climate such as women (rural women in particular), children, indigenous peoples, and ethnic minorities, scheduled castes, and persons with disabilities, Small Islands Groups among others.[188] In most of the scenario, it is the sheer loss and impact on the above-listed rights, which prompts the individuals and communities to flee and relocate to other places. The communities whose rights have already been impacted and violated are in double jeopardy since the international legal mechanisms are inadequate and do not recognize the status of these people as entitled to protective measures. An analysis of the normative foundation of human rights reveals that the idea and concept of human rights are inherent in the dignity of human being[189] and every human being is entitled[190] to

[184]The Montevideo Convention on the Rights and Duties of States (adopted 26 December 1933) 165 LNTS 19 (Montevideo Convention).

[185]Stephen Humphreys, 'Climate Change and Human Rights: A Rough Guide' (International Council on Human Rights Policy, Versoix, 2008) 3. https://www.ohchr.org/Documents/Issues/ClimateChange/Submissions/136_report.pdf; Rebecca Jacobs, 'Treading Deep Waters: Substantive Law Issues in Tuvalu's Threat to Sue the United States in the International Court of Justice' (2005) 14(1) Pac. Rim L. & Poli'y J. 103–128 https://digital.law.washington.edu/dspacelaw/bitstream/handle/1773.1/662/14PacRimLPolyJ103.pdf?sequence=1&isAllowed=y accessed 27 March 2018.

[186]Susannah Willcox, 'A Rising Tide: The Implications of Climate Change Inundation for Human Rights and State Sovereignty' (2012) 9(1) EHRR 1–19, 8 http://projects.essex.ac.uk/ehrr/V9N1/Willcox.pdf accessed 14 March 2018; United Nations Office of the High Commissioner for Human Rights (OHCHR), 'report on the Relationship Between Climate Change and Human Rights' (n 121) 40.

[187]See Sophie Pascoe, 'Sailing The Waves On Our Own: Climate Change Migration, Self-Determination and The Carteret Islands' (2015) 15(2) QUT L. Rev 72–85 https://lr.law.qut.edu.au/article/view/610/574 accessed 02 January 2018.

[188]McInerney-Lankford, Darrow and Rajamani (n 124) 30; McInerney-Lankford, 'Climate Change and Human Rights' (n 153) 431; John H. Knox, 'Linking Human Rights and Climate Change at the United Nations' (2009) 33 Harv. Envt'l L. Rev 477–498, 487 http://www.law.harvard.edu/students/orgs/elr/vol33_2/Knox.pdf accessed 15 November 2017.

[189]ICCPR (n 153) preamble.

[190]Jack Donnelly, *Universal Human Rights in Theory and Practice* (Cornell University Press, Ithaca, 2003)36.

certain fundamental rights irrespective of the extent of legal protection and yardsticks. Thus, climate refugees/displaced can claim minimum protection of standards as complementary mechanism even if they do not qualify under the restricted definition of Refugee Convention.

Several arguments are advanced for facilitating and implementing a human rights approach.

- Firstly, the concept of human rights changes the contours of climate displacement by focusing on the plight and right of individuals and communities by offering a minimum standard of protection framework in case of climate displacement. It imposes an obligation on the state to afford protection to people within their territory and offers a legal basis on which protection could be sought.[191]
- Applying human rights standard will directly promote climate justice by protecting and amplifying the concerns of the most vulnerable and marginalized communities. By focusing on the human aspect, a human rights approach has the effect of looking at climate change issue broadly by incorporating non-state actors and bringing each and every individual affected by climate displacement to the forefront of climate change discourse.[192] Generally, international negotiation though relies on consensus is marked by power asymmetry and domination of the developed north. A human rights-based approach may detangle the power asymmetry and provide an equitable platform for the victims of climate change.
- A human rights-based approach by its focus on individuals will enhance cooperation between countries and help evolve climate change discourse as true common concern of humanity.

A human rights-based approach was thought to provide moral impetus in highlighting and elevating the issue to be a matter of grave concern.[193] However, in the recent years, discourse on human rights has moved from the moral paradigm to emphasize effective legal implementation. For example, the Committee on Economic, Social and Cultural Rights under ICESCR in General Comment 3 proclaimed while specifying the legal benchmarks for the integration and implementation of international human rights covenants emphasized the duty of all the nation-states to take measures 'to realize progressively the full enjoyment of the rights contemplated in the covenants'.[194]

Under the existing legal framework, specific arguments may be extended for the protection of climate displaced/refugees. One such evidence is the principle of *non-refoulement* laid down under the Convention which prohibits states from *expelling or*

---

[191] Jane McAdam, *Climate Change Forced Migration, and International Law* (Oxford University Press, Oxford, 2012) 90–91.

[192] Scott Leckie, 'The Human Rights Implications of Climate Change Where Next?' (2008) 31 FMR 18–19 https://reliefweb.int/sites/reliefweb.int/files/resources/DB0C5ED80E7A7EE1C12574D00033E015-FMR-oct2008.pdf accessed 25 June 2018.

[193] Humphreys (n 185) 7–8.

[194] CESCR, 'The Nature of States Parties' Obligations (Art. 2, Para. 1, of the Covenant)' (General Comment No. 3, 2000) E/1991/23 http://www.refworld.org/docid/4538838e10.html accessed 17 December 2018.

returning a refugee ... to the frontiers of territories where his life or freedom would be threatened on account of his race, religion, nationality, membership of a particular social group or political opinion.[195] However, a problem arises since the Refugee Convention does not recognize the concept of climate refugees. This issue can be mitigated to a certain extent by looking at the concept of non-refoulement from a human rights perspective,[196] which goes beyond the constraints of humanitarian law and assures protection to anyone irrespective of the political refugee status as long as a real danger and risks to life involved when returned.[197] However, the cases decided show lack of consensus in expanding the scope of refoulement to non-political and civil situations. In the case of D v UK,[198] an interpretation was accepted stating that returning a patient to a country with exceptionally poor medical facility would amount to inhuman treatment. However, the same interpretations were not followed in the cases followed.[199] Though it makes legal sense to expand the scope of non-refoulement, it also sets a high threshold benchmark requiring a situation, which makes the survival of the individuals and communities to the level of privation.[200]

The most pervasive contention that advocates that the climate refugees has a right to relocate and resettle anywhere is a human right,[201] which finds its genesis in the UDHR and the two covenants.[202] However, it is questionable whether there is a clearly recognized right to relocate the across borders. In the case of a successful refugee claim, the individual is provided with a legal right to settle. In the case of the relevant human rights documents about asylum and nationality, it confers discretionary powers on the states, and state practices suggest wider discrepancy.[203] Sheri P. Rosenberg has argued that the international community owes the responsibility to take sufficient measures to protect people when a nation-state distinctly

---

195 United Nations Convention Relating to the Status of Refugees (n 137) art. 33 See generally Guy S. Goodwin-Gill and Jane Mc Adam, *The Refugee in International Law* (Second edition, Clarendon Press, Oxford, 1996).

196 Tamás Molnár, 'The Principle of Non-Refoulement Under International Law: Its Inception and Evolution in a Nutshell' (2016) 1(1) Corvinus J. Int'L Affairs 51–61, 53; see ICCPR (n 153) art 7; ECHR (n 161) art 3. These articles contain the principle of non-refoulement.

197 Ibid, Tamás Molnár, Paulina Tandiono, 'Legal Conundrum in the Plight of Pacific Island Countries: Climate Change, Displacement and Human Rights' (Master's Thesis, London School of Economics, 18 August 2016) 17–18 http://www.lse.ac.uk/humanRights/documents/Tandiono-dissertation.pdf accessed 21 December 2018.

198 *D v UK* (1997) 24 EHRR 423.

199 *GS (India) and Ors v SSHD* [2015] EWCA Civ 40.

200 Tandiono (n 198) 15.

201 Cornelis Wolfram Wouters, *International Legal Standards for the Protection from Refoulement* (Intersentia Publishers, Antwerp, 2009) 154.

202 Swapnil Tripathi, 'Climate Refugees: Acknowledging the Existence of an Imminent Threat' (2017) 4(1) NLUJ Law Review 21–34, 25 http://nlujodhpur.ac.in/uploads/16821709020334.pdf accessed 15 January 2018; Shakeel Kazmi, 'Climate Change: Human Rights in the Times of Climate Displacement' (SJD Dissertation, Pace Law School, Pace University, United States of America, 2012) 97 https://digitalcommons.pace.edu/cgi/viewcontent.cgi?article=1013&context=lawdissertations accessed 19 February 2018.

203 *Nottebohm Case (Liechtenstein v Guatemala)* [1955] ICJ Rep 18.

collapses to protect its own people'.[204] The international community owes a natural law duty to preserve humanity, and positive actions must be taken in that regard.[205] Climate change is surfacing as a tragedy that threatens and violates the human rights of the poor and the future generation.[206] It has been argued that all states have the responsibility of assisting those affected by such change out of harm's way.[207]

All UN agencies and member states have pledged to integrate human rights into their actions and to protect natural resources on which human life depends.[208] The significant causes of climate migration such as the sinking of an entire country, desertification, floods, deforestation or substantial proliferation of natural vulnerabilities are destined to affect fundamental human rights like the socioeconomic rights, third-generation human rights, the right to human security and above all the right to life.[209] In this connection, the defencelessness and susceptibility of the climate refugees are contested against the concept of human rights and contrary to the obligations and the principle of CBDR for climate change that must also be reflected in the conduct of nation-states in performing their international commitments.[210]

The discussion attempts to show the intricate relationship between climate change displacement and human rights and tries to establish that human rights regime holds extreme promise for dealing with climate change impacts. Despite the seeming obviousness of a connection in the two fields, states have been wary of developing an exclusive framework coping with climate change displacement as a human rights issue. However, recent developments point towards a shift in the approach with the incorporation of non-binding Declarations explicitly mentioning the affirmative obligations of states to cater to human rights need in climate issues. By concentrating and

---

[204] Sheri P. Rosenberg, 'Responsibility to Protect: A Framework for Prevention' (2009) 4(1) Global Responsibility to Protect 442–477, 443 http://responsibilitytoprotect.org/The%20Responsibility%20to%20Protect%20A%20Framework%20For%20Prevention%20(Rosenberg).pdf accessed 04 May 2018.

[205] J. L. Holzgrefe, 'The Humanitarian Intervention Debate' in J. L. Holzgrefe & Robert O. Keohane (eds), *Human Intervention, Ethical, Legal, And Political Dilemma* (Cambridge University Press, Cambridge, 2003) 15–52.

[206] UNDP, *Fighting Climate Change: Human Solidarity In A Divided World, Human Development Report 2007/2008* (Palgrave Macmillan, 2007)7.

[207] Bridget Lewis, 'Neighbourliness and Australia's Contribution to Regional Migration Strategies for Climate Displacement in The Pacific' (2015) 15(2) QUT Law Review 86–101,101 https://lr.law.qut.edu.au/article/view/625/575 accessed 16 January 2018.

[208] Chaloka Beyani, 'A Human Rights-Based Approach to Protection of Environmentally Displaced Persons' (*Brookings*, 6 June 2011) https://www.brookings.edu/on-the-record/a-human-rights-based-approach-to-protection-of-environmentally-displaced-persons/ accessed 27 December 2017.

[209] Benoit Mayer, 'International Law and Climate Migrants: A Human Rights Perspective' (2011) IDLO-CISDL Legal Working Paper Series on Sustainable Development Law on Climate Change, 08 http://cisdl.org/public/docs/legal/Mayer%20Benoit_International%20Law%20and%20Climate%20Migrants.pdf accessed 27 December 2017.

[210] Benoit Mayer, 'The International Legal Challenges of Climate-Induced Migration: Proposal for an International Legal Framework' (2011) 22(3) Colo. J. Int'L Envt'L L & Poli'y 357–416, 358 https://www.colorado.edu/law/sites/default/files/Mayer%20(Corrected)-S.pdf accessed 12 May 2018.

converging on human rights approach, this section by no means moderates the significance of adaptation actions as the central strategies in addressing climate change-induced displacement. The discussion also accounts for the fact that the international human rights instruments have been developed before the advent of climate change and further studies and research needs to be undertaken to appreciate the precise relationship between climate change and human rights.[211] While not substituting or diminishing the importance of the other evolving approaches in dealing with climate displacement, human rights strategies can provide fillip and play a complementary role in the existing international refugee laws and international climate change law. Increasing demand from vulnerable groups and international scholarship reiterates the requirement of protecting climate migrants, who are undoubtedly worst affected by climate degradation. Currently, it is immensely imperative to rivet upon climate change migration from a holistic manner and to account human rights in a cross-cutting manner to ensure fair and equitable international protective measures for the climate displaced.

3.7 Conclusion

The climate change-induced migration is mostly national (internal/domestic), but many displaced people are bound to cross international borders in the quest for better opportunities for their safety and livelihoods. Political and humanitarian refugees are protected under the refugee law or UNCSR, and economic migrants also enjoy some protection under the ILO conventions. But, unfortunately, environmental/climate migrants do not have any specific or specialized legal mechanism for their protection and treatment, and national governments and comity of nations have not agreed to have a unified definition or universally accepted normative understanding or guidelines for their safety and well-being. The international community is of the view that it is immensely difficult for them to incorporate the climate change displacement into the traditional framework of migration. Often, the migration is perceived regarding 'push' and 'pull' factors. However, in case of climate change migration, the 'push' and 'pull' factors become redundant as the lack of food security, vanishing land and violent conflicts drive them to migrate to safer places. These issues are not merely theoretical but involve practical geostrategic and geopolitical dimensions of far-reaching consequences.

The instant chapter has sought to provide a conceptual and normative understanding of the problem of climate change-induced displacement and migration. The chapter traced the evolution of the concept beginning with the term 'environmental refugees' to 'climate refugees. It has been seen that a definite understanding of the extent and magnitude of climate change displacement is blurred between sci-

[211]Mostafa Mahmud Naser, Tanzim Afroz, 'Human Rights Implications of Climate Change Induced Displacement' (2009) 21(3) Bond L. R. 139–153, 144 https://epublications.bond.edu.au/cgi/viewcontent.cgi?article=1379&context=blr accessed 19 May 2018.

ence, media, politics and academics. Media and politics have repeatedly projected an alarmist picture of climate displacement, whereas scholars and science have often resorted to conservative and sceptic estimate to showcase the climate-induced displacement. The lack of consensus among the international organizations, academics and policy-makers on the definition of climate change has complicated the scenario. The division between maximalists who approached the issue with a unidimensional approach and minimalist which prescribed to the idea that linkage between climate change and displacement is complex got tilted in favour of minimalists by the policy responses. The matter becomes more complicated in the absence of a comprehensive legal mechanism to protect the rights of individuals and communities, particularly in the phase of rapidly evolving nature of forced displacement in the SAARC region. Therefore, evolving a definition is significant in identifying and envisaging the level and kind of protection. However, the absence of definition should not be interpreted as lack of adequate legal mechanisms. The international community needs to evolve a consensus regarding the definition of 'climate refugees' who are in need of legal protection and humanitarian assistance.

Numerous approaches to analyse climate displacement have evolved over the years ranging from migration studies, refugees protection, security narratives and adaptation strategies. However, these approaches take a state-centric view and many a time look at the issue from a positivist point of law and fail to account for climate justice and protection of every individual affected by climate change. In this background, there has been a move to approach the climate displacement from a human rights perspective. The human rights-based approach reminds the people of the mutual dependency and needs for cooperation to solve the problems of the most marginalized and vulnerable. It also reminds us that society and law both are living organisms and requires progressive interpretation. The principles of shared responsibility must be the part of municipal legal systems of individual states and the international community to combat the climate change based on the contours of comprehensive normative lego-institutional and policy framework in the SAARC region. Epistemologically speaking, it is, indeed, critical to take into account the climate change implications, inequality and the human mobility and engraft them into an accountable, coherent, and responsible global and regional geopolitical framework for climate justice. The element of climate sustainability is equality that demands sustained action on climate change to address the continuous climate change-induced displacement of the people.

Chapter 4
Climate Change Migration: Legal Protection Under International Refugee Law and Climate Change Legal Regime

Abstract The climate change migration/displacement is a crisis moment for humanity in the contemporary world that has been impacting the geopolitical and socio-cultural trajectory of the nation-states. Countries have been reordering their geoeconomic dynamics and foreign policy semantics while taking into account the climate change and its impacts. Climate change is destined to produce an unprecedented displaced population that could be measurable to the traditional refugee-like conditions regarding vulnerability and victimhood that would suffer from natural resources and reserve crunch. The climate change-induced displacement and human migration need a response under international refugee law and international climate change law frameworks. Thus, the development of climate change law and protection of CDPs thereunder has to be located by examining the UNFCCC, the Kyoto Protocol, the Paris Agreement and appreciating the legal principles of COPs in response to the climate migration. At the same time, the legal protections under refugee law limitations thereof and application of human rights law, country of origin and host state responsibilities have also been addressed by an advocacy for an alternative regional or international legal regime on climate refugees. Therefore, the instant chapter addresses the extent of protection granted to climate-displaced people under refugee law, climate change law and human rights law. The chapter discusses the complex legal issues associated with climate change-induced displacement and analyses the existing legal protection for climate change displacement under international law. The first part of the chapter makes an attempt to identify the inherent provisions to address the issue of climate change displacement under the international climate change law. The second part of the chapter addresses the question, whether the climate change-induced displacement could be categorized as political refugees and whether the issue of climate displacement could be attended and addressed under the existing refugee law framework.

S. Jolly and N. Ahmad, *Climate Refugees in South Asia*, International Law and the Global South, https://doi.org/10.1007/978-981-13-3137-4_4

4.1 Introduction

The climate change is not just an environmental disaster in waiting; it strikes at the very existence of humanity with a multitude of social and economic consequences. To make the matters complicated, the climate change impact will not be uniformly distributed; it is expected to have a differential impact with most vulnerable and poor communities and nations facing the severe impact. The climate change[1] has caused and exacerbated human displacement[2] with an estimated 50 million refugees worldwide at present. The current estimates are bound to be multiplied in decades ahead as droughts; water shortages and natural disasters are made even more severe by global warming. The opinion of experts suggests millions of climate change-induced displaced people by 2050 or 2100.[3] Tuvalu, Kiribati, Vanuatu, Maldives and other small islands are facing the threat of total submersion of their land and have become the face of climate change.[4] The grave situation has forced many of the residents to seek refuge in other countries and demand relocation and rehabilitation measures.[5] Different solutions have been debated by experts and policy-makers to overcome this situation. The Pacific Access Category (PAC) agreement signed between New Zealand and Tuvalu whereby New Zealand has relaxed its immigration laws and stipulated a specific quota to be granted residence is one such agreement.[6] However, the moot question is on the larger role and legal response of the international community to the humanitarian crisis brought about by climate change. The problem of climate change-induced forced migration has been acknowledged in principle by the academics, UN, environmental expert groups, NGOs and other specialized agencies,

---

[1]Climate change is a permutation and transformation in global climate patterns, especially a change attributed mostly to the enlarged levels of atmospheric carbon dioxide generated by the usage of fossil fuels.

[2]Human displacement denotes forced movement of people from their home within the country called Internal Displacement or national displacement, and such people are addressed as Internally Displaced Persons (IDPs). In case of International Displacement people move across international borders from their country of origin and seek refuge in another country called refugees. Human displacement is also synonymous with the term forced migration. Human displacement is caused due to many different factors like population displacement resulting from development; there are typically two types: direct displacement, which leads to actual displacement of people from their locations, and indirect displacement which leads to a loss of livelihood.

[3]Frank Laczko and Christine Aghazarm (eds.), *Migration, Environment and Climate Change: Assessing the* Evidence (Geneva, Switzerland: International Organization for Migration, 2009), 9, Norman Myers, Environmental Refugees: A Growing Phenomenon of the 21st Century, (2002) 357 Philos. Trans. R. Soc. Lond. B. Biol. Sci. 609–613, 609, Beyond Borders: Our Changing Climate its Role in Conflict and Displacement, Environmental Justice Foundation 2017.

[4]Tiffany T.V. Duong, 'When Islands Drown: The Plight of "Climate Change Refugees" and Recourse to International Human Rights Law', [2010] 31(4) U. Pa. J. Int'l L. 1239–1266, 1239.

[5]Ibid, 1248.

[6]Helen Lee, 'Pacific Migration and Transnationalism: Historical perspectives', in Helen Lee and Steve Tupai Francis (eds) *Migration and Transnationalism: Pacific Perspectives*, (ANU Press 2009) 10; see also Angela Williams, 'Turning the Tide: Recognizing Climate Change Refugees in International Law', [2008] 30(4), Law & Poly 502–529, 515; see also Paramjit S. Jaswal and Stellina Jolly, 'Climate Refugees: Challenges and Opportunities for International Law' [2013] 55 JILI 45–58, 45.

but, unfortunately, this problem does not figure in governmental and institutional agenda of the international community.

The chapter addresses the legal protection of the climate-induced displaced population. The main focus of the chapter is not on whether to address the climate-displaced population as migrants or refugees. The idea is that climate change is a humanitarian crisis and whatever may be the complex political, economic and social factors influencing the extent of climate change, the situation will produce a large number of people who could be compared to traditional refugee-like situations in terms of vulnerability and the absolute lack of control of existence and resources. Whatever may be the name assigned to climate-induced displacement, the legal response is called for. The chapter addresses the extent of protection granted to climate-displaced people under international climate law, international refugee law, and international human rights law. The chapter discusses the intricate legal issues associated with climate displacement and analyses the available legal protection for climate displacement under international law. The first part of the chapter attempts to identify the inherent provisions to address the issue of climate displacement under the international climate change regime. The second part of the chapter discusses the question, whether the climate-induced displacement could be categorized as refugees and whether the issue of climate displacement could be attended under the international refugee law.

4.2 Protection Under International Climate Regime: Development of International Climate Regime

The issue of climate change emerged in the international sphere as an issue to be handled with some unique characteristics which previous environmental hazards and crisis lacked. Climate change threat was global and multiscalar in nature and touched upon the basic issues of human existence.[7] The multiscale nature of climate change was considered appropriate to be tackled at the domestic level, and steps were taken to address climate change.[8] However, soon the international community understood the inadequacy of solely depending on domestic measures and strived to establish concrete international cooperative measures.[9] Developing a universal consensus in the case of climate change was a herculean task with several conceptual, economic, social and legal constraints.[10] The initial complication was on account of

---

[7]Sam Barett, 'The Necessity of a Multiscalar Analysis of Climate Justice', [2012] 37(2) Progress in Human Geography 215–233; Anna Grear, *Choosing a Future: The Social and Legal Aspects of Climate Change* (Edward Elgar Publishing 2014) 2.

[8]David A. Wirth, 'The International and Domestic Law of Climate Change: A Binding International Agreement Without the Senate or Congress?' [2015] 39 (2) Harv. Envtl. L 515–566, 517.

[9]Elizabeth Burleson, 'Climate Change Consensus: Emerging International Law' [2010] 34 Wm. & Mary Envt'l. L. & Pol'y Rev. 543–588.

[10]Freidrich Solatu, *Fairness in International Climate Change Law and Policy*, (Cambridge University Press 2010) 13.

the fact that the science of climate change was not acknowledged and uncertainty was pervasive. The cause and origin of climate change were difficult to be pointed out as it touches upon all the spectre of human lives and activities ranging from power sectors, transportation, industry, agricultural, etc.[11] Any initiative to curb climate change would have the effect of touching upon the essentials of life affecting economic, social and political sphere of communities and societies.[12] The strong realization that a group of nations has mainly created climate change through their past actions makes it extremely difficult for evolving a common strategy. This also explains the fascination of the international community to concentrate on the issue of climate responsibility and developing the principle of CBDR which formed the foundation of the later climate regime.[13] The acute awareness regarding the climate change as an issue having a differential impact on different countries; communities and gender further made it a challenge for the international community to evolve a common strategy.[14] The global impact of climate change, its ensuing consequences and the enormous costs involved in mitigation presented a realization to the international community for global action.[15] The existing international legal precedents in the form of the 1979 Geneva Convention on Long-Range Trans-Boundary Air Pollution[16] as well as the 1985 Vienna Convention for the Protection of the Ozone Layer[17] and its 1987 Montreal Protocol on Substances that Deplete the Ozone Layer[18] constituted serious precedents at adopting innovative and novel mechanisms at tackling global environmental issues through multilateral cooperation.

The transition from domestic action to global action was initiated in 1979 when the scientific community highlighted the issue of rising CO2 level and organized the First World Climate conference.[19] The conference identified 'climate change' as an 'urgent world problem'.[20] The World Climate Programme was established with the

---

[11]Ibid.

[12]Ibid.

[13]Sumudu Attapattu, 'Climate Change Differentiated Responsibilities and State Responsibility: Devising Novel Legal Strategies for Climate Damage' in Benjamin J. Richardson, Yves Le Bouthillier, et al., (eds) *Climate Law and Developing Countries Legal and Policy Challenges for the World Economy New Horizons in Environmental and Energy Law series* (Edwin Elgar 2009) 37–62, 38.

[14]Joydee Gupta, 'International Law and Climate Change: The Challenges Facing Developing Countries' [2006] 16(1) YBIEL 119.

[15]Jacqueline Peel, 'Climate Change Law: The Emergence of a New Legal Discipline' [2008] 32(3) Melb U Law Rev 922–979, 926.

[16]Convention on Long-Range Transboundary Air Pollution (adopted 13 November 1979) 1302 UNTS 217.

[17]Vienna Convention for the Protection of the Ozone Layer (adopted 22 March 1985) UN Doc UNEP/IG.53/Rev.1.

[18]Montreal Protocol on Substances that Deplete the Ozone Layer, opened for signature Sept 16, 1987, S. Treaty Doc. 10, 100th Cong., 2d Sess. (1987), reprinted in 26 LL.M. 1550 (1987).

[19]Daniel Bodansky, 'The history of Climate Change' [2001] International Relations and Global Climate Change 23.

[20]Ibid, The world climate conferences are a series of international meetings, organized by the World Meteorological Organization (WMO), about global climate issues.

mandate to look for a scientific and legal solution.[21] By late 1982, a strong realization has grown that the climate change needs to be addressed through strong international cooperation.[22] The Hague Declaration on the Environment strongly affirmed the science of climate change and the dangers it presents to humanity. The Declaration stressed the need for immediate and concrete international concerted action[23] and the right to have a healthy environment as a human rights and consistent with the right to life.[24] Declaration also highlighted the issue of equity and emphasized the need to evolve a principle of responsibility based on the differential contribution of countries towards climate change, especially the industrialized nations.[25]

Following the Declaration, the growth and evolution of climate change have been rapid.[26] In 1988, the UN General Assembly, Resolution 43/53, declared that 'climate change is a common concern of mankind'.[27] In 1988, the UNEP and World Meteorological Organization (WMO) established the IPCC to provide the assessment of the issue of climate change.[28]

The IPCC consisting of scientific experts, government representatives and civil society groups provided a kind of legitimacy and increased the scientific certainty of climate change. The first assessment report of IPCC highlighted the gravity of the climate change and pointed out the environmental and socio-economic consequences of climate change.[29] The IPCC report acted as backburner for the UN Resolution 45/212, wherein the UN General Assembly (UNGA) instituted the 'Intergovernmental Negotiating Committee for a Framework Convention on Climate Change' (INC/FCCC).[30] The negotiations concluded in the adoption of a framework convention in the form of UNFCCC at the Rio summit in 1992.[31]

---

[21] Ibid.

[22] Michele M. Betsill, 'International Climate Change Policy: Complex Multilevel Governance', in Regina S. Axelrod, David Leonard et al. (eds), *The Global Environment: Institutions, Law, and Policy* (2nd ed. CQ Press 2005) 234–258.

[23] The Hague Declaration on the Environment (adopted 11 March 1989) 28 ILM 1308.

[24] Fransesco Francioni and Ottavio Quirico, 'Untying the Gordian Knot: Towards the Human Right to a Climatically Sustainable Environment' in Ottavio Quirico, Mouloud Boumghar (eds), *Climate Change and Human Rights: An International and Comparative Law Perspective*, Routledge Research in International Environmental Law (1st ed Routledge 2015) 133–156, 143.

[25] Ibid.

[26] Kevin R. Gray, Cinnamon Piñon Carlarne, et al., *The Oxford Handbook of International Climate Change Law* (Oxford University Press 2016) 4.

[27] Protection of global climate for present and future generations of mankind, UNGA A/RES/43/53, 1988 states that climate change is a common concern of mankind, since climate is an essential condition which sustains life on earth.

[28] The IPCC was established to provide the decision-makers and others interested in climate change with an objective source of information about climate change available at http://www.ipcc.ch/about/index.htm accessed on 3 March 2017.

[29] Lorraine Elliott, *The Global Politics of the Environment* (2nd ed Palgrave Macmillan 2004) 167.

[30] 'Protection of global climate for present and future generations of mankind' A/RES/45/212, 1990.

[31] United Nations Framework Convention on Climate Change (adopted 29 May 1992, entered into force 21 March 1994) 1771 UNTS 107 (hereinafter UNFCCC), see Regina (n 22) 34.

4.3 The United Nations' Framework Convention on Climate Change: Declaration of Principles

The process of negotiations under INC/FCCC resulted in a Framework Convention which was adopted at the 1992 Rio Conference. The negotiation process clearly showed the complexities associated with climate change and the differing and often conflicting position of parties.[32] The negotiation process highlighted that though the issue of climate change is based on science, the solution to it lies in evolving effective political, economic and social mechanisms.[33] The issue of assigning responsibility became the most contentious issue which brought a sharp divide between Global South and Global North countries. The answer to the issue was found in historical perspective and based on equity and pragmatism.[34] The developing nations alleged that since the industrialization, the actions of greenhouse emission by developed countries are responsible for the present global crisis. Thus, considering the polluter pays principle, it seemed that the responsibilities of countries to restore environment should be based on the proportion made to the contribution to climate change.[35] The equity also demanded differential responsibility as it would be the Global South countries that are more amenable to the hazardous impacts of climate change-induced migration in future.[36] With this background, the principle of CBDR emerged as the core fundamental aspects of the climate change regime.[37]

By approbating the Declaration and its principles, the UNFCCC shaped the framework for further measures and collaboration instead of a comprehensive and mandatory treaty to reduce and mitigate the greenhouse gas emissions (GHGs).[38] The UNFCCC intends to make the concentrations of GHG in the atmosphere stable at a stage that would curtail intimidating anthropogenic intrusion in the climate system.[39]

---

[32]Elliott (n 29) 168.

[33]James Ridge, Samuela Bassi, 'Global Cooperation and Understanding to Accelerate Climate Action' in Nicholas Stern, Alex Bowen, John Whalley eds, *The Global Development of Policy Regimes to Combat Climate Change* (World Scientific Press 2014) 1.

[34]Philippe Cullet, 'Differential Treatment in Environmental Law: Addressing Critiques and Conceptualizing the Next Steps' [2016] 5(2) Transnat'l Envt'l.L 305–328, 307.

[35]Edward A. Page, 'Distributing the Burdens of Climate Change' [2008] 17(4) Envt'l Politics 556–575, 7; see also M Coghlan, 'Prospects and Pitfalls of the Kyoto Protocol to the United Nations Framework Convention on Climate Change' [2002] 3 MJIL 165.

[36]Yoshiro Matsui, 'Some Aspects of the Principle of Common but Differentiated Responsibilities', International Environmental Agreements: Politics, Law and Economics, (2002) 2 (2) 155 quoted by Rachel Boyte, 'Common but Differentiated Responsibilities: Adjusting the Developing/Developed Dichotomy in International Environmental Law' [2010] 14 Nz. J. Envt'L Law, 63–101, 70.

[37]Lavanya Rajamani, 'Differentiation in the Emerging Climate Regime' [2013] 14(1) Theor. Inq. L 151–171, 153.

[38]Sebastian, Oberthur and Hermann E. OTT, *The Kyoto Protocol: International Climate Policy for the 21st Century* (Springer 1999) 3; see also Daniel Bodansky, 'The History of the Global Climate Change Regime' in Urs Luterbacher and Detlef F. Sprinz (eds) *International Relations and Global Climate Change*, (MIT Cambridge 2001) 32.

[39]UNFCCC, (n 31) art 2; *see* also Edith Brown Weiss, 'Climate Change, "Intergenerational Equity, and International Law"' [2008] Vt. J. Envtl. L. 615–627, 617.

The Convention is based on the principle of 'common but differentiated responsibilities and respective capabilities' and urges the developed nations to take responsibility for the present climate change crisis.[40] A key aspect of the Convention is its division of nations into 'Annex I' (forty-one industrialized countries and economies in transition, such as the former Soviet Union and Eastern European states) and 'Annex II' (24 wealthy members of the Organisation of Economic Co-operation and Development).[41]

The UNFCCC article 4.2 obliges Annex I parties to adopt national policies to mitigate climate change and strive to reduce their emissions to 1990 levels.[42] The preamble paragraphs of UNFCCC recognize the past greenhouse gasses emission of developed countries and the need for the development of developing countries.[43] The UNFCCC requires developing countries to develop in a sustainable manner and address the adverse effects of climate change through adaptation. By invoking the principle of sustainable development in the context of developing nations, the economic, social and developmental concerns of the Global South nations got a clear affirmation.[44]

The main focus of UNFCCC was on climate mitigation. The UNFCCC focused on reducing greenhouse gas emissions and enhancing sinks and reservoirs.[45] Even though the 1990 IPCC report underlined the issue of climate change migration, however, the UNFCCC is reticent and does not have any indication to the issue of climate change-induced displacement, migration and climate refugees or refugees. One of the major causes could be the fact that climate change-induced displacement and migration are primarily regarded to be part of adaptation. However, the UNFCCC is devoid of specific language pertaining to climate change adaptation and

---

[40]Ibid at art. 3.1.1 provides The parties should protect the climate system for the benefit of present and future generations of humankind, on the basis of equity and in accordance with their common but differentiated responsibilities and respective capabilities. Accordingly, the developed country parties should take the lead in combating climate change and the adverse effects; see also Lavanya Rajamani, *Differential Treatment in International Environmental law* (Oxford University 2006)176.

[41]Ibid; see also Mary J. Bortscheller, 'Equitable but ineffective: How the Principle of Common but Differentiated Responsibilities Hobbles the Global Fight against Climate Change' [2010] 10(2) SDLP 49–53, 65–68.

[42]Ibid art 4.2, *see also Malgosia* Fitzmaurice, 'Responsibility and Climate Change' [2010] 53 GYIL 89–108,106.

[43]UNFCCC (n 31) paragraph 3 to preamble.

[44]Ibid Principle 3(5) The parties should cooperate to promote a supportive and open international economic system that would lead to sustainable economic growth and development in all parties, particularly developing country parties, thus enabling them better to address the problems of climate change.

[45]Ibid, art 3.3 principles, The parties must resort to the precautionary measures to forestall, avert or reduce the factors of climate change and mitigate its negative effects. Where there are severe threats or irrevocable damage, lack of full scientific certitude should not be exhausted as a cause reason for deferring such steps, taking into consideration that policies and actions to cope with climate change must be cost-effective so as to maximize the global benefits at the lowest estimated cost. To achieve this, such policies and actions must consider multiple socio-economic approaches to address all contexts comprehensively with all relevant sources, sinks and reservoirs of greenhouse gases (GHG) and acclimatization that consist of all economic sectors.

only contemplates few obligations on the parties to propose policy-oriented actions, initiate measures and collaborate in the issues of adaptation.[46] As observed by IPCC that the 'Increasing opportunities for human mobility can lessen the vulnerability and susceptibility for such peoples. Changes in migration trends can be reactions to both severe weather events and prolonged climate unpredictability and change, and migration can be an efficacious adaptation strategy".[47] Since migration *per se* is regarded as a manner of adaptation, one of the manners of observing the issue of climate change-induced displacement is through the lens of adaptation strategy and vulnerability. Two articles under UNFCCC could be relevant here, article 4.8 which enjoins all parties to extend full consideration to the actions necessary and germane to meet the particular requirements and apprehensions of Global South countries stemming from the adverse impacts of climate change or the effects of the implementation of the response measures that consist of the actions relating to funding, insurance and the transfer of technology.[48] The UNFCCC provision looks at the issue of vulnerability through geographical indicators rather than looking at individual or community levels including gender, children. The population and communities displaced as a result of climate change from these vulnerable areas can seek measures.[49] The clear vision of the CBDR principle is visible in elaborating special provisions including technological and financial assistance for different categories of states including least developed countries,[50] vulnerable economies,[51] Small Island Nation-States; countries with low-lying shoreline areas; nation-states with arid and semi-arid zones, woodland areas and the areas disposed to forest degeneration and states with areas prone to natural disasters.[52] The broad scope of the article makes it precisely difficult to assess the differential adverse impact of climate change; however the provisions are significant in evolving adaptation strategies for the vulnerable population and increasing their resilience to reduce climate-induced displacement.[53] Though displacement is not the core focus of the UNFCCC, some of the core princi-

---

[46]Ibid art 4.b commitments.

[47]IPCC, 2014, *Climate Change 2014, Impacts, Adaptation, and Vulnerability*—Summary for Policymakers, Contribution of Working Group II to the Fifth Assessment Report of the Intergovernmental Panel on Climate Change [Field, C.B., V.R. Barros, D.J. Dokken, K.J. Mach, M.D. Mastrandrea, T.E. Bilir, M. Chatterjee, K.L. Ebi, Y.O. Estrada, R.C. Genova, B. Girma, E.S. Kissel, A.N. Levy, S. MacCracken, P.R. Mastrandrea, and L.L. White (eds.)]. World Meteorological Organization, Geneva, Switzerland, 20.

[48]UNFCCC (n 31) art 4.8.

[49]Rowena Maguire, 'Governance of Climate Displacement Regime with the UN Climate Regime' in Cadman, Rowena Maguire, Charles Sampford, *Governing the Climate Change Regime: Institutional Integrity and Integrity Systems* (Taylor and Francis 2016) 124.

[50]UNFCCC (n 31) art 4.9.

[51]Ibid art 4.10.

[52]Ibid art 4.9.

[53]Sumaya Zakieldeen, Adaptation Under UNFCCC, Working Paper prepared by ECBI partner FIELD, 2010, http://www.eurocapacity.org/downloads/Adaptation_framework_under_the_UNFCCC_June_2010.english.pdfaccessed accessed on 12 April 2017.

ples engrained under UNFCCC could be interpreted to address climate displacement concerns.

Article 3 of UNFCCC introduces the CBDR principle along with the principle of the precautionary principle and the right to sustainable development to achieve the objective of the Convention.[54] The CBDR principle under UNFCCC provides that countries shall be differentiated for their obligations towards the climate change as developed and developing country parties.[55] Differential obligations are imposed on the developed countries regarding mitigation measures and financial assistance to developing nations.[56] The principle of CBDR runs through the breadth and length of the UNFCCC.[57] Since the major cause of climate change is the result of historical GHG emissions produced by the developed nations, consequent climate displacement could also be attributed to the activities of the developed nations. Based on the CBDR principle, a differentiated responsibility could be imposed on the developed countries to protect and conserve the rights of climate-displaced population through measures aiming to reduce displacement and envisaging adequate rehabilitation and relocation measures.[58]

The principle of sustainable development attempts to balance the developmental needs and requirements of current and future generations.[59] The principle of sustainable development is echoed, in the UNFCCC. It states 'parties should protect the climate system for the benefit of present and future generations of humankind, on the basis of equity'.[60] The IPCC report confirms the view that climate change is a serious equity issue between generations. In fact, the most extreme damages are expected to occur in the far-distant future, perhaps one, two or even three centuries from now.[61] Thus, sustainable development and intergenerational equity are at the heart of climate change. Any effective climate policy should invoke the principle of intergenerational equity and sustainable development in terms of burden sharing, assistance and evolve a policy which will reduce the risk of climate displacement for the present generations and the future generations as well.

---

[54] Fitzmaurice (n 42) 106.

[55] Stellina Jolly, 'Application of Solar Energy in South Asia: Promoting Intergenerational Equity in Climate Law and Policy' [2014] 7 (1) Int. J. Private Law 20–39, 22.

[56] UNFCCC, (n 31) art. 11.5.

[57] Rowena Maguire, 'The Role of Common but Differentiated Responsibility in the 2020 (2013) 4 Climate Regime' CCLR 1–10.

[58] Sumudu Attapattu, *Human Rights Approaches to Climate Change: Challenges and Opportunities* (Routledge 2015) 123.

[59] See generally, Howard Mann, 'Comment on the Paper by Philip Sands', in Winfred Lang, *Sustainable Development and International Law* (Springer 1995) 71.

[60] Jennifer Morgan, David Waskow, 'A New look at Climate Equity in the UNFCCC' [2014] (14) 1 Clim. Policy 17–22.

[61] Jolly (n 55) 23.

4.4 Kyoto Protocol, Protecting Climate Through Obligations: An Analysis

The Kyoto Protocol supplemented the UNFCCC.[62] The background for the Kyoto Protocol was initiated immediately after the UNFCCC came into existence. The first COP meeting which took place in Berlin adopted a mandate known as Berlin mandate which called for a legally binding international response based on the principle of CBDR.[63] The Berlin COP instituted a high-level Ad-hoc Working Group on the Berlin Mandate (AGBM) to negotiate a future legal instrument.[64] The negotiations resulted in the Kyoto Protocol, which was adopted by the parties at UNFCCC COP 3 at Kyoto in 1997.[65]

The core commitment of Kyoto Protocol establishes a quantitative legally binding obligation on Annex 1 countries to reduce the greenhouse gas emissions by an average of 5% below the 1990 level for the first commitment period of 2008–2012.[66] The Kyoto Protocol divides state parties into Annex 1 consisting of developed nations and non-Annex countries representing economies considered to be underdeveloped or in the process of developing. Only Annex 1 nations have binding greenhouse gas emission targets, while non-Annex 1 countries have no emission reduction commitments. The emphasis of the Kyoto Protocol is on mitigation. The Protocol adopted an innovative concept of 'flexible market-based mechanisms' in the form of emissions trading system, Joint Implementation and the Clean Development Mechanism (CDM) for achieving the emission reduction targets.[67]

Under the Emission Trading Scheme, each Annex I party's reduction commitments translate into an 'assigned amount' of GHG emissions over the five-year commitment period.[68] At the end of the commitment period, an Annex I party's emissions cannot exceed its assigned amount. The Protocol also establishes cooperative mechanisms in the form of joint projects to reduce emissions, referred to

---

[62]Kyoto Protocol to the Framework Convention on Climate Change, 3d Sess., Agenda Item 5, U.N Doc. FCCC/CP/1997/L.7/Add.1, 37 I.L.M. 32 (1997) (hereinafter Kyoto Protocol).

[63]Wytze van der Gaast, *International Climate Negotiation Factors: Design, Process, Tactics* (Springer 2016) 56.

[64]Ibid.

[65]K. Sheeran, 'Beyond Kyoto: North-South Implications of Emissions Trading and Taxes' [2006–2007] 5 Seattle J. for Soc. Just. 697–720, 702.

[66]Kyoto Protocol (n 62) art 2; see also Edith Brown Weiss, 'Strengthening Compliance with Climate Change Commitments' in Holger P. Hestermeyer et al. (eds.), *Coexistence, Cooperation and Solidarity* Volume I, (Martinus Nijhoff 2012) 693–720.

[67]Sebastian (n 38) 340.

[68]Kyoto Protocol (n 62) art 3 provides The parties included in Annex I shall, individually or jointly, ensure that their aggregate anthropogenic carbon dioxide equivalent emissions of the greenhouse gases listed in Annex A do not exceed their assigned amounts, calculated pursuant to their quantified emission limitation and reduction commitments inscribed in Annex B and in accordance with the provisions of this article, with a view to reducing their overall emissions of such gases by at least 5 per cent below 1990 levels in the commitment period 2008 to 2012.

as Joint Implementation ('JI').[69] Based on the CBDR principle, the Kyoto Protocol establishes the CDM as a mechanism to promote sustainable development in non-Annex countries.[70] The CDM proposes that Annex I governments and companies can earn certified emission reductions ('CERs') by investing in emission reduction projects in non-Annex I countries.[71] The incorporation of CDM has been the sole mechanism for the involvement of non-Annex countries during the Kyoto Protocol's first commitment period.[72]

The Kyoto Protocol came into force on 16 February 2005. The adoption of the Protocol strengthened institutional and legal frames of the climate change regime. The Kyoto Protocol's commitment period officially began in January 2008 and was to expire on 31 December 2012; there was a clear apprehension about the future of the climate change legal regime. At COP 21 in Paris, the world community reached a historical and legally binding agreement, with the target of maintaining the global warming below 2 °C.[73]

4.5 Legal Principles of COP in Response to Climate Displacement

The UNFCCC and the Kyoto Protocol simultaneously do not focus and lack unequivocal provisions to address the menace of climate change-induced displacement and its impact on the social fabric and human rights. However, numerous COPs parleys, consultations and negotiations have underscored the issue of climate change-induced displacement although in a myopic manner. The highest decision-making body of UNFCCC is the COP and that has highlighted the specific issues of climate displacement and been evolving certain mechanisms to deal with climate displacement. The main mechanism through which the climate displacement concerns were incorporated was through the inter-linkage of adaptation and displacement.[74]

---

[69] Ibid art 6; see also Wirth (n 8) 522.

[70] Douglas R. Williams, Anita Halvorssen, J. Kevin Healy, William Pizer and Jacob Werksman, 'Rethinking the Kyoto Protocol: Are There Legal Solutions to Global Warming and Climate Change?' [2006] 5 Wash. U. Global Stud. L. Rev. 333–380, 340 http://openscholarship.wustl.edu/law_globalstudies/vol5/iss2/5 accessed 4 May 2017.

[71] Kyoto Protocol (n 62) article 12.

[72] Williams (n 70) 340; see also D.S. Olawuyi, 'From Kyoto to Copenhagen: Rethinking the Place of Flexible Mechanisms in the Kyoto Protocol's Post 2012 Commitment Period' [2010] 6 LEAD *Journal* 21–35, 23.

[73] UNFCCC. *Adoption of the Paris Agreement*. Report No. FCCC/CP/2015/L.9/Rev.1 http://unfccc.int/resource/docs/2015/cop21/eng/l09r01.pdf accessed 10 February 2017 (hereinafter Paris Agreement).

[74] Michele Klein Solomon and Koko Warner, 'Protection of Persons Displaced as a Result of Climate Change: Existing Tools and Emerging Frameworks' in Michael B Gerrard and Gregory E Wannier (eds.), *Threatened Island Nations Legal Implications of Rising Seas and a Changing Climate* (Cambridge University Press 2013) 243–298, 245.

The COP held in Marrakesh, parties, did not simply finalize the Kyoto Rulebook, but major initiative was taken towards adaptation. A work programme aimed at developing capacity building and national adaptation Plans for least developed nations were initiated.[75] The COP 8 held in Delhi defined the contours of sustainable development and was one of the initial COP where the social issue of climate change was addressed.[76] The COP emphasized the core issues important to developing nations. Poverty eradication was sought through the financial assistance and technology transfer while emphasizing adaptation and mitigation.[77]

The evolving focus on adaptation continued at the COP 9, held in Milan, Italy, in 2003. The adaption was discussed in the context of distribution of reserves, capacity building and technology transfer.[78] Based on the CBDR principle, the COP emphasized the role of developed nations in assisting Global South countries to adapt to climate change. Though the decision does not specifically mention and elaborate climate displacement, the concept of assistance, which should flow from the developed nations to developing nations regarding climate adaptation, could be interpreted to assist vulnerable communities facing displacement. The adaptation was formalized through the mandatory submission of 'National Communication' to UNFCCC.[79] The Milan COP decision also reveals that the core climate principle of CBDR was invoked to address the issue of climate displacement. This proves that the issue of climate migration was also seen as a conflict between Global North and Global South nations' issues.

The COP 11, held in Montreal, concluded and accepted a 'Five-year Programme of Work of the Subsidiary Body for the Scientific and the Technological Advice on Impacts, Vulnerability, and the Adaptation to Climate Change'[80] and 'Further

[75]Conference of the Parties, United Nations Framework on Climate Change, *Report of the Parties on Its Seventh Session, held at held at Marrakesh from 29 October to 10 November 2001, Addendum—Part 2: Action Taken by the Conference of the Parties at Its Seventh Session,* Un Doc FCCC/CP/2001/13/Add.1. (21 January 2002)10/CP 5.

[76]Earth Negotiations Bulletin, 'Summary of the Eighth Conference of the Parties (COP) to the UN Convention on Climate Change' (2002) 12 *Earth Negotiations Bulletin* 1.

[77]Conference of the Parties (COP), United Nations Framework Convention on Climate Change (UNFCCC), *Report of the Conference of the Parties on Its Eighth Session, New Delhi, From 23 October to 1 November 2002 Addendum—Part 2: Action Taken by the Conference of the Parties* (COP) *at Its Eighth Session* UN Doc FCCC/CP/2002/7/Add.1 (28 March 2002) 1/CP8.

[78]Conference of the Parties (COP), United Nations Framework Convention on Climate Change (UNFCCC), *Report of the Conference of the Parties on Its Ninth Session, Milan,* Italy, *from 1 to 12 December 2003 Addendum Part 2: Action Taken by the Conference of the Parties at Its Ninth Session,* UN Doc FCCC/CP/2003/6/Add.1 (22 April 2004) 5/CP9.

[79]Ibid 1/CP.9 and 2/CP.9.

[80]UNFCCC, Conference of the Parties (COP), United Nations Framework Convention on Climate Change (UNFCCC), *Report of the Conference of the Parties on Its Eleventh Session, Montreal, Canada from 28 November to 10 December* 2005—*Addendum—Part 2: Action Taken by the Conference of the Parties at Its Eleventh Session*, UN Doc FCCC/CP/2005/10/Add.1 (30 March 2006) 2/CP 11.

Guidance for the Proper Operation of the Least Developed Countries (LDC) Fund'.[81] Both these initiatives could be applied in the case of climate displacement.

The COP 12 saw concrete steps towards adaptation. Besides the establishment of climate change adaptation fund,[82] Subsidiary Body of Scientific and Technological Advice (SBSTA) was mandated to assess the impacts, vulnerabilities and adaptation about climate change. The Bali Action Plan emphasized adaptation and laid the elements of adaptation.[83] The adaptation measure focused on the special and immediate vulnerabilities of developing countries.[84] The conference finalized the working mechanism of the Adaptation Fund Board under the guidance of the COP/MOP.[85] It established the Ad Hoc Working Group on Long-term Cooperative Action (AWG-LCA), envisioned to negotiate and collaborate a pragmatic climate agreement to replace the Kyoto Protocol after the first commitment period.[86] The Bali Action Plan also initiated the loss and damage mechanism. The action Plan discussed in detail the disaster management and loss and damage mechanism associated with climate change, although discussion on linking loss and damage mechanism to responsibility was absent.[87] This was the first time the term loss and damage was used in a negotiated UNFCCC decision COP 14 in Poznan, while agreeing to operationalize the Adaptation Fund immediately never started any new negotiations in the area of adaptation. However, it brought the term migration and displacement to the climate law lexicon.[88] The COP 14 provided further guidance on the operationalization of the least developed fund established to assist the Least Developed Country parties (LDCs), in their preparation and implementation of national adaptation programmes of action (NAPAs).[89] The decision confirms that the issue of climate displacement

---

[81] Ibid 3/CP.11.

[82] Earth Negotiations Bulletin, *Summary of the Twelfth Conference of the Parties to the UN Convention on Climate Change* (International Institute for Sustainable Development, 2006).

[83] UNFCCC, Conference of the Parties (COP): Bali Action Plan, FCCC/CP/2007, Bali, Indonesia, 14 March 2008, www.unfccc.int/resource/docs/2007/cop13/eng/06a01.pdf accessed 12 February 2017; see also Sven Harmeling and Christoph Bals, Adaptation to Climate Change- Where do we go from Bali? An Analysis of The COP 13 and the Key Issues on the Road to a New Climate Treaty, German Watch Briefing Paper, 2008, 10 https://germanwatch.org/klima/adapt08e.pdfaccessed accessed 10 January 2017; see also Michael B. Gerrard, Gregory E. Wannier, *Threatened Island Nations: Legal Implications of Rising Seas and a Changing Climate* (Cambridge 2013) 287.

[84] Ibid.

[85] Decision 1/CMP.3 FCCC/KP/CMP/2008/11/Add.2 Adaptation Fund (Bali, December 2007).

[86] UNFCCC, Ad Hoc Working Group on Long-term Cooperative Action under the Convention (AWG-LCA), UNFCCC 2007. FCCC/CP/2007/6/Add.1, 14 March 2008, Decision 1/CP.13, para. 2 and Annex 1, http://unfccc.int/resource/docs/2007/cop13/eng/06a01.pdf#page=3 accessed 15 January 2017.

[87] See generally Andrea C. Simonelli, Graycar, *Governing Climate Induced Migration and Displacement: IGO Expansion and Global Policy Implications* (Springer 2016).

[88] Koko Warner, 'Global Environmental Change and Migration: Governance Challenges' [2010] 20(3) Global Environmental Change 1–12; see also Mizan R. Khan, *Toward a Binding Climate Change Adaptation Regime: A Proposed Framework* (Routledge 2013) 29.

[89] Evaluation of UNDP work with Least Developed Countries Fund And Special Climate Change Fund Resources, UND P, 2009, 8.

was considered a part of adaptation entirely. The primary mode of action was the financial assistance from the developed nations to developing nations. Any other independent responsibility of nations and communities towards tackling climate displacement was not discussed. No strategies specifically relating to displacement were adopted.

COP 15 in Copenhagen is credited with directly bringing the issue of climate displacement as an agenda of negotiation through AWG-LCA. The AWG-LCA Report at COP 15 has stated:

Affirms that all parties (are) to augment action threshold on acclimatization and it should be resorted to in accordance with the Convention and the provisions thereunder; follow a country-propelled, gender- sympathetic, participatory approach to handle the adaptation; … taking into consideration their CBDR and corresponding capabilities, and dedicated national and regional development precedences, priorities, targets, and circumstances, …. With the support from developed nations envisage …. Steps to enhance the understanding and cooperation relating to national, regional, and international climate change-induced displacement, migration and estimated relocation, where they lay claim and it is perceived appropriate.[90]

Although it is a positive development, the idea of looking at climate displacement only from the prism of compensation to the developing nations is a one-sided approach, which masks the multiple scenarios associated with climate displacement. Further even regarding compensating the developing nations, there exists a lack of clarity in terms of granting and calculating compensation.[91] National Adaptation Plans (NAPs) are a useful planning tool for parties to prevent and minimize displacement. It should be emphasized that in Resolution 10/4, adopted in March 2009, the UN Human Rights Council recognized the core inter-linkage between human rights and climate change.[92] The issue of inter-linking human rights and climate change has been part of LCA negotiations. The preamble to LCA negotiating text included reference to human rights.

> Noting the resolution 10/4 of the UN Human Rights Council (UNHRC) on human rights and climate change-induced displacement that acknowledges the existence of human beings and their rights are at the centre of concerns for the sustainable development, and the significance of reverend Mother earth and its ecosystems and all its natural creatures. Mindful of the fact that the negative effects of climate change have a wide range of direct and indirect consequences for the full enjoyment of the human rights, including living well, and that the impacts of climate change will be experienced most intensely by those parts of the global population that are already at vulnerable stage due to age, gender, or disability.[93]

---

[90]UNFCCC, *Work Undertaken by the COP at its Fifteenth Session on the Basis of the report of the* Ad Hoc *Working Group on Long-term Cooperative Action under the Convention*, UN Doc FCCC/CP/2010/2 (11 February 2010) Annex II, para 4(f).

[91]Sébastien Jodoin, From Copenhagen to Cancun: A Changing Climate For Human Rights in The UNFCCC? (2011) 3 http://www.preventionweb.net/files/17552_fromcopenhagentocancun20110110.pdf accessed 10 March 2017.

[92]U.N. Human Rights Council [UNHRC] Res. 7/23 (28 March 2008) UN Doc. A/HRC/7/78.

[93]Jodoin (n 91) 2.

However, the preamble lacked clarity regarding the terms including the meaning of 'living well' and the interrelationship between climate change, human rights and sustainable development.[94] Further in spite of the strong demands from the Small Island Developing States (SIDS), COP 15 could not reach an understanding and agreement on the issue of linkage of human rights, climate change and climate migration and the topic does not find any place as part of the Copenhagen accord which elevated the issue of climate change to the political summit level.[95]

At the Cancun COP, the SIDS highlighted the issue of climate displacement; the Environmental Minister of Ghana underscored the internal climate-induced displacement and the more than 300,000 lakh death as a result of climate change.[96] A major victory was celebrated when the Cancun Agreements mentioned the climate change-induced displacement and migration issues for the first time—in particular, ascertaining them as plausible adaptation schemes.[97] The conference emphasized on giving the same priority to adaptation as mitigation strategies under paragraph 11 of the agreement that recognizes adaptation as a challenge faced by all parties and calls for heightened action and international collaboration on adaptation directed at plummeting vulnerability and fostering resilience in Global South country parties, taking into consideration the urgent and immediate requirements of those Global South countries that are exceptionally vulnerable.[98] Paragraph 14(f) of the Cancun Agreement called on the parties to adopt measures to enhance the understanding, harmonization and collaboration about climate change-induced displacement, migration and planned relocation, where it is suitable at the domestic, regional and global levels.[99] By explicitly incorporating the concern of displacement, the conference brought into focus the need for policy research on the core relationship between climate change, migration and displacement and planned relocation at national and international levels. The COP not only recognized climate displacement as a real issue but also gave the legal foundation to the UN regime to address the issue.[100] The nomenclature preferred by the decision was 'climate change-induced displacement', migration and planned resettlement.[101] This was a preliminary step aimed at enhancing awareness and understanding of climate displacement. However, the emphasis on cooperation points to the direction where states are expected to initiate cooperative measures to tackle climate displacement issues. Importantly, the mandate of UNFCCC to look into the climate displacement issue was officially acknowledged.

---

[94] Ibid, 2.

[95] Ibid.

[96] Andrea (87) 145.

[97] The Cancun Agreements: Outcome of the Work of the Ad Hoc Working Group on Long-Term Cooperative Action under the Convention, The *Report of the Conference of the Parties* (COP) *on its Sixteenth Session—Addendum—Part Two: Action taken by the Conference of the Parties*, UN Doc FCCC/CP/2010/7/Add.1 (15 March 2011) para 14(f).

[98] Ibid, see para 11 Cancun Agreement; see also Nina Hall, *Displacement, Development, and Climate Change: International organizations moving beyond their mandates* (Routledge 2016) 33.

[99] Ibid Cancun Agreement.

[100] Maguire (n 49) 124.

[101] Koko Warner, (n 88) 12.

Cancun Agreement also concentrated on the programme 'loss and damage' associated with compensating the people for their loss and damages and enhancing the adaptive capacity of the vulnerable population.[102]

At the UN Climate Change Conference (UNCCC) at Durban in 2011, adaptation remained a central focal theme and climate displacement, or migration was part of it. The conference emphasized balancing the allocation of the green climate fund between mitigation and adaptation.[103] The conference operationalized the Adaptation Board, the primary advising body on adaptation under the UNFCCC.[104] Conference also emphasized the assistance to developing nations in developing national adaptation Plans and on a work Plan to consider tissues of loss and damage.[105] The conference agreed to extend the Kyoto commitment period and negotiate a legally binding climate agreement by 2015.[106]

COP 18 which took place in Doha while linking climate change-induced displacement, migration and human movement with deprivation and damage calls on the parties to undertake enhanced work to advance the understanding of and knowledge on destruction and damage and onset of events including adverse effects of climate change are disturbing patterns of migration, displacement and human movement, climate change-induced displacement is recognized as a non-economic damage.[107] This could make the claim procedure for compensation complex. COP 18 also directed the parties to work on enhancing the understanding of the "patterns of migration, displacement and human mobility as a part of Loss and damage implementation."[108] At Warsaw in Poland, the COP 19 decided to take the formal execution of loss and damage mechanism forward and created the Warsaw International Mechanism (WIM) for the purpose.[109] The WIM mechanism is not compensation awarding body, it does not create any liability, and stress is on the capacity building and institutional development.[110] The loss and damage mechanism creates an institutional structure to provide technical assistance, capacity building to address

---

[102]Lavanya Rajmani, 'Addressing Loss and Damage' [2015] 1(30) *EPW* 17.

[103]Nina Hal (n 98) 33.

[104]Koko Warner and S. Zakieldeen, 'Loss and Damage Due to Climate Change: An Overview of the UNFCCC Negotiations' European Capacity Building Initiative (ECBI) 3, http://www.eurocapacity.org/downloads/LossandDamage.pdf accessed on 12 June 2018.

[105]Nina Hall (n 98) 33.

[106]United Nations Framework Convention on Climate Change (UNFCCC), *Durban Climate Change Conference—November/December 2011* (5 January 2012) http://unfccc.int/2860.php accessed on 12 June 2018.

[107]Decision 3/CP.18, 'Approaches to address loss and damage associated with climate change impacts in developing countries that are particularly vulnerable to the adverse effects of climate change to enhance adaptive capacity', FCCC/CP/2012/8/Add.1 (28 February 2013), para 26, Non-economic Losses in the Context of the Work Programme on Loss and Damage, Technical paper, FCCC/TP/2013/2 (2013).

[108]Ibid.

[109]Decision 2/CP.19, 'Warsaw International Mechanism for Loss and Damage Associated with Climate Change Impacts', FCCC/CP/ 2013/10/Add.1 (2014).

[110]Moguire (n 49) 124.

the onset of extreme weather events disasters and consequent displacement.[111] At COP 20 in Lima, Peru referred to loss and damage mechanism only in the preamble and two-year work agenda was agreed on relating to the loss and damage apparatus.[112] The work agenda seeks information from international organizations to the anticipated migration and the vulnerable communities. The COP also refers to the need for establishing a climate coordination facility to look into and assist organized migration issues of relocation/resettlement and compensation modalities.[113] The steps initiated at these COPs are at the preliminary stage and lack explicit and specific protective programmes on climate displacement. At the COP 21 that was held in Paris, international community negotiated a new agreement to address the concerns of climate change.

4.6 Paris Agreement: Consensus of the International Community

The Kyoto Protocol envisaged the top-down strategy, whereunder specific obligations have been imposed on the nation-states to reduce and mitigate their GHG emissions. However, the Paris Agreement offers flexibility and distinct obligation to each nation-state to proclaim their Intended Nationality Determined Contributions (INDC) for mitigating the climate change and its adverse impacts. The 153 nation-states have submitted and declared their GHG emissions reduction assurances through INDC as of now.[114] However, the fact that these emission reduction objectives will consume years to produce the tangible results and the previous emissions of carbon dioxide would remain in the atmosphere for a more considerable time and the repercussions of climate change would perpetuate over the next some decades to numerous years while pandering to an outpouring of climate displacement.[115]

The Paris Agreement has recognized the climate change migration at various stages. In the way up to Paris negotiations, the draft consultations to the text on the loss and damage had required the instituting of a 'climate change-induced displacement coordination and collaboration facility'.[116] Precisely, the paragraph 50 of the COP decision called the Executive Committee of the WIM to constitute a task force on human displacement that will concentrate on developing recommendations for

---

[111] Ibid.

[112] Decision 1/CP.20, 'Lima Call for Climate Action', FCCC/CP/2014/10/Add.2 (2 February 2015).

[113] Sumudu Attapattu(n 58) 29.

[114] UNFCCC. *INDCs as communicated by parties*. http://www4.unfccc.int/submissions/indc/Submission%20Pages/submissions.aspx.

[115] The Future of Climate Change, https://www.epa.gov/climate-change-science/future-climate-changeaccessed (3 February 2017).

[116] Jessica Wentz and Michael Burge, Designing a Climate Change Displacement Coordination Facility: Key Issues for COP 21, 2015 1–19, 2 http://columbiaclimatelaw.com/files/2016/06/Wentz-and-Burger-2015-09-Displacement-Coordination-Facility.pdf.

the combined approaches to avoid, minimalize, and attend displacement related to the negative impacts of climate change displacement.[117] Therefore, two items attain importance; first, the term 'refugees' has not been subscribed by the Paris Agreement in its schematization. Second, the Paris Agreement makes reference to the issue of preventing or reducing the climate change-induced displacement by underscoring the resilience of populations and securing their livelihoods opportunities. The Paris Agreement has simply impressed upon the nation-states to respect, advance and reflect upon their corresponding obligations in the context of climate change-induced displaced people or migrants. What precisely are these commitments and obligations? What are the rights of CDPs or people or migrants which have not been delineated under the Paris Agreement? The Paris Agreement disappointed the affected communities in recognizing any particular legal protection about prevention, sustenance and clarification of their rights. The Paris Agreement does not stipulate any institutional framework that would take into account the rights of migrants or CDPs. However, solid measures are endorsed to address the loss and damage mechanism comprising the all-inclusive risk assessment and management, emergency preparedness, redressal of non-economic losses, the creation of risk insurance facilities and strategies to increase the resilience of humanity and its institutions.[118]

4.7 Limitations of Climate Law Regime in Dealing with Climate Displacement

Koko Warner sums up the normative evolution and approaches of UNFCCC regarding climate displacement. In the COP held in Poznan, the text referred to only 'migration and displacement'. In June 2009, AWG-LCA draft of the term 'climate refugees' was used.[119] By Copenhagen, the terminology underwent slight modification to 'climate change induced displacement, migration and planned relocation.'[120] Cancun made a deviation from the past development and differentiated the climate migration based on the kinds of mobility (climate change-induced displacement, human migration, planned relocation) and levels (national, regional and global). It also articulated the kind of efforts required (research, coordination, cooperation).[121]

Among the 22 COPs, which have taken place till today, the only actual text, which addressed climate migration, displacement and the consequences of climate displace-

[117]Paris Agreement, (n 71), UNFCCC. (2015) 'Draft decision CP.21 on Agenda item 4(b): Adoption of the Paris Agreement'. Proposal by the President. (https://unfccc.int/resource/docs/2015/cop21/eng/l09r01.pdf).

[118]Ibid.

[119]Koko Warner, 'Climate Change Induced Displacement: Adaptation Policy in the Context of the UNFCCC Climate Negotiations'. Legal and Protection Policy Research Series, UN High Commissioner for Refugees (UNHCR), May 2011, 1–17, 12 http://www.unhcr.org/4df9cc309.pdf accessed 10 May 2017.

[120]Ibid.

[121]Ibid.

ment, was the Cancun Adaptation Framework. The steps are only in the infancy and preliminary, where they are attempting to take stock of climate displacement scenario. Specific institutional and legal protective mechanisms delineating the rights of climate-displaced people are missing. In this scenario, a question arises whether international climate law with its foundation in UNFCCC is the right forum to address the legal challenges arising out of climate displacement. On the positive side, UNFCCC with its near universal membership with a focus on mitigation and adaptation is the right institutional mechanism to address climate change and its multitude of consequences including displacement. Recent times UNFCCC and the various COPs have initiated specific focus on climate displacement.[122] The universality of UNFCCC has forced bodies such as 'Equity and Justice Working Group Bangladesh' (EquityBD), to strive for negotiation for a new Protocol to the UNFCCC for ensuring the security and rehabilitation of climate-induced displaced communities and acknowledging them as 'Universal Natural Persons'.[123] However, these suggestions have not gained much acceptance mainly because there is a consistent view that UNFCCC may not be the most appropriate legal institution to specifically address climate displacement due to its lack of enforcement powers.[124] UNFCCC has approached climate displacement mainly from adaptation. Climate migration may be one aspect of adaptation strategy; however, looking the entire issue of climate displacement from the prism of adaptation is an oversimplification and a restrictive approach to the complicated scenario which goes on to affect a host of human rights.

4.8 Application of International Customary Law

The development of international climate change law suggests that the treaty and convention mechanism are the primary methods of resolving issues. However, most of the treaty mechanisms under the environmental law are couched in the language of soft law lacking legal enforceability, suffer from slow evolving negotiations, require strong compliance mechanisms and have been left to the states to implement through national legislations. In this background, some of the experts and states have started debating the feasibility of applying CIL to effectively address the issue of climate change and the ensuing consequences.

One of the core CIL is expressed as the principle of no trans-boundary harm. The principle could be traced back to the well-known *Trail Smelter* case: wherein the Court observed that:

---

[122] Christina Hibb, James Ford, 'Should the United Nations Framework Convention on Climate Change Recognize Climate Migrants?' [2012] 7(1) *Environ. Res. Lett.* 1.

[123] Jane McAdam, 'Swimming against the Tide: Why a Climate Change Displacement Treaty is Not the Answer' [2001] 23(1) Int'l J. Refugee L. 2–27, 19, see Equity and Justice Working Group Bangladesh (EquityBD), 'Climate Change Induced Forced Migrants: in need of a dignified recognition under a new Protocol, 2009, 2 http://www.equitybd.net/wp-content/uploads/2015/10/climate-refugee-en-bg.pdf accessed 13 June 2018.

[124] Hibb & Ford (n 122) 4.

> no State has the right to utilize or permit the utilization of its territory in such a method as to inflict injury by emission of fumes in or to the territory or land of another or the properties, individuals or persons therein when the situation or case is of severe, consequence, and the injury is recognized and established by clear and cogent evidence.[125]

Subsequently, the principle has been adopted under Principle 21 of the 1972 Stockholm Declaration[126] and Principle 2 of the 1992 Rio Declaration and has also been relied upon as the CIL.[127] The principle was invoked by ICJ in the case of Corfu Channel case.[128] In the case of the ICJ's 1996 Advisory Opinion on *The Legality of the Threat or Use of Nuclear Weapon*,[129] the Court reiterated the general obligation of the states to prevent activities causing harm to other countries and further extended the obligation to areas beyond national jurisdiction as part of the *magnum opus* of international environmental law.[130] The principle of no harm this entitles a state whose rights has been breached by the action of other countries to seek legal remedies.[131] The contours of the responses have been elaborated by the International Law Commission (ILC) draft articles on state responsibility. article 2 of the draft elaborates the requirement for establishing an International Wrongful Act.

- The impugned act or commission or omission must be attributable to the state.

---

[125]*Trail Smelter Arbitration, United States v. Canada*, Arbitral Trib., 3 U.N. Rep. Int'l Arb. Awards 1905 (1941).

[126]Declaration of the United Nations Conference on the Human Environment, U.N. Doc A/Conf.48/14/Rev. 1(1973); 11 ILM 1416 (1972), Principle 21 states: 'States have, in accordance with the Charter of the United Nations and the principles of international law, the sovereign right to exploit their own resources pursuant to their own environmental policies, and the responsibility to ensure that activities within their jurisdiction or control do not cause damage to the environment of other states or of areas beyond the limits of national jurisdiction'.

[127]Rio Declaration on Environment and Development, UN Doc A/CONF.151/26 (vol. I); 31 ILM 874 (1992) Principle 2 states: 'States have, in accordance with the Charter of the United Nations and the principles of international law, the sovereign right to exploit their own resources pursuant to their own environmental and developmental policies, and the responsibility to ensure that activities within their jurisdiction or control do not cause damage to the environment of other States or of areas beyond the limits of national jurisdiction'. See also the preamble of the UNFCCC recalls that 'States have, …, the sovereign right to exploit their own resources pursuant to their own environmental and developmental policies, and the responsibility to ensure that activities within their jurisdiction or control do not cause damage to the environment of other States or of areas beyond the limits of national jurisdiction'. It validates that the no-harm rule forms part of the international law encompassing climate change, and has some significance to the relationship between parties to the Convention.

[128]*Corfu Channel (U.K. v. Alb.)*, Merits, 1949 I.C.J. Rep. 4, 22.

[129]*Legality of the Threat or Use of Nuclear Weapons*, Advisory Opinion, 1996 I.C.J. 266.

[130]Louka, Elli, *International Environmental Law: Fairness, Effectiveness, and World Order*, (Cambridge: Cambridge University Press 2006) 45.

[131]Richard S.J. Tol, Roda Verheyen, State Responsibility and Compensation for Climate Change Damages—A Legal and Economic Assessment, [2004] 32 *Energy Policy* 1109–1130, 1111.

- The act or commission or omission must form the violation of an international legal norm in force for the nation-state at the time.[132]

The fore, practice of international law reveals that attributing the activities of private entities to a state is questionable unless and until clear evidence can be adduced for the effective control of state over the entities.[133] In the case of climate change, most of the GHG emissions emanate from the private entities and state attributability can be raised only when the states fail to ensure compliance with international obligations under national legislation to mitigate GHG emissions.[134] Further establishing a specific breach of international obligation may be difficult in the case of climate change as climate change is an ongoing process and not a result of the specific and single incident. Furthermore, the international law does not always rely on the absolute or strict liability as a yardstick of responsibility, but the emphasis is on fault liability based on negligence.[135] Thus, in case of responsibility for climate change, particularly in case of responsibility towards people displaced by climate change, the establishment of those factors is highly challenging as evidenced from the multitude of climate litigations witnessed in multiple jurisdictions.[136] Further, in the context of environmental law, the progress about liability and compensation has been very slow to come. Principle 22 of the Stockholm Declaration states that 'the states shall collaborate to evolve and develop further the international law regarding fiscal encumbrances, liability, and damages for the victims of environmental pollution and other environmental injuries and damages caused due to the actions within the jurisdiction or control of such nation-states to regions beyond their jurisdiction'.[137] However, the lack of consensus prevented the international community from evolving any concrete steps. In the 1992 Rio Declaration, Principle 23 emphasized developing national law regarding liability than the international law for harmful effects of environmental degradation caused by the actions within their jurisdiction or control to areas outside their jurisdiction.[138] The current prospects of employing the principle of no trans-boundary harm in the context of climate displacement appear bleak.

---

[132]International Law Commission, Draft articles on Prevention of Trans-boundary Harm from Hazardous Activities—with commentaries—report of the ILC on the Work of its Fifty-Third Session (Yearbook of the International Law Commission, vol. II, Part Two, 200). See also, United States Diplomatic and Consular Staff in Tehran case (USA v. Iran), Judgment, I.C.J. reports 1980, p. 3, para. 56; *Military and Paramilitary Activities in and against Nicaragua (Nicaragua v. United States of America),* Merits, I.C.J. reports 1986, p. 14, para. 226; *Gabčikovo-Nagymaros Project (Hungary/Slovakia),* I.C.J. reports 1997, p. 7, para. 78.

[133]Lana Gorell, Climate Change and State Responsibility—Migration as a Remedy? 33 http://lup.lub.lu.se/luur/download?func=downloadFile&recordOId=4905454&fileOId=5010730 (thesis submitted to Lund university, 2014).

[134]Ibid, 33.

[135]Ibid, 34.

[136]Jaquilene Peel, 'Issues in Climate Change Litigation' [2011] 5 *Carbon and Climate Law Review, U of Melbourne Legal Studies Research Paper No. 575*, 1–19, 11 https://ssrn.com/abstract=2060696 accessed 4 March 2017.

[137]Stockholm Declaration (n 126) Principle 22.

[138]Rio Declaration (n 127) Principle 13.

In this background, the concept of human rights obligates states to safeguard the life and property of those within a state's territory. The obligation operates in the case of environmental hazards and climate change consequences and mandates the states to initiate appropriate legal and institutional mechanisms to safeguard the existing population.[139] In the absence of clear-cut legal avenues and rights of the displaced victims under international climate law, it is imperative to look into the potential channel of protection under international refugee law and international human rights law for several reasons. Firstly, the problem of climate law displacement is primarily a humanitarian crisis, and the plight of many categories of climate-displaced people is comparable to the plight and vulnerabilities of refugees. As pointed out earlier, it makes practical sense to employ the term refugees to denote climate-induced displaced population if they can fulfil the criteria advanced by the maximalists. The further part of the chapter employs the term 'climate refugees' primarily for its appeal and also due to lack of a compelling alternative existing at the international level. Secondly, the refugee framework is the closest legal mechanism existing at the international level with decades of experience to tackle the issue of people who are forced to flee from the country of their origin and seek refuge in other jurisdictions.

4.9 The Legal Protection Under International Refugee Law

The International Refugee Law (IRL) can be conceptualized and contemplated as a global human good like a clean environment. The idea of public human good pervades the refugee law that is uncompetitive and inalienable from its cores. These core values are available to each citizen of the world including the refugees as their rights. Tribal people were the first migrants in quest of food reserves, water sources and resources, but they were not yet recognized and designated as asylum seekers and refugees. However, they were only acknowledged as collectors or huntsmen in the pursuit for new habitation and settlements. The new territories of their settlements fulfilled their needs and, subsequently, got amalgamated and united as their inherent assets.[140] Consequently, the struggles and hostilities followed to capture or protect the lands conquered by the new owners or habitants. The new occupiers of land prepared themselves as per official model by protecting and administering the new territories. In the early time, governments formulated the rules and regulations to possess and protect their natural reserves and resources and environmental assets due to the security factors. The scenario remains the same even today as the existing climate change migration legal regime or migrations regulations in the modern nation-states also attend the national security narratives and sovereignty implica-

[139]Dinah Shelton, 'Human Rights and Climate Change' Buffet Center for International and Comparative Studies, Working Paper no 09, (2009) 6.

[140]Daniel Warner, 'Migration and Refugees: A Challenge for the 21st century' in Jean-Yves Carlier, Dirk Vanheule, *Europe and Refugees: a challenge?* (The Hague: Kluwer Law International 1997) 58.

tions.[141] Therefore, it is axiomatic that the tale of migrants is depicted as a narrative of abuse, hostility and racism in the host societies, which has made the position of migrants vulnerable and imperilled.

Thus, movement refugees are coterminous with the history of human existence. However, the understanding of international community to consider the protection and humane treatment and to formulate pragmatic solutions for refugees commenced only under the patronages of the League of Nations.[142] Dr. Fridtjof Nansen organized an inter-governmental conference on 3–5 July 1922, in Geneva, Switzerland. The concept of *Identity Certificates* in the form of Nansen Passport materialized at the conference that made the human mobility easy who otherwise were divested of the national protection of their home countries or countries of nationality. Though, the League of Nations had restricted and defined the refugees by categorizing them according to their deprived, divested or former nationalities of their countries of origin, the Nansen Mandate was successively widened to include other categories of the refugees such as the Armenians refugees (1924), the Assyro-Chaldean refugees and the Turkish refugees (1928). In 1947, the UN has founded the International Refugee Organization[143] (IRO), particularly for the European refugees. However, the IRO was soon replaced by the 'Statute of the Office of United Nations High Commissioner for Refugees'[144] (UNHCR) in 1949. Subsequently, the UNHCR has concluded and adopted the '1951 UN Convention[145] relating to the Status of Refugees' under the mandate of the 'Office of Statute of UNHCR' for an ephemeral period of three years only. Therefore, in June 1950, it was resolved to wind up the same. However, the international community instantly realized that the refugee problem or crisis is not limited to Europe only; rather, it had trickled beyond the European region. Consequently, in 1967, an Additional Protocol was concluded and adopted that abolished the geopolitical demarcation about the applicability of the UNCSR and established it as an instrument of universal application. Thereafter, UNO had created and established

---

[141] The Protection of Refugees and their Right to Seek Asylum in the European Union, Collection Europya Vol 70, 2011, 4 https://www.unige.ch/gsi/files/6614/0351/6348/Bacaian.pdf accessed 13 June 2018.

[142] 1919 Covenant of The League of Nations adopted in Paris on 29 April 1919.

[143] The International Refugee Organization (IRO) was an intergovernmental organization founded on 20 April 1946 to cope with the massive refugee problem in the wake of World War II.

[144] The Statute of the Office of the United Nations High Commissioner for Refugees was adopted by the UN General Assembly on 14 December 1950 vide Resolution 319 (IV) of 3 December 1949, and the United Nations General Assembly decided to establish a High Commissioner's Office for Refugees as of 1 January 1951. The UN General Assembly also called upon the governments to collaborate and cooperate with the High Commissioner in the performance of his or her functions relating to refugees as per the mandate of the Office of the UNHCR. In accordance with the Statute of the UNHCR, the work of the High Commissioner is humanitarian and social and of completely non-political character.

[145] 1951 UN Convention relating to the Status of Refugees was adopted on 28 July 1951 by the UN Conference of Plenipotentiaries on the Status of Refugees and Stateless Persons convened under UNGA Res. 429 (V) (14 December 1950), 189 U.N.T.S. 150 (entered into force 22 April 1954); Protocol Relating to the Status of Refugees adopted by U.N.G.A. Res. 2198 (XXI) (adopted 16 December 16 1966, entered into force 31 January 1967) 606 UNTS. 267.

many lego-institutional frameworks and international arrangements to address the plight of refugees, forced migrants and stateless persons and concluded and adopted the UN Convention on Statelessness[146] and the UN Convention on the Reduction[147] of Statelessness. In addition to these global measures, there have been a plenty of UN conventions, international treaties[148] and customary international law to make available for them legal protection under all circumstances. In substance, the refugee protection regime (RPR) is founded on the UNCSR whereunder UNHCR fulfils its delegated functions and derives its origins from the customary international law principles of human rights. The RPR is conclusively produced by the treaty and customary law standards, especially those stemming from the 1951 UNCSR with its 1967 Additional Protocol.

As per the UNHCR, the 1951 UNCSR has an ethical, moral, legal and political significance and relevance that transcends its particular expression:

> "legal in that it advances the basic norms and standards on which righteous action can be established and substantiated; political in that it affords a really worldwide framework within which nation-states can conjoin, collaborate and share the responsibility stemming from the forced displacement; and ethical in that it is an exceptional instrument acceded to by the 146 states parties of their obligation to preserve and protect the rights of some of the world's most defenceless, vulnerable and disadvantaged people."[149]

Thus, the definition provided under the Refugee Convention outlines the central principal of 'risk of persecution' and *non-refoulement* in providing a clear and certain legal yardstick for the humanity in approaching the concerns of refugees in a shared manner. Hence, 1951 UNCSR defines a refugee as:

> 'Any person who is owing to well-founded fear of being persecuted for reasons of race, religion, nationality, membership of a particular social group or political opinion, is outside the country of his nationality and is unable or, owing to such fear, is unwilling to avail himself of the protection of that country; or who, not having a citizenship or nationality and being outside the country of his former habitual residence as a result of such events, is unable or, owing to such fear, is unwilling to return to it.'[150]

---

[146] 1954 UN Convention relating to the Status of Stateless Persons (adopted 28 September 1954) by a Conference of Plenipotentiaries convened by Economic and Social Council Resolution 526 A (XVII) of 26 April 1954 (Entry into force: 6 June 1960, in accordance with article 39).

[147] 1961 UN Convention on the Reduction of Statelessness adopted on 30 August 1961 by a Conference of Plenipotentiaries which met in 1959 and reconvened in 1961 in pursuance of General Assembly Resolution 896 (IX) of 4 December 1954 (Entry into force: 13 December 1975, in accordance with article 18).

[148] International Human Rights Law (IHRL), International Humanitarian Law (IHL), International Criminal Law (ICL) and Customary International Law (CIL).

[149] UN High Commissioner for Refugees (UNHCR), Refugee Protection: A Guide to International Refugee Law, 1 December 2011, available at http://www.refworld.org/docid/3cd6a8444.html accessed 20 June 2018, see also Erica Feller, International refugee protection 50 years on: The protection challenges of the past, present and future' (2001) 83 RICR, 581–606, 582.

[150] Article 1 (A) 2 of Convention relating to the Status of Refugees Adopted on 28 July 1951 by the United Nations Conference of Plenipotentiaries on the Status of Refugees and Stateless Persons convened under General Assembly Resolution 429 (V) of 14 December 1950.

The refugee definition hereinabove has been phrased comprehensively in general expressions, and it appears ostensibly to incorporate all categories of refugees, but it is a fallacious impression. The 1951 UNCSR stipulates that a person to be eligible as a refugee has to provide that (1) the person has been considered or recognized as a refugee under the previous treaty[151] or (2) the person who 'is outside the homeland or country of his or her nationality' (or not having any nationality) and 'is not able to or not willing to avail himself or herself of the protection of the country of origin or nationality due to a "well-founded fear" of persecution' based on the grounds of nationality affiliation, political beliefs, social origin, religious consideration, and racial roots.[152] However, the persecution grounds specified in the refugee definition have not been elucidated or defined by the framers of the 1951 UNCSR. Moreover, among the common people, the term 'refugee' is understood as relating to people fleeing from ethnic cleansing civil strife, genocide and political persecution deviant to legal contextualization of the 1951 UNCSR refugee definition in the strict sense. Moreover, the most significant element of the refugee definition happens to be 'persecution', but it still remains undefined in the existing refugee discourse due to it being highly politicized in global politics, international relations domestic political priorities. While the 1951 UNCSR definition did offer legal protection to persons with a '*well-founded fear of being persecuted for reasons of race, religion, nationality, membership in a particular social group or political opinion*'[153] during the middle of the twentieth century, but it does not address the contemporary problems such as climate change-induced human displacement and migration that drive people to seek shelter or refuge outside the country of their nationality or origin by crossing international borders,[154] and unfortunately, the legal protection of *climate refugees* or *territorially dispossessed persons* (TDPs) remains flagrantly unaddressed under the existing refugee law or climate change law.

The only way out for incorporating the climate change-induced displacement in the existing refugee definition is by perceiving and treating state mishandling of climate change concerns or negligence in environmental degradation matters as amounting to persecution.[155] However, persecution necessitates a definite intent which is conspicuously absent in the case of general environmental supervision and management and evidencing *mala fide* practices on the part of the governmental steps are

---

[151] Article 1 A (1), 1951 UN Refugee Convention (a person who has been considered a refugee under the arrangements of 12 May 1926 and 30 June 1928, or under the Conventions of 28 October 1933 and 10 February 1938, the Protocol of 14 September 1939, or the Constitution of the International Refugee Organization).

[152] UN Refugee Convention Relating to the Status of Refugees (n 145) art 1A (2).

[153] Brittan J. Bush, 'Redefining Environmental Refugees' (2013) 27 Georgetown. Immigr. L. J., 554.

[154] Bonnie Docherty and Tyler Giannini, 'Confronting a Rising Tide: A Proposal or a Convention on Climate Change Refugees' (2009) 33 Harv. Envt'l. L. Rev. 349–402,402; *see also* Aurelie Lopez, 'The Protection of Environmentally Displaced Persons in International Law' (2007) 37(2) Environmental Law, 365–409.

[155] Jessica Cooper, 'Environmental Refugees: Meeting the Requirements of the Refugee Convention', N.Y.U. Envt'l. L. J. (1988) 6 480–529, 502.

immensely problematic. However, the general environment maladministration policies may have an effect on the population; the policies are not directed at a specific group of persons which is a compulsory requirement under the 1951 UNCSR.[156]

There are no protection clauses in the 1951 UNCSR coping with climate change-induced displacement and climate change migration, and the same is confined to only those people who cross international borders.[157] In case of internal displacement, there is another legal framework called the '1998 UN Guiding Principles on Internal Displacement'[158] which were prepared for the UN Commission on Human Rights that has normative provisions for protection and assistance to[159] IDPs and defines them as under:

> Persons or groups of persons who have been obligated or forced to flee or to leave their homes or places of habitual residence, in particular as a result of or in order to avert the effects or impacts of armed conflict, situations of generalized violence, violations of human rights or natural or human-made disasters, and who have not crossed an internationally recognized State border.[160]

Internal displacement[161], acknowledged through the late 1980s and sprouted into prominence on global priorities in the 1990s. However, the dynamics and undercurrents of internal displacement have shifted globally as aid, assistance and protection are no more limited to refugees as per the 1951 UNCSR, but aid and assistance have also become the necessity of the people, population and persons displaced within their geopolitical boundaries.[162] It is obvious that the IDPs definition provided hereinabove alludes to a concept of protection for climate refugees; however, it also suffers from defects and inadequacies. The IDPs definition is tremendously constrained, lacks universal applicability and also excludes those climate refugees who cross international boundaries. This definition protects only those IDPs who got displaced or deracinated, and their security has been entrusted to the municipal jurisdictions and their instrumentalities. On the other hand, duty-bearers are not performing their obligations that are stuck in external aid and assistance for capacity extension, etc. However, climate change-induced displacement and migration is global problem with global implications in the existing circumstances and encumbered with international obligations. Furthermore, the IDPs framework is not a legally implementable mechanism, and its execution exclusively depends upon the national governments.

---

[156]Ibid, UN Refugee Convention Relating to the Status of Refugees (n 145) art 1A (2).

[157]Angela Williams (n 6) 502.

[158]1998 UNHCR Guiding Principles on Internal Displacement for the UN Commission on Human Rights (UNHCR).

[159]Francis M. Deng, Foreword, in Sorensen, B. and Vincent, M., (eds.), *Caught Between Borders: Response Strategies of the Internally Displaced* (London: Pluto Press 2001) xiii.

[160]Ibid.

[161]See generally Roberta Cohen and Francis M Deng, *Masses in Flight: The Global Crisis of Internal Displacement* (Washington DC: Brookings Institute 1998).

[162]David Korn, *Exodus Within Borders: An Introduction to the Crisis of Internal Displacement* (Washington DC: Brookings Institute 1999) 63.

The 50th anniversary of the 1951 UNCSR offered the international community an opportunity to have a process of global consultations[163] to map the wide-ranging dimensions, ramifications and facets refugee issues and to facilitate the pragmatic and realistic implementation of the UNCSR with its 1967 Additional Protocol. But the question of climate change-induced human displacement did not figure in their trajectory of priorities that has produced the new class of uprooted people called 'climate refugees'.

Under the 1951 UNCSR, a very limited[164] group of people is protected that pandered to a dichotomy in refugee protection project as states incline to coddle with ethical and humanitarian narratives to expand protection to people and persons seeking for refuge, but, unfortunately, states are reluctant and recalcitrant to take the responsibility of such persons. Though regional arrangements in Africa and Latin America have taken several initiatives in expanding the definition of refugees, they do not provide clear and explicit legal protection to 'climate refugees', and domestic, legal systems do not make sufficient arrangements for their well-being.[165] Particularly, South Asia is a region that has not developed any regionally acceptable principle of protecting climate refugees, and there are no conspicuous state practices of protection compatible with global standards of human rights. South Asian countries have not participated in the initial international legal developments about refugees and have also not become a party to the established refugee framework. According to Prof B.S. Chimni, 'passage of national legislation would allow states in the South Asian region to identify and debate their individual concerns, both at the level of security and resources and thereby bring to the fore the divergent perceptions to the refugee problem. They would also accumulate critical experience in their implementation'.[166] Although, he national governments in South Asia have been coping with the problem on *ad hoc* basis with no lego-institutional transparency, commitment, conviction, and framework, there are multiple policy predilections and institutional priorities offered for diverse refugee groups from different nation-states in South Asia devoid of national non-discrimination *juris corpus* and global convention law as contemplated under article 3 of UN Torture Convention of 1984 and 33 (1) of the UNCSR under which the principles of *non-refoulement* envisioned the obliteration of the *forced deportation*. It must be noted that the principle of *non-refoulement* has been acquiring the position of *jus cogens* under international law. The Hindu refugees from Pakistan are extended wide range of rights by the government of India such as

---

[163]Global Consultations on International Protection/Third Track: Asylum Processes (Fair and Efficient Asylum Procedures), Access to Procedures, EC/GC/01/12 (2001) *UN High Commissioner for Refugees (UNHCR).*

[164]William Thomas Worster, 'The Evolving Definition of the Refugee in Contemporary International Law' (2012) Berkeley. J. Int'l. L. Law 30(1) 94–160, 100 available at: http://scholarship.law.berkeley.edu/bjil/vol30/iss1/3 accessed 12 June 2018.

[165]Benoît Mayer, 'Environmental Migration: Prospects for a Regional Governance in the Asia-Pacific Region' [2013] 16 (4) Asia Pac. J. Envt'l. L. 77–103, 92.

[166]B.S. Chimni, 'The Law and Politics of Regional Solution of the Refugee Problem: The Case of South Asia' RCSS Policy Studies 4 https://www.rcss.org/publication/policy_paper/RCSS%20Policy%20Studies%204.pdf accessed 10 June 2018.

education, medical care and employment in the unorganized sector; even the government of India has carried out significant changes in the rules in July 2012, whereby these refugees from Pakistan can now aspire and seek long-term visas that guarantees them employment in the organized sector also. The Tibetan refugees are considered at par citizens of India with all kinds of benefits, and Tamil refugees enjoy political clout in the corridors of power at the state and union government levels. Therefore, the safety and security of the Tamil refugees are advanced at the highest political and government level in India, whereas refugees from Bangladesh and Myanmar (Burma) get a discriminatory treatment and raw deal only on the ground of religion that too at the lego-institutional level copiously supported by the government of India.

4.10 Protection Under International Human Rights Law (IHRL)

Of late, there have been a number of overwhelming natural disasters: the tsunami in South Asia, Hurricane Katrina in the USA, earthquakes in Afghanistan, India, Iran, Nepal, Pakistan and elsewhere, and the list is endless. The natural disasters, coupled with mounting concern for the adverse effect of climate change, have contributed to the comity of nations prioritizing the forced displacement caused due to environmental and climate reasons on its agenda. Consequently, a vortex of issues has been mentioned by the international and non-governmental organizations (NGOs), policy-makers, scholars and the communities who have been affected by the multiple aspects of climate change and environmental and hexicological imbalances. Therefore, there is an emerging need to reconceptualize the idea of forced migration concerning the climate change while evaluating the significance of the existing protection formulations in offering the legal solutions, the possibility of embracing a treaty controlling this category of migrants and evolving the lego-institutional governance response architecture to the climate change migration based on the human rights paradigm.

The international lawyers in their life legitimately assume what Louis Henkin remarkably stated: 'almost all the nations observe and heed to almost all the principles of international law and almost all of their obligations and onuses almost all of the time'.[167] Though international lawyers contest the theory of observations and efficacy of international law, in positive terms, international human rights law has the capability to alter the state behaviour that bewilders the nation-states as human rights conventions impact the core areas of state sovereignty devoid of any geostrategic benefits. Therefore, human rights law makes a case for its central role in shaping and providing the impeccable testimony to the human rights community of practitioners and defenders to invoke the pervasive power of legitimate legal obligations against the state actions.

[167]Louis Henkin, How Nations Behave: Law and Foreign policy (2nd ed. Columbia University Press, 1979) 47; *see also* Abram Chayes and Antonia Handler Chayes, *The New Sovereignty: Compliance with International Regulatory Agreements* (Harvard University Press 1995).

The legal protection of migrants/refugees has evolved from the notion of a minimum standard based on the state responsibility to fundamental rights consecrated in the international human rights law and as such available to every person.[168] Therefore, the ascendance of international human rights law on the global protection scenario has come to stay as the complementary source of protection. Today, international human rights law has been profoundly reshaping and recalibrating the contours of general international law that has progressively diluted the traditional notion of state responsibility and replaced it with the majesty of the human rights paradigm. There is a normative synthesis between the traditional conceptions of treatment of aliens in the host states and contents of fundamental rights contemplated under the rubrics of international human rights law popularly known as minimum standards of protection of refugees.

The human rights approach is attended by a host of positive elements in the context of refugee protection. States have conventionally been perceived as the primary actors in international law. The nation-states are parties to international conventions and treaties including the UNCSR,[169] and they are defined as having specific attributes, obligations and responsibilities, including the responsibility to protect (R2P) their citizens.[170] In particular, the present existence of state protection is a vital component of the definition of refugee status under article 1A (2) of the UNCSR.[171] In the context of refugees, countries of asylum generally do not bear the same degree of responsibility as the countries of origin of refugees in the context understood in the refugee framework.[172] Thus, countries of origin have direct responsibility for refugee flights: it is their responsibility not to construct issues of crisis dimensions for other states, and they should have a greater threshold of responsibility than the refugee-receiving states under the international refugee law. However, the international human right law discourse emphasizes both host states and the country of origin responsibility. The human rights discourse envisions the responsibility of the countries to protect people in the territory from all sorts of harms that may have an adverse impact upon the peaceful enjoyment of the full range of fundamental human rights stipulated in the national constitutions and international human rights instruments. Therefore, if human rights get violated by climate change, such rights are also protected under

168 Vincent Chetail, 'The Human Rights of Migrants in General International Law: From Minimum Standards to Fundamental Rights' [2013] 28 Geo. Immigr. L. J 225–255, 231.

169 189 UNTS 150, supplemented by the 1967 Protocol relating to the Status of Refugees, 606 UNTS 267, Art 39 (2) and (3) provides that signature and accession to the Convention is open only to states (specifically, those states who participated in the Convention drafting process or who are member states of the United Nations).

170 Matthew J. Gibney, 'Liberal Democratic States and Responsibilities to Refugees' [1999] 93 American Political Science Review 169–181, 170; see also Andrew. E. Shacknove, 'Who is a Refugee?' [1985] 95(2) Ethics 274–284.

171 1951 UN Refugee Convention Relating to the Status of Refugees (UNCSR) (n 145) art 1A (2) of the 1951 UN Convention relating to the Status of Refugees that refers to an applicant's capacity to seek the protection of the country of his nationality.

172 Nafees Ahmad, 'Refugees: State Responsibility, Country of Origin and Human Rights' (2009) 2 Asia-Pacific Journal on Human Rights and the Law 1 1–22.

the national as well as international law. However, it is the notion of human dignity occupying the central place in the UN Charter and the UDHR[173] that has emerged as a core value throughout regional and global human rights regimes.[174] Thus, negative impacts of climate change would cripple the capacity of nation-states to meet the requirements of human rights and the protection of human dignity[175] under their legal obligations.

Since the UNCSR was drafted in 1951, the key role of nation-states under international law has been impugned. While states remain to be the predominant actors in the international arena, some scholars[176] record the increasing importance of non-state actors, such as the United Nations Organization, other international organizations, transnational corporations, non-governmental organizations (NGOs) and other groups. The human rights approach assists in comprehensively evaluating the scenario of elevating the individual to the centre stage. Today, the individual has acquired a centrality in the entire scheme of human rights discourse and is now considered as a subject of international law. Moreover, national boundaries are diminishing their significance as international barriers. Consequently, a new world human order has been emerging. The human rights of all people, persons and individuals, including those of refugees, have become a multidimensional discourse heralding a novel premise, whereas the human rights and state concerns are at hammerhead with each other.

Therefore, international human rights law is germane to climate change that causes human rights transgressions. But the prism of human rights is equally important in visiting the consequences of climate change. Therefore, the relationship between human rights[177] and climate change protection[178] under the international law is multidimensional and complicated to understand its socio-legal context, especially to the 'climate refugees', but, unfortunately, the environmental dimensions are rarely cogitated upon in the theoretical, intellectual and scholarly treatments of international human rights law.[179] There is practically no debate on the relationship between human rights and the environment[180]; therefore, there is an emergent need to integrate the

---

[173]Universal Declaration of Human Rights (UDHR), Preamble, GA Res 217A (III), UN Doc A/810 (10 December 1948) ('whereas recognition of the inherent dignity and of the equal and inalienable rights of all members of the human family is the foundation of freedom, justice and peace in the world …').

[174]UN Charter, 26 June 1945, Preamble (determining to 'reaffirm faith in fundamental human rights, in the dignity and worth of the human person …').

[175]Christopher McCrudden, 'Human Dignity and the Judicial Interpretation of Human Rights' (2008) 19 Eur. J. Int'l L. 655, 655, 668.

[176]R Mc Corquodale, 'An Inclusive International Legal System' [2004] 17 Leiden J. Int. L 477–504; see also P Alston (ed), *Non-State Actors and Human Rights* (OUP 2005); see also A Bianchi (ed), Non-state Actors and International Law (Ashgate 2009).

[177]Francesco Francioni, 'International Human Rights in an Environmental Horizon', (2010) 21 EJIL 41–55.

[178]Loukis Loucaides, 'Environmental Protection through the Jurisprudence of the ECHR' (2004) 75 BYBIL 249–267.

[179]Philip Alston, Henry Steiner, and Ryan Goodman, *International Human Rights in Context* (3rd edn, 2008 Oxford University Press) 1433.

[180]Oliver De Schutter, *International Human Rights Law* (2010 Cambridge) 192.

environmental values prominently into the international human rights law framework.[181] In 2011, the UNHCR attempted to develop and codify international law on 'climate refugees' by following the existing corpus of international human rights law. The UNHCR has considered the various possibilities such as recognizing the procedural environmental rights[182] as the most significant addition to the human rights since the 1992 Rio Declaration on Environment and Development,[183] adopting a Declaration or Additional Protocol as a relevant mechanism to advocate the right to a decent environment and addressing the extra-territorial application of international human rights law to trans-boundary pollution.[184] Therefore, the response of the international human rights law to global climate change with a sense of environmental democracy[185] has to be of global application and standard that covers the entire humanity and not merely by emphasizing on just host state or country of origin responsibility. For instance, in the context of 'climate refugees', the flight of population is not due to the direct activity of country of origin; in fact, the climate displacement is directly proportional to the vulnerability and marginalization of nations. Thus, it makes a case for the greater responsibility of host states and also for the role of international human rights law to provide protection and grant rights to climate refugees.

The study of human rights framework establishes the fact that climate change causes human suffering, human misery and human deprivation of large scale and magnitude directly percolated from the human-made imbalances in the natural ecology. The human rights framework must be the ground and criterion to determine future planning for climate-displaced population. However, in the context of climate change displacement, there are many fundamental grounds, that climate change is only one of the provocations why people decide to relocate, move, and that the interrelationships of environmental, socio-economic and political reasons make it insurmountable to determine whether climate change is the prevalent cause. Since climate change migration is demonstrated in multiple and different ways, a one-size-fits-all response is not correct but should look at the human rights dimensions in a holistic manner accounting for regional and local dynamics.

In such a scenario of human rights preservation, promotion and protection, climate human rights cannot be left in a vacuum as they constitute the centrality of human existence. Therefore, addressing the human rights also amounts to recognizing the emerging menace of climate change that has been transgressing the human rights and has made the significant population in many parts of world vulnerable and susceptible to climate injustice. Moreover, there is a need to design with dexterity and pragmatism, international and regional climate programmes. At each stage,

---

[181]Desgagné, 'Integrating Environmental Values into the ECHR', (1995) 89 AJIL 263–294.

[182]Sumudu Atapattu (n 58) 161–162.

[183]Menno T. Kamminga (ed.), Challenges in International Human Rights Law, Ashgate, 2014, p. xvi–xxii.

[184]A. E. Boyle, 'Human Rights or Environmental Rights? A Reassessment' (2007) 18 Fordham Envt'l L Rev 471–508, 498.

[185]Principle 10, Rio Declaration (n 127).

there is a need to assess the substantial burden that incurs and whether it should be shared or attended otherwise.[186] The human rights do not recognize the geopolitical boundaries and make them translucent that the obligations of the national government do not stop at their boundaries.[187] For example, nation-states have an individual responsibility to supervise and, where necessary, regulate the conduct of private entities within their jurisdiction, including those functioning overseas. It is specifically crucial in the context of climate change, where the reasons are generally located in individual actions. Therefore, the big emitters and polluters must not fall through the network of a global system that enforces various obligations on Global North and Global South countries. To a certain extent, entrepreneurial imagination and creativity must be utilized towards reasonable solutions that can mitigate and forestall undesirable human harms presented by the climate change. The climate change-induced displacement and migration jeopardizes the lives and livelihoods of people in different nook and corners of the world. The emergence of climate change has been causing to increasing prices of grains and staples that have been destabilizing the food security for millions of people particularly in the Global South countries where weather patterns are unstable and volatile.[188] Thus, climate change poses an acute danger to development in the Global South countries that presently do not have the sufficient resources to enjoy the fundamental human rights.

In this regard, it is substantially relevant that the pertinent human rights treaties, by and large, continue to be silent on the consequences flowing from the non-compliance of commitments by a state.[189] The human rights basis always requires that a state must satisfy its international commitments to the extent of its obligations.[190] Thus, to that extent, one can presume the presence of an individual right under international law based on the right to specific performance on the part of the government of the country of origin or country of nationality. Therefore, an individual has the right to claim, in case of his or her being in the confrontation with the government or its instrumentalities, humane treatment and behaviour as enshrined in the relevant provisions of the international human rights law. However, as far as human rights transgressions of perennial nature are concerned, an individual may invoke the right to cessation of refugees if they are duly found involved in war crimes, crimes against peace, crimes against humanity and acts committed against the purposes and principles of United Nations. However, it is not completely clear rather to a great extent remain shrouded in obscurity as to the determination whether the right to

[186]Mary Robinson, 'Climate Change and Human Rights: A Rough Guide' (2008) International Council on Human Rights policy, 2 https://www.ohchr.org/Documents/Issues/ClimateChange/Submissions/136_summary.pdf, accessed 20 June 2018.

[187]Frank Newman et al. 'International Human Rights Law, policy, and Process' (4th ed. 2009 Lexus Nexis) 18.

[188]Andrew Borrell, 'As Global Food Demand Rises, Climate Change is Hitting our Staple Crops' (2017) the Conversation https://theconversation.com/as-global-food-demand-rises-climate-change-is-hitting-our-staple-crops-73360, accessed 5 June 2018.

[189]Charles Lipson, 'Why are Some International Agreements Informal?' (1991) 45 INT'L ORG. 495, 506.

[190]Andrew T. Guzman, 'A Compliance Based Theory of International Law' (2018) 56–76.

compensation genuinely comes into existence that would stipulate to annihilate the repercussions of the commission of the criminal or illegal acts by the refugees.

There is a problem that exists between rights and obligations warranting a solution based on the doctrine of state responsibility, climate change law and the laws of humanitarianism and interstate cooperation. However, there are many more concerns as to an obligation for states to help the victims of natural and environmental disasters in general, and therefore, assistance and security should not be segregated. There is an obligation to assist the climate refugees based on humanitarian norms and the same could be transformed into an obligation to accept persons into the host territories. These are the issues, which could be ruminated and rummaged further while keeping in mind the different areas of law, the various methodologies and goals. These capacities of law do not certainly envision obliging states to transgress on their sovereignty by permitting admission grounded on global protection requirements. It is, indeed, pressing to note that there are many types of migration that pandered to human displacement. The climate change and rising sea levels attracted the state responsibility, but it is not possible to construe it into a state obligation to legitimize the trans-boundary migration. In the wake of the abrupt inception of migration due to natural disasters, the acknowledgement of international protection is easily established along with an inherent duty to help on the basis of core values of humanitarian understandings and global collaboration. The impugned duty could elongate as far as conceiving an obligation *erga omnes*, as delineated in the *Barcelona Traction* Case[191] by the International Court of Justice. Thus, over the years, a number of other categories of persons have also been cogitated upon who are in need of international protection with time and space, and their rights have also been codified in the international treaties on the basis of constraining humanitarian considerations.

Therefore, the contemporary migration influxes scenarios have shot the issue of migration into unprecedented prominence that has also found its legitimate place in the international agenda. The international migration, off late, has attained the crisis proportions and mind-boggling magnitude that cannot merely be confined to the trajectory of development models transcending country of origin, transit or reception or a conflation of all. The rhetoric of refugee protection still awaits a comprehensive concept of refugee rights including the rights of CDPs, territorially dispossessed persons (TDPs), climate change-induced displaced people (CCiDPs) or climate refugees. There is an urgent need of having a liberal and cosmopolitan vision of broader refugee protection inclusive of climate change.

4.11 Protection Under International Refugee Law: Kinds?

The most substantive concerns were asked regarding the rights of persons, groups and communities displaced because of climate change and the obligations of indi-

[191] *Case Concerning the Barcelona Traction, Light and Power Company Limited (Second Phase) (Belgium v Spain)* [1970] ICJ Rep 3.

vidual states in South Asia and the global community to offer protection and help. Therefore, here arises a pertinent question: is an individual entitled to seek protection from states other than his/her state of nationality and origin and are such states obligated to provide such protection? As evident from the earlier discussion, for being recognized as a refugee under the legal framework, a person who is outside of his/her country of origin has to establish *inter alia* that he/she is unable or, owing to well-founded fear of being persecuted for the reasons adumbrated in the definition clause of UNCSR, are unwilling to avail themselves the protection of their country of nationality. As per the existing refugee law framework, there is no space to bestow protection to persons compelled to migrate due to climate change that has not even been covered by the UNCSR codifying the principle of *non-refoulement*.[192] In Europe, refugee law is based on the European Convention on Human Rights[193] (ECHR) which encompasses human rights protection regime and the UN Convention against Torture[194] (CAT). Therefore, it can be understood that the word 'protection' conveys the admission to the territory of the host country, protection against forced return to the circumstances where fundamental rights may be imperilled, extending the temporary or permanent legal status and facilitating the accessibility to rights in the reception country. The current discourse on the refugee protection is flawed and fallible due to the misunderstanding of the mandate of the refugee law. The idea of protection that has been contemplated in the refugee law is internal protection which has to be made available by the state within its territories to victims and potential victims of persecution.

But, unfortunately, authors do not find this view in the drafting history of the refugee instruments. In the current circumstances of human displacement of high magnitude, this view must be ideal in interpreting the mandate of UNCSR in conformity with global human rights standards. Hence, the idea of 'protection' must be interpreted as a much broader concept while taking into consideration the mandate of human rights law, refugee law and humantarian law, whereunder the protection of individuals and the human security as a whole has been appreciated. Consequently, the 'protection broadly incorporates the actions intended at ensuring full respect for the fundamental rights of all people and individuals in accordance with international law, deviant to their frailties like gender, age, nationality, socio-economic viability, ethnic, religious or another backdrop.'[195] Therefore, a human rights perspective would underscore non-discrimination, the protection of indigenous groups and ethnic, racial and religious minorities, and the right to preserve one's culture which goes beyond just the question of possible admission to the host country. Therefore, protection concerns of climate refugees displaced because climate change-induced

[192]UN Refugee Convention Relating to the Status of Refugees (n 145) article 33 (1).

[193]Ibid article 2 and 3, Convention for the Protection of Human Rights and Fundamental Freedoms (European Convention on Human Rights, as amended) (ECHR).

[194]UN Convention against Torture and Other Cruel, Inhuman or Degrading Treatment or Punishment (CAT) (adopted 10 December 1984, entered into force 26 June 1987) 1465 UNTS 85 article 3.

[195]Inter-Agency Standing Committee, 'Protection of Internally Displaced Persons: Inter-Agency Standing Committee Policy Paper' (December 1999) 4 http://www.unicef.org/emerg/files/IDPPolicy.pdf accessed 16 August 2017.

migration must be addressed in an adequate manner by adopting a holistic view of their displacement and flight. The contents of protection, justifications and sources must also be addressed on the anvil of political, economic and social conditions as the negative effect impact of climate change for diverse groups and communities and countries depends upon the wider interpretation of safety and protection to the requirement of addressing the defencelessness and vulnerabilities while advancing the capabilities and enhancing the competences.

4.12 Extended Approaches Along the Contours of the Proposed Solutions to Remedy the Plight of Climate Change Refugees

In the context of the SAARC region, the evolutionary peregrination of the 1951 UNCSR and the role of Europe in its development could be the basis of South Asian and Asian extended approaches to regionalize its mandate as per regional aspirations for the protection of the climate refugees. The SAARC reaction to the contemporary refugee law framework in dealing with climate refugee predicaments while challenging the limitations of the Westphalian model for protecting the rights of climate refugees needs to evolve the regional spirit and cooperation. The concerns of climate refugees, responses of refugees and future of refugees must form the part of these extended approaches in the SAARC jurisdictions. Thus, understanding the pathology, progressive evolution and codification of the refugee law and climate change law frameworks under the SAARC geopolitical peculiarities and sociocultural particularities are the critical issues in the contemporary climate change imperatives and refugee regime narratives. Therefore, these issues must be ruminated and delineated in a regional perspective based on the national and transnational participation of stakeholders, climate change experts, refugee defenders, refugee policymakers, CSOs and NGOs, etc.

In essence, such an approach should explore how endeavours to differentiate between forced relocation and voluntary migration have influenced and fashioned international norms, obligations, standards and lego-institutional responses, as well as South Asian states practices and localized strategies and tactics taking into account the broader human rights interpretation developed in oriental and occidental regions.

The South Asian perspective must delineate the role of human rights surrounded by the pre-flight and post-flight gyrations of the refugee know-how in a climate change context. In addition, the socio-economic and political context of deterrence, promotion, safety and resolution stratagems must also be cajoled and delved into. A fundamental appraisal of jurisprudence within global, regional and national courts will offer an insight into how international refugee law should be interpreted and implemented that might also examine the obstacles, disadvantages and advantages of other forms of protection, e.g. temporary protection, subsidiary protection, protection city sanctuaries and safe havens. Therefore, the patterns vis-a-vis the enlargement of the refugee definition should be considered by studying regional perceptions and approaches to climate change refugees, mass exoduses, mass escapes, gender persecution, the emergence of non-state actors and 'development/environmental or CDPs' within South Asia, Africa, Europe, Latin America and rest of the Asian regions. The increasing issues such as IDPs, restitution rights and emergence of sanctuary cities for refugees should also be focused and offered in any model to be adopted for regional refugee order in future. The attempt to modify the gaps in the refugee law framework will, in turn, guide to a comparative research and study of the existing refugee definitions as a foundation for the refugee status determination (RSD), the concern of the safe third country option and the problem of responsibility determination for an asylum entitlement, and other contested problems relating to RSD must be addressed on priority basis in any regional and global understanding. The evolution of the regional refugee protection framework should not rivet on the laws of a single country in particular; rather, the approach must be regional, cosmopolitan and transnational in identifying, arriving and providing the commonalities, cohesion and struggles within the global system so as to achieve exhaustive and comprehensive understanding of the 1951 UNCSR as well as the core regional treaties and conventions for a better South Asian vision that may embrace the necessary UN norms for the accomplishment, deprivation and recantation of refugee status, the evolution of the *non-refoulement* code and the standard and principles of reception and treatment of refugees.

In this context, the reformulation of international refugee instruments and arrangements with all incidental and vicarious refugee protection pacts may have become more imperative now than ever before for all South Asian nations. However, evolving, emplacing and institutionalizing the national refugee protection regime in SAARC nation-states must attend the regional and national requirements, objectives and understandings and this can no more be misjudged and undervalued under international human rights law, whereunder even a person is now considered as a subject matter of general international law, national borders have been losing their utility, and human rights take precedence over sovereignty concerns. In India and somewhere else, the asylum seekers from Bangladesh and Rohingya refugees from Myanmar have become something of a political campaign issue in South Asia in the name of homeland security concerns at the hustings with incumbent governments and opposition parties rivalling with each other to appear toughest on the 'bogus' asylum seekers 'flooding' into India and other SAARC jurisdictions while forgetting the fact they intriguing human sufferings and pain of Himalayan enormity that is iniquitous, impervious and invidious.

The lawmakers must, thus, take the initiative in promoting deference, coexistence and broadmindedness for refugees and in encouraging local people to perceive climate refugees not only as people necessitating and justifying global protection but also as human beings with a real and enduring contribution to make to the advancement and diversification of their societies. The parliamentarians have a major and central role to play in making sure that national debates in South Asia take proper consideration of this important reality and the many positive experienced countries that have enjoyed in welcoming, integrating and assimilating refugees into their communities and societies. The SAARC countries have yet to disconnect refugee problems from their national security narratives and do share the holistic global view of construing them as human rights and humanitarian apprehensions. Therefore, it is most improbable that a regional refugee protection regime, either in the shape of a Declaration, Additional Protocol or a convention, is possible to materialize. Even if it does, in the absence of national laws such a legally binding instrument is expected to be constrained by some reasons and the rights of climate refugees are liable to be compromised. This leaves us with the alternative of having inclusive national law on refugees including climate change refugees and change-induced or forced displacement. In the long run, it may also be rummaged *de nova* to have a SAARC Model law on refugees for stability, progress, and peace in South Asia.

4.13 Advocacy for an Alternative Regional/International Legal Regime on Climate Refugees

Off late, the repercussions of climate change—from storm escalation to saltwater incursion to desertification—have triggered an increase in climate refugees, or climate migrants, and galvanized the scholarship on the disputed issue. Therefore, climate change-induced migration has become one of the fundamental thrusts of climate justice movements.[196] The policy-makers, scholars, advocates and stakeholders have developed a considerate perception of climate change-induced migration.[197] In the absence of international consensus on augmenting climate refugees to the contemporary refugee law framework, one section of scholarship has argued for creating a new, comprehensive and consolidated architecture for climate change migration. This could possibly appear like an 'Additional Protocol to UNCSR' adopted by the UNHCR on climate refugees or a completely new and separate UN Convention or Treaty on the climate refugees as there are many common features in these proposals in terms of rights and definitions of climate refugees at international level.

However, the debate on expanding the notion of refugees to incorporate climate refugees is conspicuously divided. There are academics and scholars who are of the view and perceive the climate change-induced displacement as 'climate migrants',[198] and there is a debate whereunder they have impugned the validity of the concept of 'climate refugee' expression and it being advocated to be the right reference and connotation[199] as climate refugees are forced to acclimatize in the absence of

---

[196]Matthew Gauthier, 'Climate Refugees and International Law: Legal Frameworks and Proposals in the US and Abroad', 1 http://studentorgs.law.unc.edu/documents/elp/2016/m_gauthier.pdf accessed 10 March 2018. *See* generally, Shaun McCullough, 'In a Rising Sea of Uncertainty: A Call for a New International Convention to Safeguard the Human Rights of Citizens of Deterritorialized Asia-Pacific Small Island-States', (2015) 26 Colo. Nat. Res. Energy & Envt. L. Rev. 109–137, 120.

[197]Erik Haites et al., Possible Elements of a 2015 Legal Agreement on Climate Change, Working Paper 16, (IDDRI, 2013) 1–24, 16.

[198]Francois Gemenne, '*One Good Reason to Speak of Climate Refugees*', (2015) FMR. 6 (1) 22–24.

[199]Ibid.

any legal protection. But some scholars[200] advocate that the 1951 UNCSR refugee definition has the capacity to address environmental displacement too. However, such an approach is facing serious challenge and obstruction from national governments concerned as they contend that such a measure would open the 'refugee flood gates' in the wake of sheer magnitude of the problem. Moreover, it is considered that such an enlargement of the existing refugee definition might result in the possible weakening of the contemporary protection for refugees.[201] This is because the 1951 UNCSR refugee definition creates two paradigms of interpretation; firstly, it is the '*human rights concern-oriented interpretation*', and secondly, it is the '*state concern-oriented interpretation*'; however, the later standard limits its ambit and purview for persons seeking protection and instead deprives them.

The experts who contend against the expansion of the refugee status to climatically displaced people underscore the purpose and object of the 1951 UNCSR to be the protection of civil and political rights of people from the persecution perpetrated by the oppressive governments.[202] The environmental hazards were never intended to be accounted for by the 1951 UNCSR. However, all conventions and the legal mechanism have to be dynamic in addressing new emerging situations of human displacement and mobility. The 1951 UNCSR would not have been able to provide for environmental concern as the apprehension for the protection of environment never existed as a priority at the time of its inception. The trepidation that such an enlargement of the current refugee definition could result in the likelihood of dwindling of the existing model of protection for refugees is unsubstantiated and speculative as any efficient legal mechanism would evolve to cover new challenges and situations.[203] The judgement handed down in the case of *Teitiota v. The Chief Executive of the Ministry of Business Innovation and Employment*[204] makes a brilliant example of this evolving notion of 'climate refugees'. The instant case involved Teitiota seeking refugee status in New Zealand for himself and for the members of his family in the wake of the threat of submergence of Kiribati Island due to climate change. Justice priestly empathized with the idea that the legal definition of refugee is not confined to the one incorporated in the 1951 UNCSR, but it can be expanded, extended and defined while incorporating 'a person or individual driven from his or her residence, home to seek shelter, refuge, especially in an alien or foreign country, from conflicts, war, racial discrimination, religious or faith-based persecution, social origin, political troubles, natural disaster, etc.'[205] The expansion of human rights is a good precedent to follow here as it would ensure the safety and security of the cli-

---

[200] Jessica B. (n 155) 480; see also: Christopher M. Kozoll, 'Poisoning the Well: Persecution, the Environment, and Refugee Status' (2004) 15 Colo. J. Int'l Envtl. L. & Pol'y 271–307.

[201] Keane, David, 'The Environmental Causes and Consequences of Migration: A Search for the Meaning of 'Environmental Refugees' [2004] 16 Geo. Int'l Envtl. L. Rev. 209–223, 218.

[202] Suzette Brooks Masters, 'Environmentally Induced Migration: Beyond a Culture of Reaction' [2000] 14 Geo. Immigr. L. J. 855, 864.

[203] Keane, David (n 204).

[204] In the High Court of New Zealand Auckland Registry between Ioane Teitiota and the Chief Executive of the Business Innovation and Employment NZHC 3125 [26 November 2013].

[205] Ibid.

mate refugees. Therefore, the international human rights law which was exclusively concerned with civil and political rights gradually expanded to incorporate economic and cultural rights under its realm.[206] Further, the human rights bodies have evolved a mechanism to provide individual representations for the redressal of grievances as a paradigmatic shift from merely a state-centric lego-institutional system for solutions for human rights violations.[207]

Mayer points that the climate change-induced displacement or migration has been construed as a challenge to international governance[208] and the climate justice does not find any credible place in the ongoing refugee discourse as enunciated in the refugee law. In this scenario, the question arises should we explore new legal alternatives to have a new connotation for them by adopting a new Global Additional Protocol to 1951 UNCSR, whereunder the current mandate of UNHCR is expanded to embrace the climate refugees. The climate refugees must be defined in an additional global instrument, whereunder their civil and political rights should be crystallized and concretized as a distinct compendium of rights while taking into account temporary and permanent nature of legal protection. Because there are a significant number of climate refugees who are not in a position to return to their country of nationality or origin, therefore, reception countries should also extend citizenship with its entire incidental and accompanying benefits to them at par their nationals. Further, in a situation of non-availability of all rights in receiving countries, climate refugees must receive fundamental civilian freedoms and human rights necessities without sociopolitical discrimination and persecution. Such kind of aid and assistance should be extended in the form of social initiatives and incentives, particularly to those climate refugees who permanently migrated and are willing to be assimilated into the host countries.

A dynamic law is required to deal with the rising problem of climate refugees which is why the need for an Additional Protocol to existing refugee regime is being advocated since the existing regime already has experience in delaying refugee situation. Secondly, it is also an opportunity to bring synergy and hybrid integration from other branches of law as climate change displacement is cross-disciplinary. Additionally, this protocol is not meant as a top-down but can be a supplement and complement to regional mechanisms that take into account the hybrid law.

---

[206]ICESCR 1966 (the International Covenant on Economic, Social and Cultural Rights stipulates obligations on State parties to take actions, individually and through international assistance and cooperation, to the maximum of their available resources, with a vision of attaining gradually the full accomplishment of economic, social and cultural rights. The Covenant's complaint mechanism is established by its Optional Protocol of 10 December 2008.

[207]'Complaint Procedure 1503' has been improved now by the Human Rights Council on 18 June 2007 by adopting a new Resolution 5/1 that entitled 'Institution Building of the United Nations Human Rights Council'.

[208]Benoit Mayer, 'Environmental Refugees? A Critical Perspective on the Normative Discourse' (18 July 2012) 1–22, 1 available at SSRN: http://dx.doi.org/10.2139/ssrn.2111825.

The international hybrid law has been developing since 2007 as a rights-oriented approach with the liberal, reformative and progressive interpretation[209] of the law. Such an approach has been utilized in many court decisions,[210] whereunder the purview of municipal and international law systems has been expanded to extend the protection ambit to vulnerable populations affected by the climate change. The hybrid law embodies a tool of legal research that examines contemporaneously and inseparably, a climate change situation from threefold perspectives such as international environmental law, international human rights law and international refugee law.[211] Thus, its utilization is streamlined by employing a specific benchmark as an alternative to a threefold assessment.

The decision to migrate due to environmental factors is not only is based on environmental reasoning but also is dictated by the human rights violations such as economic trauma, non-accessibility to the healthcare system and unreasonable governmental limitations. Therefore, such an interwoven approach to environmental degradation, forced migration/displacement and human rights transgressions characterizes a paradigmatic shift and a substantial reconceptualization of duties of nation-states to protect based on CBDR application premise for both climate refugee generating and climate refugee hosting states. Therefore, it is a fundamental understanding that hybrid approach based on this simple premise would ensure the protection, preservation and promotion of human rights and in turn that would further safeguard the human mobility as a hybrid concept incorporating the climate change-induced human displacement, migration and premeditated relocation. The approbation of migration as an adaptive tool by the affected communities in the face of environmental degradation gets support under the hybrid approach and ensures state responsibility to protect. Therefore, the nation-states are obligated to address the existing climate change migration crises and to avert the future human mobility impacts caused by the climate human rights violations under the hybrid law approach. Recently, even the climate talks[212] in Bonn proactively relied upon the human rights language of the Paris Agreement that talks of migrants' rights in its preamble as an integrated approach. Therefore, in an environmentally inhospitable world, climate change-induced human displacement, migration and relocation require binding legal protection arrangements at national, regional and international levels.[213] It is also evident that the hybrid law approach has to be further strengthened to address the protection needs of the climate refugees by conceiving and adopting an Additional Protocol or new hybrid law mechansim. Such a legal arrangement would emplace the

---

[209]Bradley W. Miller, 'Beguiled by Metaphors: The 'Living Tree' and Originalist Constitutional Interpretation in Canada' [2009] 22 The Can JL & Jur 331–354.

[210]*AD (Tuvalu) v. C Curtis* [2014] NZIPT 501370-371; see also Siegeo Alesana v. New Zealand [2014].

[211]Tiffany T.V. Duong (n 4) 1254.

[212]The Bonn Climate Change Conference, (30 April–10 May 2018) Earth Negotiations Bulletin (ENB) 12 (726) 2018 http://enb.iisd.org/climate/sb48/ accessed on 29 May 29 2018.

[213]Frank Biermann and Ingrid Boas, 'Preparing for a Warmer World: Towards a Global Governance System to Protect Climate Refugees' [2010] 10(1) Global Envi'l Politics 60–88.

robust human rights-oriented solutions, and the idea of protection would have natural preventative conduct while neutralizing the negative impacts of climate change.

Off late, international law has been witnessing some sort of hybrid developments as seen under the Nansen Protection Agenda and the UN Global Compact on Migration (GCM) stemmed from the New York Declaration, 2016, on Migrants and Refugees based on the rule of law, transparency and inclusion. The GCM Framework would have an actionable programme for implementation of UN Global Goals Agenda by 2030 and international migration in all its manifestations. It will also ensure the enforcement of Addis Ababa Action Agenda and the 2013 Declaration of High-Level Dialogue on International Migration and Development. Although, these legal initiates have integrated principles from human rights, humanitarian law and refugee law to deal with the situations of international migration and climate change-induced displacement. However, these are soft law initiatives, nevertheless, they constitute a very significant hybrid development in international law. Today, the climate-induced displacement and climate change migration is threatening to blow into a crisis proportions affecting the human rights of all, and it is further destined to attain worst forms in future. Thus, the global community must evolve the GCR Framework Agenda as an international hybrid law approach for safeguarding the climate-displaced population. Therefore, the GCR as an Additional Protocol must implement its mandate under the existing UNHCR as a principal organ along with cooperation and collaboration with other international bodies like UNEP and human rights bodies along with the mandate of UNFCCC, Paris Agreement, relevant COPs, etc. But the most significant challenge is there in the hands of states or international community as for how to execute and implement the new GCR framework. It has been axiomatic from the experience of UNHCR that protection under 1951 UNCSR did not trickle down to the most deserving refugees despite the UNHCR's best efforts endeavours. Therefore, new GCR legal architecture must not be allowed to succumb and suffer from such inadequacies and bottlenecks in the context of climate refugees. The concept should develop that refugee problem should not be attended as an issue of 'illegal migrants' or 'immigrants' and demographic assault on particular geopolitical entity. It is the political landscape and characteristics of the persecution that primarily perpetuates the deprivation, denial and denunciation of national protection to the refugees by the governments, their country of nationality or origin.

Notwithstanding, the emergence of 'climate refugees' should not be perceived as adversity or hostility, but it has posed an unprecedented opportunity to the comity of nations to meet, formulate and innovate the novel strategies, good practices and ideas to provide legal protection to climate change-induced displaced people or migrants or climate refugees.

4.14 Conclusion

It is increasingly been a recognized fact of contemporary human life that climate change-induced displacement and migration will blatantly and manifestly affects population, people and their human rights across the geopolitical continuum. The myopic and prismatic lenses of environmentalism are incapable of addressing a complex issue as climate change. The repercussions and negative impact of climate change are calamitous and terrible and will wobble the very foundation of human existence. Among the short-term and long-term after-effects of climate change, mass migration/climate change displacement of people, communities and societies attains profound threats. The differential impact of climate displacement will be on those people and the societies who have the least acclimatizing capacity, and they are prone to enhanced vulnerability threshold. It is only after the Bali COP that the international community has initiated paying attention to the issue of climate adaptation and climate change-induced displacement and migration. The Paris Agreement has constructively elevated the threshold of issue of human mobility and its consequences and analysed the issue with a focus on interrelationship with human rights and developmental law. However, these endeavours of international community are at embryonic stage and, therefore, it requires well-calibrated measures to emplace a concrete, robust and pragmatic system specifying the modalities of recognizing the climate change-induced displacement while determining the rights of people who got climatically displaced and the obligations of nation-states and other stakeholders towards climate change-induced displacement and migration.

The contemporary international refugee regime has been concluded and adopted as an integral measure of a process whereunder specific classes of people have been identified and recognized by the comity of nations who are in need of international protection. These instruments have been addressing and focusing the situation of these classes of persons while their interpretation is continued to be static. The present refugee law based on legal pluralisms and fragmentation has to develop and expand to understand the significance of climate change discourse. Most of the impugned discourse on climate change-induced displacement is focused on verbalizing the appropriate terminology in the case of climate change-induced displacement and concentrates on whether the climate change-induced displaced people could be eligible for seeking the status of refugees. Not disregarding the significance of a correct expression, the instant chapter advocates that climate change-induced displacement debate should address more on the legal niceties and remedies regarding the prevention, protection and promotion of climate justice. To using or employing the term 'climate refugees' generically may be a misleading one for some scholars. However, the fact is that the majority of the people remain displaced, are living in 'refugee-like situations' and are not presently protected under the multiple fields of international law. Whatever may be the name or title assigned to this displaced group of people, the legal steps have to be calibrated.

The adoption of a treaty or an addition Protocol to UNCSR may be perceived by few as a weak proposition to address the current situations of climate change migration and forced displacement, but adopting the international and regional arrangement treating all possible forms of climate change movements remains the most viable option before the international community. The potential arrangement on climate refugees must ground itself on the principle of global solidarity, international burden-sharing arrangements and evolving the best state practices incommensurate with global international human rights standards based on hybrid law approach. These arrangements can be advanced with the codification of rights of climate refugees having identified their particular protection requirements by evolving a global consensus within the comity of nations including South Asia. However, there is a lack of political will to have a new international or regional convention or treaty because lawmaking in the field of international law requires various stages, wherein international soft law instruments and documents like UN Declarations, UNGA resolutions, Guidelines of UN Agencies and regional understandings and recommendations of different organizations are adopted that conceive and confirm norms and obligations until unanimity and consensus are achieved on specific matters stemming into a treaty and the same process could also be resorted to in addressing the climate change migration. Thus, there is state responsibility obligation to help climate refugees and to ensure the protection of their rights on the grounds of humanitarianism, existentialism, human rights, the international rule of law, world constitutionalism, global pluralism and multiculturalism.

Climate change-induced displacement or environmental migration is not a disconnected phenomenon but affects every continent, region and country sans any distinction or discrimination. The present century is an epoch of migration, connecting it with securitization would entail a mass conceptual dimension and emergence of meta-politics, and that would be a disservice to the cause of rights of the climate refugees. However, climate change migration is impregnated with infinite portents hooked up to the material mobility of the people. Therefore, the demonization of "climate change migrants," "climate refugees" or "refugees" *per se* as a potential picture of terrorists that generates fear and ontological perception of intimidation among the architects of national security narratives devoid of ground reality is a new political narrative. Unfortunately, the climate change-induced displacement and migration have been hogging the limelight in the media, but the image that is being portrayed does not convey the human rights paradigms of security narratives. Thus, the climate change and issues germane to migration and displacement have to be recavilled and reconsidered by the comity of nations by adopting a consolidated new Additional Global Protocol based on the hybrid approach of law. Therefore, the present chapter makes a case for an Additional Global Protocol on Climate Refugees to the 1951 UNCSR.

Chapter 5
Climate Refugees: South Asian States' Legal Protection Practices

Abstract The grimmest sordid state of affairs of being a refugee in the contemporary world has attained a new terrain of crisis depicted by the emergence of 'climate refugees' produced by climate change. People do not move on a whim or megrim rather sociopolitical, economic and environmental factors compel them to migrate from their homelands. Almost seventy years ago, European Jews were deprived of asylum and hounded by the Nazis. Similarly, the climate change and 'climate refugees' are making news daily due to the calamities and catastrophes surrounding them with unprecedented visibility. South Asia is one of the regions where states do not afford effective legal protection to refugees or climate refugees. However, the international protection to political refugees was institutionalized in the 1951 UN Convention relating to the Status of Refugees (UNCSR) with its 1967 Additional Protocol and its lodger the UN High Commissioner on Refugees (UNHCR). In South Asia, no country has acceded to UNCSR except Afghanistan, and there are no national refugee laws and no legal guarantees. Refugees, as well as 'climate refugees', have been enduring insecurity and condemnation in the SAARC region. The present chapter tries to examine the South Asian state practices to deal with the climate refugees. Unfortunately, climate refugees are completely deprived of any legal protection due to the gaps in the national laws in South Asia. The chapter further evaluates the reception of international climate change law in South Asia that is flagrantly ineffective, and climate change migration governance in South Asia that fallibly revolves around national security narratives.

5.1 Introduction

South Asia is fast emerging as the new climate refugee flash point in the world. There are umpteen instances of climate change-induced displacement and migration in South Asia. In 2009, half a million Bangladesh population was displaced after the submergence of the Bhola Island. Many of the displaced population have found its way to India through its porous border with Bangladesh. Those migrants/refugees do not enjoy any legal protection and constantly live in fear of deportation to their

S. Jolly and N. Ahmad, *Climate Refugees in South Asia*, International Law and the Global South, https://doi.org/10.1007/978-981-13-3137-4_5

home land which does not exist. Studies have identified major migration corridors in South Asia between Bangladesh–India, Afghanistan–Pakistan, India–Pakistan and Nepal–India. Experts predict a future scenario characterised by increasing migration/displacement ranging from internal to external, temporary to permanent which could be part of adaptation, planned relocation or involuntary forced displacement.

The eight municipal jurisdictions of South Asia constitute the fastest growing region of the world that has been witnessing remarkable metamorphosis in the fields of health, education and infrastructure. But studies and reports point out to the dangerous threat of climate change-induced human displacement and climate change migration that requires being tackled regionally and institutionally.[1] Historically, South Asia as a region has been receiving the refugees as well as producing the refugees owing to a multitude of factors such as conflicts, wars, violence, persecution, social origin, political proclivities, religion, caste, communalism, sub-regionalism and geophysical events. The conceivable connection between climate change and migration in South Asia has been ruminated in the academic discourse for the last many years. However, South Asian states do not have uniformity in addressing the emergence of this new class of people called 'climate refugees' out of this crisis.

In this scenario, the determination of climate change as a cause of migration and drivers of internal displacements have to be identified and linked in South Asia. Therefore, it poses few profound questions. What are the South Asian states' legal practices regarding protection of CDPs? How should South Asian states assist those in need of asylum? This chapter examines the existing domestic laws and policies regarding climate change-induced displacement in South Asia. It also emphasizes the range of legal and policy options to deal with multifarious challenges posed by climate change-induced displacement and climate refugees thereunder. Particular emphasis will be on the specific climate legislation, national refugee laws, human rights laws, disaster management laws and other environmental laws, which are relevant in the sphere of climate change displacement and migration.

5.2 Reception of International Refugee Law in South Asia

The comity of nations has been dithering over the refugee migration since its discernible nascency, and it got flustered with the issue. Today, the 1951 UN Convention relating to the Status of Refugees (UNCSR) is 'one of the most widely accepted international norms, and remains the sole legally binding international instrument that provides specific protection to refugees'.[2] On the one hand, this is a prodigious accomplishment which marks the recognition and fulfilment of the human

---

[1] Koko Warner, 'Climate Change Induced Displacement: Adaptation Policy in the Context of the UNFCCC Climate Negotiations', Legal And Protection Policy Research Series, Division of International Protection UNHCR, (2011) 1–22 http://www.unhcr.org/4df9cc309.pdf accessed 22 March, 2018.

[2] Saadia M. Pekkanen, John Ravenhill, and Rosemary Foot, *The Oxford Handbook of the International Relations of Asia* (Oxford University Press, 2014) 625–626.

rights to asylum in multinational law.[3] On the other side, it is not functioning for the non-western nation-states who were excluded from the first institutionalization of international refugee regime and who now deal with the majority of the world's refugee flows.[4]

South Asia currently hosts substantially large per cent of the world's refugee population.[5] Historically, South Asia has witnessed substantial intra-regional movement and displacement of regional groups fleeing religious or ethnic persecution and sociopolitical instability.[6] The empirical experience of the South Asian region shows countries can be both refugee generating and refugee hosting. Bangladesh, India, Pakistan and Nepal are countries that receive refugees, while Afghanistan, Bhutan and Sri Lanka have produced refugees.[7] But South Asian countries have not tried to formulate any regional refugee protection legal regime despite the fact it has witnessed one of the largest refugee movements and population transfers between India and Pakistan during partition of the subcontinent in 1947 and creation of a new country in 1971 on the eastern border of India called Bangladesh that has also generated refugee exodus and flooded India with Bangla refugees. These refugee exoduses were sufficient to have a regional regime for refugee protection in South Asia. Unfortunately, South Asian Association of Regional Cooperation (SAARC) governments could not attend the plight of political refugees, and now another class of refugees has emerged in the region in search for climate justice[8] called climate refugees and CDPs climate IDPs. Unfortunately, UNCSR does not even remotely talk of climate refugees.

Though the climate change displacement has been emerging at a pace that has not been witnessed[9] the problem of climate change migration and displacement has seen little diplomatic engagement in the international arena for its resolution.[10] The climate change displacement poses challenges to the regional and international

---

[3]U N General Assembly, Convention on the Status Relating to the *Refugees,* 28 July 1951, United Nations Treaty Series, Vol 189, 137, See also Roman Boed, 'The State of the Right of Asylum in International Law', (1994) 5(1) Duke J. Comp & Intl. L, 1–33, 8–9.

[4]Andrzej Bolesta (ed), *Forced Migration and the Contemporary World: Challenges to the International System*, (Libra, 2003) See generally Daniel Kanstroom, 'The "Right to Remain Here" as an Evolving Component of Global Refugee Protection: Current Initiatives and Critical Questions', (2017) 5(3) JMHS 614–644 (The discussion of the article pertains to non-refoulement and resettlement.

[5]United Nations Human Rights Commission, (2017) 'Report on Global Trends: Forced Displacement in 2016' 14 http://www.unhcr.org/5943e8a34.pdf. Accessed on 25th March 2018, Aung Phyro and Tapan Kumar Bose, *Refugees in South Asia: An overview in The Fleeing People of South Asia: Selections from Refugee Watch* (Anthem Press, 2009) 143–149.

[6]Sugata Bose and Ayesha Jalal, *Modern South Asia: History, Culture, Political Economy*, (Routledge, 1998) 165–167.

[7]Partha S. Ghosh, *Migrants, Refugees and the Stateless in South Asia* (Sage India, 2016) 21–22.

[8]Paul Martin and others, *The Search for Environmental Justice*, (Edward Elgar Publishing, 2015) 11–12.

[9]Fernández, María, 'Refugees, Climate Change and International Law'. (2015) 49 FMR 42–43.

[10]Vikram Kolmannskog, 'Climate Change, Environmental Displacement and International Law' (2012) 24 J. Int. Dev. 1071–81, 1072.

community regarding protection under the multiple fields of international law that requires immediate attention and resolution. Climate change migration has often been considered an effective adaptation strategy,[11] but it should never be construed as a permanent and sole solution. Therefore, climate change migration consultations must incorporate a pragmatic mechanism to implement the adaptation and mitigation strategies and reframe the contours of climate change discourse[12] in the SAARC region for addressing the protection of the climate refugees or climate IDPs.

European nation-states concluded much arrangement[13] and ultimately evolved and conceived a legally binding refugee protection regime[14] for addressing the intra-European refugee migration and displacement in post-World War II and finally adopted the 1951 UNCSR. The UNCSR is historically European[15] and morally Global North, and the concept of human rights is legally international, philosophically universal and historically Western,[16] but such an occidental understanding of human rights does not fit into the traditionally 'Global South' and ethically South Asian duty-oriented jurisprudence.[17] The story of UNCSR is the history of human migration and displacements in Europe after the conclusion of World War II.[18] Therefore, the South Asian nation-states did not participate in the drafting of the UNCSR, as it was limited to address only the intra-European refugee migration and displacement.

All the original member states of the SAARC (except the newly joined[19] Afghanistan) did not ratify the UNCSR and its Additional Protocol of 1967. The majority of the nation-states in South Asia did not provide formal reasons for their

[11]Muhammad Abid, Janpeter Schilling, Jurgan Scheffran and Farhad Zulfiqar, 'Climate Change Vulnerability, Adaptation and Risk Perceptions at Farm Level in Punjab, Pakistan' (2016) 547 Sci Total Environ 447–60.

[12]Uma Kothari, 'Political Discourses of Climate Change and Migration: Resettlement Policies in the Maldives' (2014) 180(2) Geogr. J. 130–140.

[13]Erika Feller, 'The Evolution of the International Refugee Protection Regime' (2001) 5 Journal of Law & Policy 130–132.

[14]Laura Barnett, 'Global Governance and the Evolution of the International Refugee Regime' (2002) 14(2) IJRL 238–262.

[15]Lucy Mayblin, 'Historically European, Morally Universal? The 1951 Geneva Convention on the Status of Refugees' (*e-International Relations Students,* 28 March 2010) http://www.e-ir.info/2010/03/28/historically-european-morally-universal-the-1951-geneva-convention-on-the-status-of-refugees/ accessed 28 April 2018.

[16]Michael Freeman, 'Human Rights: Asia and the West', in James Tuck-Hong Tang (ed.) *Human Rights and International Relations in Asia-Pacific* (1995) 13, 17.

[17]Emma Haddad, 'Refugee Protection: A Clash of Values' (2003) 7(3) IJHR 1–26.

[18]Eduardo Arboleda and Ian Hoy, 'The Convention Refugee Definition in the West: Disharmony of Interpretation and Application' (1993) 5(1) IJRL 66–90.

[19]Afghanistan joined as the eighth member state of SAARC at the 14th SAARC summit at New Delhi in April 2007; Ministry of Foreign Affairs of the Islamic Republic of Afghanistan, 'SAARC' http://mfa.gov.af/en/page/6546/6565/6566, accessed on 24 April 2018.

opposition to the UNCSR.[20] There are many reasons attributable by India for not signing or ratifying the UNCSR as under:

- That the UNCSR had a constricted, restricted and conservative definition of the term 'refugee' and branded it as Eurocentric[21] since it does not address the nature of refugee migration in South Asia;
- That in 2003 at the 54th session of the EXCOM Meeting[22] of the UNHCR, India stated that the UNCSR definition does not recognize the primary actors who contribute to proliferating refugee mobility;
- That India contended at the same meeting that majority of the refugee migrations had been caused due to prevalent utter poverty and deprivation around the world particularly in the developing world including South Asia where refugees find even survival migration[23] most difficult;
- That there are different classes of displaced persons who have not been covered by the UNCSR and India objects to article 35 of the UNCSR, whereunder the supervisory responsibility has been vested in the UNHCR[24] alone that undermines the sovereignty of India;
- That India objects to UNCHR perception of India as a fit country for local integration and does not promote refugee repatriation that cast the heavy burden upon India and India cannot afford it due to its growing population, and it is also unfair to treat India as a refugee destination on a permanent basis as it also a developing country;

Thus, India is opposed to UNCSR as it is Eurocentric,[25]and it does not address the concerns of developing nation-states[26] and problems of economic migrants in South Asia.[27] Further, restricted scope of refugee definition and its incapacity to address the situations of mass exoduses, concerns of sovereignty, internal security[28] and climate change as a security threat[29] have posed irreconcilable grounds.

---

[20] Paula Banergee, 'Forced Migration in India', in Elena Fiddian-Qasmiyeh, Gil Loescher, Katy Long, NandoSigona (eds) *The Oxford Handbook*, (OUP 2014), 613–626, 614.

[21] Bimal Patel, *The State Practice of India and the Development of International Law: Dynamic Interplay Between Foreign Policy and Jurisprudence*, (Martinus Nijoff, 2016) 118, see also Alex Balch, Immigration *and the State: Fear, Greed and Hospitality* (Springer, 2016) 44–46, Andrzej Bolesta (n 4) 21.

[22] UNHCR's Executive Committee 54th Session contained in United Nations General Assembly document A/AC.96/987 and document no. 12A (A/58/12/Add.1).

[23] Alexander Betts, *Survival Migration: Failed Governance and the Crisis of Displacement* (Cornell University Press, 2013) 10.

[24] See for discussion, Hoi Trinh, UNHCR and Accountability: The Non Renewability of UNHCR Decisions, in Andrzej Bolesta (ed), *Forced Migration and the Contemporary World: Challenges to the International System*, (Libra, 2003) 51.

[25] Patel (n 21) 118.

[26] Ibid, 136.

[27] Ibid, 138.

[28] Ibid.

[29] Hans Joachim Schellnhuber, *Climate Change as a Security Risk*, (Earthscan Publication, 2008) 19–20.

The refugee problem is grave in South Asia, as countries concerned have not developed any formal structure to deal with the issue. In the absence of refugee-specific law, refugees and asylum seekers are addressed under ad hoc and volatile administrative apparatus that is arbitrary, discriminatory and preferential in their treatment.[30] The South Asian countries have followed the policy based on political justifications and underscoring the centrality of sovereign jurisdiction[31] in their bilateral parleys while addressing the refugee migration between refugee producing and refugee receiving countries that have turned out to be a significant obstruction in evolving a viable refugee protection framework[32] in South Asia.

Despite many inadequacies, the South Asian countries have been magnanimous in extending protection to a large number of refugees since time immemorial without having a refugee protection legal regime. The reception of millions of Afghan refugees in Pakistan and Tibetan refugees in India since March 1959 is notable examples of South Asian historical traditions of refugee protection in this region.[33] The lego-institutional frameworks of countries in South Asia have recognized the importance of IRL principle of *non-refoulement* even without becoming the party to UNCSR. India, Pakistan and Bangladesh have been members of the Executive Committee of the UNHCR, since 1995, the highest decision-making body of the UNHCR.[34] Human Rights instruments and the role played by international politics in this region have also committed to these measures taken by the states in protecting the refugees within the limited resources they possess. Presently, the SAARC countries are being assisted by the UNHCR which includes ensuring the protection of asylum seekers and refugees and promotion of international refugee law standards vis-a-vis promotional activities for adoption of the national legal framework in line with international standards.[35] The reluctance of respective governments in ratifying the Refugee Convention or adopting national legal framework and complexities of refugee issues with regional dimensions, needs to be evaluated and there is a need to look into the prospect of a regional approach to address forced migration and refugee protection issues in South Asian region in the wake of the looming crisis of climate change.

---

[30] R. Abrar Chowdhury, 'Legal Protection of Refugees in South Asia', (2001) 10 FMR 22–23.

[31] Galina Cornelisse, *Immigration Detention and Human Rights: Rethinking Territorial Sovereignty*, (Martinus Nijhoff Publishers, 2010) 33–44.

[32] *Veerabhadran* Vijayakumar, 'Developing A Regional Approach to Refugee Problems in South Asia', (1997) *Draft Paper Presented for the Fourth Regional Consultations on Refugee and Migratory Movements in South Asia, Dhaka*, published in Refuge, (2001) 19 (2) 6–16.

[33] Narayan Sharma, 'Refugee Situation in South Asia: Need of a Regional Mechanism', (2008) 1,KLR 103–122,104.

[34] N.L. Mitra, 'Refugee Law at Cross-roads', (1996) 1(2) Bulletin on IHL and Refugee Law, 265.

[35] Rosa da Costa, 'Rights of Refugees in the Context of Integration: Legal Standards and Recommendations', (2006) *Legal and Protection Policy Research Series, Division of International Protection Services, UNHCR Geneva*. 17–18.

5.3 Domestic Legal and Institutional Mechanism to Protect Climate Refugees Under Refugees Laws in South Asia

5.3.1 Afghanistan

For Afghanistan, the first large-scale flow of refugees from Afghanistan was a result of the Soviet invasion in 1979 and the ensuing ten-year war.[36] The second mass movement was triggered by the conflict between the various parties and Mujahideen factions in the wake of the Soviet withdrawal (1992–1994).[37] A third mass exodus of the refugees out of Afghanistan was caused by the brutal Taliban regime (1996–2001).[38] The climate change migration will further aggravate these circumstances as Afghanistan has started experiencing migration patterns[39] due to desertification and lack of water, or floods and failed crops, ailing livestock and localized conflicts over resources.[40]

Regarding the legal framework, Islamic Republic of Afghanistan has become the 146th state party to UNCSR and its 1967 Additional Protocol in August 2005.[41] Afghanistan's accession to UNCSR is significant, and domestic measures are envisaged for the protection of refugees. However, the domestic implementation of international refugee law framework will not be able to protect CDPs in Afghanistan as the refugee protection regime in Afghanistan follows the criteria laid down under the international framework for the determination of refugees which do not account for the climate-induced refugee problem. The blend of human rights guaranteed by the international human rights Conventions and the Islamic tradition is a unique feature of the Afghan democracy.[42] The Constitution of the Islamic Republic of Afghanistan incorporates the mandate of UDHR 1948 in the Afghan Constitution along with core human rights treaties.[43] The provision which could be claimed by the climate refugees is under article 7 of the Constitution of Afghanistan which protects

[36]Hiram A Ruiz, 'Afghanistan: Conflict and Displacement 1978 to 2001', (2002) 13 FMR 9–10. See also Rafael Reuveny and Aseem Prakash, 'The Afghanistan War and the Breakdown of the Soviet Union' (1999) 25 Rev Int'l Stud 693–708.

[37]Larry Goodson, *Afghanistan's Endless War: State Failure, Regional Politics, and the Rise of Taliban*, (Seattle: University of Washington Press, 2001) 114.

[38]Hiram A Ruiz (n 36) 9–10.

[39]Khalid Koser, 'Transition, Crisis and Mobility in Afghanistan: Rhetoric and Reality' (2014), *Geneva: IOM* 19–20.

[40]M. Alimullah Miyan, 'Droughts in Asian Least Developed Countries: Vulnerability and sustainability, Weather and Climate Extremes' 7 (2015) Sci. Direct 8–23, 9 See also, Ramesha Chandrappa and others, *Coping with Climate Change: Principles and Asian Context*, (Springer, 2011) 310.

[41]n (19) Islamic Republic of Afghanistan has become party to both 1951 UN Convention and Additional Protocol 1967 on 30 August 2005.

[42]Torunn Wimpelmann, *The Pitfalls of Protection: Gender, Violence, and Power in Afghanistan*, (University of California Press, 2017) 42–43.

[43]Gordon Brown, *The Universal Declaration of Human Rights in the 21st Century: A Living Document in a Changing World, NYU Global Institute for Advanced Study*, (Open Book Publishers, 2016), 29–32.

all persons under international human rights treaties.[44] However, for a systematic approach to climate refugee issues, Afghanistan must formulate holistic legal norms and policies that are culturally sensitive in this context while strengthening its agricultural infrastructure and water conservation system during high water availability, and diversify its livelihood options to address the extreme events of climate change.

5.3.2 Bangladesh

Bangladesh currently is hosting massive population of Rohingya refugees from Myanmar and has also been acting as a host country for refugees from Myanmar since its inception. As per the inter-agency Humanitarian Response Plan (HRP), for the year 2017–2018, the Rohingya population in Bangladesh is highly vulnerable and many having experienced severe trauma.[45] These complexities of a humanitarian crisis are expected to assume dangerous proportions once we account for the climate displacement, which has already been manifested in the social and political milieu of Bangladesh. The reports show massive-scale displacement by major natural events like flood and cyclone over 40 years (1970–2009).[46]

Bangladesh has not acceded to the 1951 UNCSR and its 1967 Additional Protocol along with other South Asian countries except Afghanistan. However, Bangladesh is privy to several international human rights treaties whereunder rights of refugees have been indirectly received protection and other benefits. However, being a dualist legal system, the international human rights are not enforceable in courts of law unless specific provisions are incorporated into existing municipal statutes or given effect through separate legislation.[47] Nour Mohammed avered that under the municipal legal system of Bangladesh, the refugees are regarded and treated as foreigners governed under the provisions of the Bangladesh Foreigners Act of 1946[48] and several other regulations governing the aliens and non-nationals in Bangladesh under

---

[44]The Constitution of Islamic Republic of Afghanistan, 2004 where under article-7 states; 'The state shall observe the United Nations Charter, inter-state agreements, as well as international treaties to which Afghanistan has joined, and the Universal Declaration of Human Rights. The state shall prevent all kinds of terrorist activities, cultivation and smuggling of narcotics, and production and use of intoxicants'.

[45]UN Children's Fund Report, 'Bangladesh: Humanitarian Situation Report No. 11 (Rohingya influx)', (*Relief web*, 21 January 2018) 1–10 https://reliefweb.int/report/bangladesh/bangladesh-humanitarian-situation-report-no19-rohingya-influx-21-january-2018 accessed on 1 February 2018.

[46]Harun Rasid and Bimal Paul, *Climate Change in Bangladesh: Confronting Impending Disasters*, (Lexington Books, 2013) 79.

[47]Abul Hasanat, 'Using International Law in National Courts: Bangladesh Perspective', (2013) 13(1) and (2) BJL 50–72, 52.

[48]Nour Mohammad, Refugee Protection Under the Constitution of Bangladesh: A Brief Overview (MCRG) 141–156, 148–149 http://www.mcrg.ac.in/rw%20files/RW39_40/12.pdf accessed 2 February 2018

the Bangladesh Registration of Foreigners Act 1939,[49] the Bangladesh Passport Act 1920,[50] the Bangladesh Citizenship (Temporary Provision) Order 1972,[51] the Extradition Act 1974,[52] the Naturalization Act 1926,[53] etc.[54] In the absence of national refugee protection law or specific legal framework to deal with the protection of refugees, Bangladesh depends upon these legislations and some of them are of colonial era to regulate the entry, stay and exit of aliens and foreigners in Bangladesh. However, these legislations, like the Indian legislations, do not mention expression 'refugees' or do not accord them any specific contemplations.[55]

Further, Nour Mohammed exemplifies[56] that the constitutional provisions that could be invoked to ensure the refugee protection.[57] The article 31 of the Constitution of Bangladesh has bestowed the right to them and states that *To enjoy the protection of the national law, and to be treated in accordance with law, and only in accordance with law, that is the inalienable and non-derogable right of every citizen, wherever he or she may be, and of every other person for the time being within Bangladesh, and in particular no action detrimental or harmful to the life, liberty, reputation, body or property of any individual or person shall be taken except in accordance with the law*.[58] Further, article 32 of the Constitution of Bangladesh grants that '*no person shall be deprived of life or personal liberty save by law*',[59] and article 33 of the Constitution of Bangladesh expounds the safeguards against arrest and detention for the citizens and non-citizens alike by explicating that '*no person who is arrested shall be detained in custody without being informed, as soon as possible, of the grounds for his arrest*'. Moreover, 'his right to consult and be defended by a legal practitioner of his choice' cannot be denied. However, these constitutional provisions legally and legitimately may be made available to the climate refugees by having a holistic interpretation as they are or may be non-citizens, aliens, foreigners and other

---

Under section 2 (a) 'foreigner' means a person who is not a citizen of Bangladesh as per Foreigners Act, 1946 (Bangladesh), XXXI of 1946, 23 November 1946 as amended by Act LIII of 1974, Sec. 8, 2nd Sch.

[49] Bangladesh: The Registration of Foreigners Act, 1939 [Bangladesh], Act No. XVI of 1939, 8 April 1939.

[50] Bangladesh: The Passport Act, 1920 (Bangladesh), Act No. XXXIV of 1920, 9 September 1920.

[51] National Legislative Bodies/National Authorities, Bangladesh: Bangladesh Citizenship (Temporary Provisions) Order, 1972, 26 March 1971, Order No. 149 of 1972.

[52] Bangladesh: The Extradition Act, 1974 (Bangladesh), Act No. LVIII of 1974, 30 July 1974.

[53] Naturalization Act, 1926 (Bangladesh), See Also: Ridwanul Hoque, Report on Citizenship Law: Bangladesh RSCAS/EUDO-CIT-CR 2016/14 December 2016.

[54] Mohammad(n 48).

[55] Navine Murshid, *The Politics of Refugees in South Asia: Identity, Resistance, Manipulation,* (Routledge, 2013) 7.

[56] Mohammad (n 48).

[57] These rights are civil and political in nature. These 18 fundamental rights have been primarily divided into two groups such as a) Rights granted to all persons- citizens and non-citizens alike. These rights are enumerated in articles 32, 33, 34, 35, 41 and 44 of the Constitution of Bangladesh.

[58] The Constitution of the People's Republic of Bangladesh, 1972 article 31.

[59] Ibid, article 32.

persons. Therefore, they are constitutionally entitled to be protected against all sorts of discrimination and deprivation in Bangladesh. There has to be an interpretation based on constitutional liberalism and environmental democracy that allows and envisages the expansion of constitutional protection, preservation of human dignity in conformity with constitutional morality and global human rights obligations.

Bangladesh has been regarded as one of the most susceptible countries at risk due to climate change, but it does not have any protection framework to protect the climate change refugees who cross its border and enter into Bangladesh. In the context of climate-induced displacement, current reports suggest to the serious crisis of internal displacement. To address the IDPs in Bangladesh, the enumerated principles of fundamental rights under articles 26, 27, 28 and 43 of the Constitution are of relevance, whereunder the general policy states that no laws of the state should be in contravention of the general principles of the fundamental rights.[60] Article 28 says that no law or policy of the state should discriminate against its citizens by religion, caste, creed or gender.[61] Moreover, Bangladesh has acceded to numerous legally binding international human rights conventions.[62] Since 1995, Bangladesh is there on the EXCOM of the UNHCR, and it is well versed with the 1998 United Nations Guiding Principles on Internal Displacement but, unfortunately, the government of Bangladesh does not utilize them to protect the IDPs. Therefore, the determined efforts are required from the government as well as civil society and media to understand the problems of IDPs and to find alternative ways to provide adequate humanitarian assistance to the most affected. It is not only a humanitarian duty but also a constitutional obligation for the state of Bangladesh.

5.3.3 Bhutan

Bhutan is the youngest democracy in the South Asian region that has adopted a new Constitution in 2008 clearing the way from a hereditary monarchy to a 'constitutional one' under a parliamentary system.[63] In the 1990s, around 95,000 people of Nepali origin arrived in Bhutan and petitioned for Bhutanese citizenship but were expelled from the country on cultural grounds in contravention of international human rights

---

[60] Ibid, article 8(2).

[61] Meghna Guhathakurta and Suraiya Begum, 'Bangladesh: Displaced and Dispossessed', in Paula Banerjee, Sabyasachi Basu, Ray Chaudhury and Samir Kumar Das (eds) *Internal Displacement in South Asia: The Relevance of the UN's Guiding Principles*, (Sage Publications, 2005) 178.

[62] Mohammad Shahabuddin, 'Human Rights and the Law', in Ali Riaz, Mohammad Sajjadur Rahman (eds) *Routledge Handbook of Contemporary Bangladesh* (2016) 283–292, 284. See generally Lassa Oppenheim, Sir Robert Jennings and Sir Arthur Watts, *Oppenheim's International Law* (Peace, Longman, 1992) 1261.

[63] Marian Gallenkamp, 'Democracy in Bhutan An Analysis of Constitutional Change in a Buddhist Monarchy', (2010) IPCS Research Papers, *New Delhi* 3–7.

instruments.[64] The fundamental rights under article 7 of the Constitution of the Kingdom of Bhutan states that the '*All persons or individuals shall have the right to life, liberty, and right to security and safety of the person and shall not be deprived of such rights except in accordance with the due process of law*'[65] and it is a right which is also available to all the persons whether they are aliens, foreigners, non-citizens, political refugees or climate refugees and CDPs, etc. But, unfortunately, the government of Bhutan does not extend such constitutional protection to a significant number of refugees in Bhutan.

There are many fundamental rights under the Constitution of the Kingdom of Bhutan that is equally available and extendable to the political as well as climate refugees such as right to intellectual property,[66] right not to be deprived or divested of property,[67] right to equality before the law[68] and right not to be arrested or detained arbitrarily.[69] The Constitution of Bhutan reflects upon the mandate of international human rights instruments, but it has acceded to very limited international human rights treaties such as the UN Convention on the Elimination of All Forms of Discrimination Against the Women (CEDAW),[70] the UN Convention on the Rights of the Child (CRC)[71] along with the First Two Optional Protocols[72] to CRC. However, Bhutan has signed but not ratified the UN Convention on the Elimination of All Forms of Racial Discrimination (CERD)[73] and the UN Convention on the Rights of Persons with Disabilities (CRPD).[74] The UNCSR and other international human

[64]Michael Hutt, *Unbecoming Citizens: Culture, Nationhood, and the Flight of Refugees from Bhutan*, (Oxford University Press, 2005) 14–18.

[65]The Constitution of the Kingdom of Bhutan, 2008, article 7 (1).

[66]Ibid, article 7 (13), states that 'Every person in Bhutan shall have the right to material interests resulting from any scientific, literary or artistic production of which he or she is the author or creator'.

[67]Ibid, article 7 (14), states that 'A person shall not be deprived of property by acquisition or requisition, except for public purpose and on payment of fair compensation in accordance with the provisions of the law'.

[68]Ibid, article 7 (15), states that 'All persons are equal before the law and are entitled to equal and effective protection of the law and shall not be discriminated against on the grounds of race, sex, language, religion, politics or other status'.

[69]Ibid, article 7 (20), states that 'A person shall not be subjected to arbitrary arrest or detention'.

[70]UN Convention on the Elimination of All Forms of Discrimination Against Women (CEDAW), 1979, was ratified on 31 August 1981.

[71]UN Convention on the Rights of the Child (CRC), 1989, was ratified on 01 August 1990.

[72]Optional Protocol to the UN Convention on the Rights of the Child on the Involvement of Children in Armed Conflict was signed on 15 September 2005 and was ratified on 09 December 2009, and Optional Protocol to the Convention on the Rights of the Child on the sale of children child prostitution and child pornography was signed on 15 September 2005 and was ratified on 26 October 2009.

[73]UN Convention on the Elimination of All Forms of Racial Discrimination (CERD), 1965, was signed on 26 March 1973.

[74]UN Convention on the Rights of Persons with Disabilities (CRPD), 2009, was signed on 21 September 2010.

rights treaties[75] have not been acceded to and, therefore, Bhutan maintains a limited relationship with global human rights legal arrangements. Thus, the refugees of all classes are discriminated by the Bhutanese state officials and agencies in violation of constitutional safeguards and its international human rights commitments.

5.3.4 India

India has been a destination for refugees and asylum seekers of all faiths[76] and sects since antiquity.[77] There are refugees in India from countries like Afghanistan, Iran, Iraq, Somalia, Sudan and Uganda.[78] India has been trying to evolve a consensus on asylum[79] and its mode of granting as per global human rights norms more as an administrative policy rather than as a legal requirement.[80] But, unfortunately, India has not signed the UNCSR and 1967 Bellagio Protocol.[81] Further, India has not ratified the 1954 Convention on the Statelessness and 1961 Convention on the Reduction of Statelessness.[82] India is a party to many international human rights Conventions such as the UDHR 1948,[83] the ICCPR1966,[84] the ICESCR 1966,[85] the

---

[75]Bhutan is not a party to the International Covenant on Civil and Political Rights (ICCPR), 1966, CCPR-OP2-DP—Second Optional Protocol to the International Covenant on Civil and Political Rights aiming to the abolition of the death penalty, International Covenant on Economic, Social and Cultural Rights (ICESCR), International Convention on the Protection of the Rights of All Migrant Workers and Members of Their Families (CMW), UN Convention against Torture and Other Cruel Inhuman or Degrading Treatment or Punishment (CAT), 1984, UN CAT-OP—Optional Protocol of the Convention against Torture, and UN Convention for the Protection of All Persons from Enforced Disappearance (CED), etc.

[76]Jenny Yang, 'Chins in Mizoram State, India: A Faith-Based Response', (2014) 48 FMR 57–58.

[77]Roger Zetter and Héloïse Ruaudel, 'Refugees' Right to Work and Access to Labor Markets—An Assessment,' (2016) KNOMAD Study, 40–42, See generally Kim Knott, 'Living Religious Practices' in Saunders Jennifer, Fiddian-Qasmiyeh Elena Snyder Susanna. (eds), *Intersections of Religion and Migration. Religion and Global Migrations*, (Palgrave Macmillan, New York, 2016) 71–90.

[78]Ananthachari, T., 'Refugees In India: Legal Framework, Law Enforcement And Security,' (2001) 7 ISIL Year Book of International Humanitarian and Refugee Law, http://www.worldlii.org/int/journals/ISILYBIHRL/2001/7.html. See generally Alexander Betts and Gil Loescher, *Refugees in International Relations*, (Oxford University Press, 2011) 268.

[79]Smrithi Talwar, 'Building a Regional Consensus on Asylum: The Indian Perspectives', (2000) 1(2) *Bulletin, IHL and Refugee Law* 251.

[80]Ibid, see also Saumitra Mohan, *Indian Policy and Development*, (McGraw Hill, 2017).

[81]Omar N. Chaudhary, 'Turning Back: An Assessment of Non-Refoulement under Indian Law' (2000) 39 (24) EPW 3257.

[82]Haq Centre for Child Rights, 'Twenty Years of CRC: A Balance Sheet' (HAQ Centre for Child Rights, 2011) 3,346 http://haqcrc.org/wp-content/uploads/2016/07/twenty-years-of-crc-a-balance-sheet-volume-II.pdf, accessed 20 February 2018.

[83]Ratification by India on 10 December 1948.

[84]Ratification by India on 10 April 1979 (a).

[85]Ratification by India on 10 April 1979.

CERD 1963, [86] the CEDAW 1979,[87] the CAT 1984[88] and the CRC 1989.[89] However, India does recognize the right to asylum under article 14 (1) of the UDHR.[90]

The government determines the status of refugees by *ad hoc* administrative decisions under the Foreigner Registration Act 1939 and the Foreigner Act 1946 to regulate the entry, stay and departure of all aliens in India. Other applicable laws are the Passport (entry into India) Act 1920, Passport Act 1967, and Extradition Act 1962. India does not have any national agency except Foreigner Regional Registration Office (FRRO) under the Bureau of Immigration of India to handle the refugees.[91] However, the Constitution of India under the Part III on fundamental rights prohibits discrimination on the grounds of race, religion, caste, creed, sex or place of birth, and other fundamental freedoms with reasonable restrictions are equally available to refugees.[92] Several members of Parliament across the party lines presented private members' Bills in the Parliament for enacting a National Refugee Law such as Asylum Bill 2015, National Asylum Bill 2015, the Protection of Refugees and Asylum Seekers Bill 2015, but these Bills are still pending before the Parliament.[93]

Refugees and asylum seekers have been enjoying protection under the Constitution of India.[94] The Supreme Court (SC) of India has done exceptional service to the cause of refugee rights. SC has interpreted the word 'person' also includes non-citizens. In the case of *Chairman, Railway Board & Others v. Mrs. Chandrima Das & Others*,[95] the SC necessitated the Indian state to follow the international human rights covenants and the UN Declarations and also adheres to the provisions of the UDHR into domestic jurisprudence.[96] International human rights core treaties can also be referred to in interpreting and understanding the national law.[97] Therefore, SC has directed and appreciated the predicament of refugees in a plethora cases such

[86]Ratification by India on 03 December 1968 with certain reservations.

[87]Ratification by India on 09 July 1993.

[88]India has signed on 14 October 1997 but it has not ratified yet.

[89]Ratification by India on 11 December 1992.

[90]U.N. Gupta, *The Human Rights: Conventions and Indian Law*, (Atlantic Publishers & Distributors, 2004) 259–260.

[91]Canada: Immigration and Refugee Board of Canada, India: Ability of Tibetan refugees to exit and re-enter India, including the requirements and procedures for Tibetan refugees to obtain and renew an Identity Certificate (IC) and a '*No Objection to Return to India' stamp; whether the IC can be renewed from abroad; Visa requirements for Tibetans returning to India*, 12 December 2014, IND105024.E available at http://refworld.org/docid/55825a44.html accessed 10 May 2018.

[92]Article 14 and 15 of the Constitution of India, 1950.

[93]Hamsa Vijayaraghavan and others, 'It's Time India Had a Refugee Law' (*The Wire*, 19 December 2015) https://thewire.in/law/its-time-india-had-a-refugee-law accessed on 25 February 2018.

[94]T Ananthachari, 'Refugees in India: Legal Framework, Law Enforcement and Security', (2001)7 ISILYBIHRL, 118–144. See: Constitution of India, 1950 (n 92) articles, 14, 20 and 21.

[95](2000) 2 SCC 465.

[96]Ibid.

[97]*Nilabati v. State of Orissa* (1993) 2 SCC 746; *D. K. Basu v. State of W.B*, (1997) 1 SCC 416; *PUCL v. Union of India* (1997) 3 SCC 433; *Githa Hariharan v. RBI*, (1999) 2 SCC 228; *Apparel Export Promotion Council* v. *A. K. Chopra* (1999) 1 SCC 759.

as *Khudiram Chakma v. State of Arunachal Pradesh and Ors,*[98] and *National Human Rights Commission v. State of Arunachal Pradesh,*[99] in which the SC observed that 'all the refugees presently residing in India are entitled to have the right to life and the personal liberty' as enunciated under article 21 of the Constitution of India. The 'Indian state is beholden to protect the life and freedom of each individual, be a citizen or otherwise, and it cannot permit individual, non-state actors or group of persons to intimidate the refugees or jeopardize their lives or compel them to leave involuntarily'. The SC further cautioned and directed that the state is constitutionally obligated to protect and safeguard the life, liberty, dignity, health, human rights and overall well-being of the Chakma refugees in India. Consequently, there is a chain of judgements handed down by the SC and many High Courts in India on the rights of Chakma refugees, Sri Lankan, and all other refugees that the rights of refugees must be safeguarded at the pedestal of procedure established by law in a liberal and secular democracy.[100] But, regrettably, the refugee jurisprudence evolved finds itself at war with the ordinary law relating to the foreigners that empower the government to deport them to their countries of origin arbitrarily. Unfortunately, the Law Commission of India in 2000 recommended in its 175th report that government should enact a more rigorous law to deal with the 'illegal entrants'.[101] However, the SC in *Louis De Raedt v. Union of India*[102] and Ors held that article 21 of the Constitution protects the life and personal liberty of all persons including aliens and foreigners present in India. Therefore, judicially created refugee rights under the Constitution of India has successfully been protecting the life and liberty of vulnerable persons. The climate-displaced population can be benefitted from the protection evolved by the Indian judiciary[103]; however, in the absence of an understanding on the concept of refugees, the climate-induced displaced population may not be legally entitled to claim the human rights protection[104] extended by the Indian judiciary.

---

[98] *KhudiramChakma v. State of Arunachal Pradesh and Ors*, 1994 SC AIR 1461.

[99] *National Human Rights Commission v. State of Arunachal Pradesh*, 1996 SCC (1) 742.

[100] *Francis C. Mullin v. Administrator*, AIR 1981 SC 746, *Govind v. State of MP*, AIR 1975 SC 1378, *Satwant Singh v. Assistant Passport Officer, New Delhi*, AIR 1967 SC 1834, *M. H. Hoskat v. State of Maharastra*, AIR 1978 SC 1548, *A. R. Antulay v. R. S. Nayak*, AIR 1992 SC 1701, *Charles Sobhraj v. Superintendent Central Jail*, AIR 1978 SC 1574, *Sunil Batra v. Delhi Administration*, AIR 1978 SC 1675, *Sheela Base v. State of Maharastra*, AIR 1983 SC 378, Parmanand Katara v. UOI, AIR 1989 SC 2039, Shantistar Builders v. N. K. Totame, AIR 1996 SC 786, *Unnikaishnan v. State of A. P.*, AIR 1993 SC 2178, *Olga Tellis v. Bombay Municipal Corporation*, AIR 1986 SC 180, *Francis C. Mullin v. Administrator*, AIR 1987 SC 746, *D. K. Basu v. State of West Bengal*, AIR 1997 SC 610.

[101] 'Law Commission of India, One Hundred and Seventy Fifth Report,' The Foreigners (amendment) Bill 2000, (Law Commission, India 2000) 15–16).

[102] 1991 3 SCC 554.

[103] Deepa Badrinarayana, 'The Emerging Constitutional Challenge of Climate Change: India in Perspective', (2009) 19(1) Fordham Int'l L J. 1–39, 23–27.

[104] Roger Zetter, 'Protecting Environmentally Displaced People: Developing the Capacity of Legal and Normative Frameworks, Research Report', (2011) *Refugee Studies Centre, Oxford Department of International Development University of Oxford* 13–14.

The government of India (GoI) presented the Citizenship (amendment) Bill 2016 in Lok Sabha on 19 July 2016. The impugned Bill aims to extend citizenship to an individual who belongs to the six religions of minorities such as Buddhists, Christians, Hindus, Jains, Parsis and Sikhs belonging to from Afghanistan, Bangladesh and Pakistan who enters into India without valid visa or travel documents and such persons shall not be considered as illegal migrants.[105] Thus, the proposed amendment would permit eligible them for Indian citizenship by the process of naturalization. The Bill aims to reduce the cumulative period of residential qualification from eleven years to six years for getting the Indian citizenship by naturalization.[106] The impugned Bill signifies a positive change in the refugee policy of India, and it would be beneficial to the refugees from Afghanistan, Bangladesh and Pakistan. In India, there are as many as 9,200 refugees from Afghanistan, and out of which, 8,500 are the Hindus and there are more than 400 Pakistani Hindu refugee localities in Indian cities like Bikaner, Jaisalmer, Jodhpur and Jaipur in Rajasthan and Surat and Ahmedabad in Gujarat.[107] The GoI has made religious persecution as the sole criterion to confer citizenship.[108] But, the controversial Bill has left other minorities and groups from these countries who are also facing the refugee-like situation based on religious persecution like Ahmadiyya Muslims in Pakistan, Rohingya Muslims in Myanmar, Afghan Hazaras and Tamil Muslims in Sri Lanka who have taken refuge in India.[109] But, unfortunately, new amendment Bill excludes[110] the Muslim refugees on the ground of religion from these countries.

The new Bill deliberately uses the term ‘migrant’ to deprive refugees in the sense of international refugee law. The Bill makes the illegal migrants eligible for Indian citizenship on the ground of all the six religions excluding Islam. The present Bill is

---

[105] In the Citizenship (amendment) Bill, 2016, wherein section 2, in sub-section (1), after clause (b), the following proviso shall be inserted, namely: ‘Provided that persons belonging to minority communities, namely, Hindus, Sikhs, Buddhists, Jains, Parsis and Christians from Afghanistan, Bangladesh and Pakistan, who have been exempted by the Central Government by or under clause (c) of sub-section (2) of section 3 of the Passport (Entry into India) Act, 1920 or from the application of the provisions of the Foreigners Act, 1946 or any order made thereunder, shall not be treated as illegal migrants for the purposes of this Act’.

[106] Amendment of Third Schedule by the Citizenship (amendment) Bill, 2016 where under in the principal Act, in the Third Schedule, in clause (d), the following proviso shall be inserted, namely: ‘Provided that for the persons belonging to minority communities, namely, Hindus, Sikhs, Buddhists, Jains, Parsis and Christians from Afghanistan, Bangladesh and Pakistan, the aggregate period of residence or service of a government in India as required under this clause shall be read as ‘not less than six years’ in place of ‘not less than eleven years’.

[107] PTI, ‘Indian citizenship for Hindus from Pakistan to be made Easier: government’, *The Indian Express* (New Delhi, 17 April 2016).

[108] Ashna Ashesh and Arun Thiruvengadam, ‘Report on Citizenship Law: India’, (European University Institute, Robert Schuman Centre for Advanced Studies’, July 2017) 19 http://cadmus.eui.eu/bitstream/handle/1814/47124/GLOBALCIT_CR_2017_12.pdf accessed on 26 February 2018.

[109] Lovish Garg, ‘If India Wants to Remain Secular, the New Citizenship Bill Isn’t the Way to Go’, (*The Wire*, 21 September 2017) https://thewire.in/communalism/citizenship-amendment-bill-2016. Accessed on 25 February 2018.

[110] Sanjib Baruah, ‘Who can become an Indian Citizen?’, *The Indian Express*, (New Delhi, 28 November 2016).

unconstitutional, and it must include the religion of Islam along with other faiths in its scope, or it must have a religion-free migrant or refugee definition.[111] It infringes the right to equality guaranteed under article 14 of the Constitution of India. The absence of a national law on refugees has placed the refugee rights in a vacuum, and India has to proactively go for a refugee law for maintaining the territorial integrity, and for establishing our high benchmarks of respecting the international human rights.

5.3.5 *Maldives*

The Maldives is the most vulnerable geopolitical entity in South Asia that finds itself at the gallows of climate change persecution. The Maldives has been unsuccessfully bracing itself for climate change displacement and migration to a destination that remains undecided.[112] However, the Maldives is bound to have climate refugees due to fast emerging climate change caused by the rising sea levels, extreme weather forecasts, permanent flooding, changes to the island and marine ecosystems caused by the rise in sea and land temperatures[113] that might engulf the existence of Maldives as a nation-state.[114] The three-quarter land area of the Maldives is one metre above of the sea level, and the smallest amount of rising in the mean sea level would jeopardize the human rights and create a humanitarian crisis[115] for the existential survival of the Maldives.

There are huge gaps[116] in the legal protection of refugees in the Maldives, and it does not host any refugee, and there is no NGO(s) or any civil society institution that claims of providing any legal aid or assistance to refugees in the Maldives. Therefore, there is no existing legal framework for climate refugees' protection in the Maldives and UNCSR does not mention at all the terms like 'climate change refugee', 'climate refugee', 'climate displacement' or 'environmental refugee'. Moreover, non-binding international human rights treaties and Conventions do not provide any protection

[111]Nafees Ahmad, 'The Status of Refugees in India', (*The Fair Observer*, 22 September 2017) https://www.fairobserver.com/region/central_south_asia/refugees-rights-india-south-asian-world-news-headlines-97021/ accessed on 12 January 2018.

[112]Chloe Anne Vlassopoulos, 'Defining Environmental Migration in the Climate Change Era: Problem, Consequence or Solution', in Faist Thomas, Jeanette Schade (eds.) *Disentangling Migration and Climate Change: Methodologies, Political Discourses and Human Rights*, (Springer, 2013) 145–163.

[113]Sumudu Atapattu, 'Climate Change: Disappearing states, Migration, and Challenges for International Law' (2014) 4. Wash. J. Envtl. L. & Pol'y, 1–34, 9.

[114]Jane Mc Adam, (ed.), *Climate Change and Displacement: Multidisciplinary Perspectives*, (Oxford, Hart Publishing, 2010) 1 introduction.

[115]Donald K. Anton and Dinah L. Shelton, *Environmental Protection and Human Rights*, Cambridge University Press, (June 2011) 767.

[116]G. Chomette, 'Indian Ocean: Maldives, an Archipelago in Peril', (2010) Massachusetts Institute of Technology and Dominique Carrééditeur, 124–163.

to climate refugees. Further, the existing legal regime on climate change does not provide any protection to climate refugees.

5.3.6 Nepal

In Nepal, the government has been hosting Tibetan and Bhutanese refugees without any regard to humanitarian considerations, and it is not a party to the UNCSR and its Additional Protocol of 1967, and it has not acceded to the 1954 UN Convention on the Statelessness and the 1961 UN Convention on the Reduction of Statelessness.[117] Recently, the government of Nepal has declined to act upon the recommendations of the UN Human Rights Council (HRC) on the adoption of refugee law and Conventions incidental thereto or connected therewith due to the misplaced national security narrative.[118] However, the human rights defenders and stakeholders have prepared a Draft Human Rights Bill without the protection of climate refugees that got infructuous due to the dissolution of Nepali Parliament.[119]

On the other hand, the municipal law of Nepal does not make any reference of the term ‘refugee’ and domestic law of Nepal treats all non-citizens as foreigners and does not have any middle category of non-citizen inhabitants of Nepal, and there is no national policy on the treatment of refugees in Nepal.[120] The all judicial facets of foreigners in Nepal are dealt with under the Immigration Act of 1992, and it suffers from lack of competence in protecting the refugees and asylum seekers. The overall responsibility for refugee protection programme has been entrusted to the National Unit for the Coordination of Refugee Affairs (NUCRA) that is responsible for the National Refugee Policy under the Ministry of Home Affairs (MoHA) of Nepal.[121]

---

[117] Submission by the United Nations High Commissioner for Refugees For the Office of the High Commissioner for Human Rights’ Compilation Report—Universal Periodic Review: 2nd Cycle, 23rd Session NEPAL, 1 https://www.upr-info.org/sites/default/files/document/nepal/session_23_-_november_2015/unhcr_upr23_npl_e_main.pdf accessed 1 March 2018.

[118] Kathmandu Post, ‘Nepal refuses to adopt refugee Convention citing national interest’, *Kathmandu Post*, (3 December 2015) http://kathmandupost.ekantipur.com/news/2015-12-03/nepal-refuses-to-adopt-refugee-convention-citing-national-interest.html accessed 1 March 2018.

[119] Inhured International, ‘National Roundtable on Refugee Protection Proceedings at-a-Glance’, (Inhured International, March 25, 2016) 1 http://inhuredinternational.org/source/National%20Roundtable%20on%20Refugee%20Protection-Proceedings%20(4).pdf accessed on 1 March 2018.

[120] Barun Ghimire, ‘Lawful Limitation or Northern Influence? Restricting the Freedom of Expression of Tibetan Refugees in Nepal, A dissertation submitted in partial fullfilment for the degree: Master in Human Rights Practice School of Global Studies, University of Gothenburg School of Business and Social Sciences, University of Roehampton Department of Archaeology and Social Anthropology’, (University of Tromso, 2013), 4 https://munin.uit.no/bitstream/handle/10037/5319/thesis.pdf?sequence=2 accessed on 5 March 2018.

[121] Himalayan News Service, ‘Government Urged to Frame Refugee Policy’, *The Himalayan Times* (Nepal, 4 May 2016).

There are plenty of human rights mechanisms and instruments[122] that have been acceded by Nepal, and there are number of provisions whereunder the entire gamut of human rights has made available to Nepalese such as the right to live with dignity under article 16, the right to equality under article 18, the right to justice under article 20, the right against torture under article 22 and the right against preventive detention[123] under article 23 whereunder protection to all refugees in Nepal can be extended as per the new Constitution of Nepal. However, the government of Nepal has been clamping restrictions[124] on refugees in violation of its international and constitutional obligations to respect and ensure the peaceful enjoyment of fundamental human freedoms in Nepal. However, these provisions are not adequate due to their legal volatility and susceptibility to multiple interpretations that are bound to undermine the rights of political refugees and climate refugees in the absence of a national refugee law and global standards of refugee protection. Therefore, there is need to have a participatory national refugee regime in Nepal wherein the concerns of all stakeholders are addressed for once and all.

5.3.7 *Pakistan*

Pakistan is not a signatory state to the UNCSR and its 1967 Additional Protocol, but it has been hosting refugee population from all corners of the world.[125] Pakistan is the second largest country in the world hosting refugee population of 1,505,525 including 1.5 million registered Afghan refugees.[126] The refugee status in Pakistan is regulated under the tripartite agreement between the Afghanistan, Pakistan and the UNHCR whereunder non-refoulement and voluntary repatriation have been included.[127] Article 9 of the Constitution of Islamic Republic of Pakistan provides the security of person[128] that includes refugees and other non-nationals in Pakistan. By being a member of UNO, Pakistan is obligated to cooperate with the UNHCR and facili-

[122]Sarah Paoletti, Eleanor Taylor-Nicholson, Bandita Sijapati and Bassina Farbenblum, 'Migrant Workers' Access to Justice at Home: Nepal (2014) Faculty Scholarship. Paper 1326, 85 (Open Societies Foundation, 2014) https://ceslam.org/docs/publicationManagement/Migrant%20Workers%20Access%20to%20Justice%20Nepal.pdf accessed on 5 March 2018.

[123]Ghimire (n 121) See The Constitution of Nepal, 2015.

[124]Ibid. *See* also James C. Hathaway, *The Rights of Refugees under International Law,* (Cambridge University Press, 2005) 280.

[125]Ali Nawaz Chowhan, 'Signing the Refugee Convention', (*The Express Tribune*, (Pakistan 5 July 2011), https://tribune.com.pk/story/202214/signing-the-refugee-convention/ accessed on April 10, 2018.

[126]The Nation, 'Second largest refugee population hosted by Pakistan according to UN' (The Nation, 20 June 2017) https://nation.com.pk/20-Jun-2017/second-largest-refugee-population-hosted-by-pakistan-according-to-un accessed on April 15, 2018.

[127]Elizabeth Mavroudi, Ben Page and, Anastasia Christou, *Timespace and International Migration*, (Edward Elgar Publishing, 2017) 68–70.

[128]The Constitution of Islamic Republic of Pakistan, 1973, article 9, states that 'Security of person. No person shall be deprived of life or liberty save in accordance with law'.

tate its operations in Pakistan. Nevertheless, such agreements do fulfil gaps in the refugee protection but only to a limited extent and these ephemeral arrangements cannot be a substitute for specific national refugee law in Pakistan. The principle of *non-refoulement*[129] is a part of customary international law [130] wherefrom no derogation is possible unless there is a treaty contrary thereto that is the '*evidence of a general practice accepted by states*'[131] determined by the general practice of states and *Opinio Juris* (states' acceptance as law or legal opinion or opinion of law) and, therefore, all countries are bound to follow the same in protecting the refugees.[132] In this context, it would, indeed, be unfortunate to think of forced repatriation of Afghan refugees on ethical and political grounds from Pakistan. The government of Pakistan must abide by the customary international law norms and refugee protection commitments[133] while observing the *non-refoulement* principle along with its tripartite agreements about the Afghan refugees. However, Pakistan cannot escape from its national and international responsibility to have a consolidated and comprehensive national law based on participatory protection architecture that includes the entire range of human displacement; refugees, asylum seekers, IDPs and climate refugees.

5.3.8 Sri Lanka

In Sri Lanka, there is no refugee law of its own, and it has neither acceded to the UNCSR nor signed the 1967 Additional Protocol thereto. However, the constitution of Sri Lanka protects all persons including refugees and asylum seekers from the '*torture or cruel, inhuman or degrading treatment, or punishment*'[134] and provides to all persons '*equality before law and equal protection of law*'[135] and '*no person shall, on the grounds of religion, race, caste, sex, language or any one of such grounds, be subject to any disability, liability, restriction*'[136] or arrested,[137] detained,[138] charged[139] or punished.[140] Therefore, it is evident from these constitutional safeguards that politi-

---

[129]Refugee Convention (n 3) article 33 (1).

[130]Hathaway (n 126) 364.

[131]A Mark Qeisburd, 'The International Court of Justice and the Concept of State practice', (2009) 31(2) U. Pa. J. Int'l L. 296–298.

[132]Jörg Kammerhofer, 'Uncertainty in the Formal Sources of International Law: Customary International Law and Some of Its Problems', (2004) 15 EJIL 532–536.

[133]Organization of the Islamic Conference (OIC), Resolution No. 32/7-P(IS) on the Problem of Refugees in the Muslim World, 15 December 1994, No. 32/7-P(IS), available at: http://www.refworld.org/docid/48c7b4e12.html accessed 30 May 2018.

[134]The Constitution of the Democratic Republic of Sri Lanka, 1978, article 11.

[135]Ibid, article 12 (1).

[136]Ibid, article 12 (3).

[137]Ibid, article 13 (1).

[138]Ibid, article 13 (2).

[139]Ibid, article 13 (3).

[140]Ibid, article 13 (4).

cal refugees, as well as climate refugees, can also be protected, but the government of Sri Lanka has not respected the constitutional mandate. However, the UNHCR has been undertaking the refugee status determination (RSD) procedure of refugees and asylum seekers in Sri Lanka under an agreement with the government that provides education and health care to refugees and asylum seekers, but they are not permitted to work and earn their livelihood. But, unfortunately, the impugned agreement does not mention the plight of climate refugees or human displacement and migration caused by the climate change.

It has been seen that Sri Lanka violated the international human rights law norms[141] and *non-refoulement* principles[142] while forcefully deporting Pakistani refugees back to their country of origin on the ground that '*a state's responsibility under international law has to be nuanced and balanced in the context of domestic compulsions*'.[143] Therefore, the government of Sri Lanka must enact national legislation on all displaced population that incorporates political and climate refugees and CDPs while adhering to the global norms on refugee protection. However, the existing national and international provisions are incapable of extending any protection to climate refugees or CDPs as object and context of their protection are substantially different.

5.3.9 Reception of International Climate Change Law in South Asia

The climate change is the most dynamic development in the anatomy of human existentialism that has presented a compendium of challenges to the international community to delineate its ramifications and repercussions of across-the-board consequences. In the words of Prof. Anand, 'law is an outgrowth of the man's needs in society'.[144] Climate change affects the social, economic and guaranteed human rights of the individual and the existence of society and nations.[145] It engulfs upon matters ranging from health, food security, access to water, and livelihood and migration patterns.[146] The international community has responded to the grave and complex

---

[141]Universal Declaration of Human Rights, G.A. res. 217A (III), U.N. Doc A/810 at 71 (1948) article 14.

[142]Refugee Convention (n 3) article 33 (1), and CAT, 1984 (n 75).

[143]Adrian Edwards, Sri Lanka Violates International Refugee Law: UN, Aug 12 2014, https://www.livemint.com/Politics/Fb90rgQRVLYktzZ5esc3nJ/Sri-Lanka-violates-international-refugee-law-UN.html, accessed 15 April 2018.

[144]R.P. Anand, 'The Influence of History on the Literature of International Law', in Ronald St. J. Mac Donald and Douglas M. Johnston (eds) *The Structure and Process of International Law: Essays in Legal Philosophy, Doctrine and Theory*', (Brill 1986) 341.

[145]Randall Abate, *Climate Justice: Case Studies in Global and Regional Governance Challenges*, (West Academic Press, 2016) 513.

[146]Maxmillan Martin, Yi Hyun Kang, Motasim Billah, Tasneem Siddiqui, Richard Black and Dominic Kniveton, 'Policy Analysis: Climate Change and Migration Bangladesh'

economic, social and environmental challenges posed by climate change and has constructed an edifice of the robust legal framework in the visage of the UNFCCC,[147] Kyoto Protocol[148] and the Paris Agreement.[149]

The international legal framework recognizes climate change as a 'common concern of mankind' lays down the obligation of state parties to mitigate and adapt to climate change and emphasize on state cooperation, dialogue and participation in good faith.[150] The construction of legal framework governing climate change has been nurtured with the universal, voluntary, vigorous and active involvement of states.[151] The successful and efficient tackling of climate change requires the implementation of the legal obligations ascribed under the international climate instruments through the comprehensive strategies and policies.[152] Regarding climate change responses, South Asian countries have been guided by the international legal instruments and have been executing its international obligations.[153] However, the definite reception and responses to climate change at national levels depend to a great extent at the level of participation and influence countries has exerted at the international law-making process.[154] Before deliberating on the reception of International position on climate displacement with regard to the legal and institutional responses initiated by South Asian jurisdictions at the domestic level, it is pertinent to assess the role and part played by South Asian countries and developing countries in general in the development and evolution of climate change legal structure and International Environmental Law in general. The assessment of the role and perspectives of South

---

Working Paper, (Refugee and Migratory Movements Research Unit (RMMRU), University of Dhaka, and Sussex Centre for Migration Research (SCMR), University of Sussex, 2013) 3–4 http://admin.indiaenvironmentportal.org.in/files/file/climate%20change%20and%20migration%20bangladesh.pdf accessed on 2 April 2018.

[147] United Nations Framework Convention on Climate Change (adopted 29 May 1992, entered into force 21 March 1994) 1771 UNTS 107 (hereinafter UNFCCC).

[148] Kyoto Protocol to the Framework Convention on Climate Change, 3d Sess., Agenda Item 5, U.N Doc. FCCC/CP/1997/L.7/Add.1, 37 I.L.M. 32 (1997) (hereinafter Kyoto Protocol).

[149] UNFCCC. *Adoption of the Paris Agreement*. Report No. FCCC/CP/2015/L.9/Rev.1, http://unfccc.int/resource/docs/2015/cop21/eng/l09r01.pdf accessed 10 February 2018 (Hereinafter Paris Agreement).

[150] Volker Roeben, 'Responsibility in International Law', (2012) 16 Max Plank, YBIL, 99–158,116.

[151] Karin Bäckstrand and Eva Lövbrand, 'The Road to Paris: Contending Climate Governance Discourses in the Post-Copenhagen Era', (2016) J. Environ. Policy Plan, 1–19.

[152] Robert N. Stavins, 'Policy Instruments for Climate Change: How Can National Governments Address a Global Problem?' (University of Chicago Law School, 1997) 293–329, 325 https://ageconsearch.umn.edu/bitstream/10757/1/dp970011.pdf accessed on 20 March 2018.

[153] Sumudu Atapattu, 'Climate Change in South Asia: Towards Climate Change in South Asia: Towards an Equitable Legal Response Within a Frame work of Sustainable Development and Human Security', (2011) *IDLO Sustainable Development Law on Climate Change Working Paper Series*, 1–14, 11.

[154] Ibid.

Asian jurisdictions in the evolution of climate change jurisprudence is significant in assessing the overall question of legitimacy and fairness of the legal structure.[155]

The existing international law has been primarily originated in the practices of the European community and has developed without the participation of the Asian-African, Latin American and other developing countries.[156] Prof. Anand cites Oppenheim and points out that for the development of international law, the Christian state of Western Europe constituted the international community.[157] He states 'when modem international law was developing in big strides, international society or the family of nations was confined to a "selective community with a provincial outlook"'.[158] The marginalization and exclusion of these nations were deliberate and was defended by reference to culture, religion, civilization and biology.[159] The role of Asian and African countries in the development of international law has thus been kept negligent since they were colonized and was considered incapable of bearing the torch of sovereignty following the industrial revolution.[160]

The establishment and inauguration of the United Nations after the World War II represents a scenario marked by the formal and substantial equality of states and the legitimate period for the creation of equitable global order.[161] The United Nations was established with the major objective of promoting 'cooperation in solving international problems of an economic, social, cultural or humanitarian character'.[162] The period witnessed the liberation and independence of colonized nations who attempted to create a discernible influence on the development of international law and challenged the very foundation of Eurocentrism and colonial orientation of international law and focused on the vast lack of participation and legitimacy in the lawmaking process.[163] Their resentment and challenge became palpable in the two landmark General

---

155 Hayley Stevenson and Johan S. Dryzek 'The Legitimacy of Multilateral Climate Governance: A Deliberative Democratic Approach' (2012) 6(1) Crit'l. Poli'y. Stud. 1–18.

156 Simon Chesterman, 'Asia's Ambivalence about International Law and Institutions: Past, Present and Futures', (2016) 27(4) EJIL 945–978.

157 R P Anand, *Toward a New International Legal order, Changing Dimensions of International Law: An Asian Perspective* (Martinus Nijhoff Publisher, September 2006) 8.

158 R P Anand, *Confrontation or Cooperation? International Law and the Developing Countries*, (Hope India, 2011) 28.

159 Simon Chesterman (n 158) 946, see Antony Anghie, *Imperialism, Sovereignty, and the Making of International Law* (Cambridge University Press, 2005) 33–34.

160 Surya Prakash Slnha, *Legal Polycentricity and International Law* (Carolina Academic Press, 1996) 6 &15, R.P. Anand, 'Role of the 'New' Asian African Countries in the Present International Legal Order' (1962) 5 AJIL, 384–406.

161 Preuss, Ulrich K. 'Equality of States—Its Meaning in a Constitutionalized Global Order', (2008) 9(1) Chi. J. Int'l L.17-49. See Quincy Wright, 'The Equality of States', (1970) 3(1) Cornell Int'l L. J. 1–7.

162 U.N. United Nations, Charter of the United Nations, 1945, 1 UNTS XVI, article 1, Para 3 http://www.un.org/en/sections/un-charter/chapter-i/index.html (accessed on 22 April 2018); See Charles W. Kegley and Gregory A. Raymond, *The Global Future: A Brief Introduction to World Politics*, (Cengage Learning, 29 January 2013), 112.

163 Anand (n 162) 385.

Assembly Resolutions (UNGA) adopted in the early phase of the United Nations.[164] United Nations was utilized as a forum to debate and campaign for economic rights, greater participation and enhanced protection of sovereignty for developing countries as part of the development of international law of natural resources.[165] It is not to mention that the newly independent states dramatically rejected the whole of current dynamics of international law based on political social and economic exclusion, they believed and desired a system, which is dynamic, diversified, inclusive, equitable and responsive to the change with the changing circumstances.[166]

Having convinced that the continued colonialism threatens the establishment and achievement of international cooperation, participation, development and collective pursuit of the international community, the developing nations clamoured for the adoption of Declaration on the Granting of Independence to the Colonial Countries and the Peoples[167] and vociferously argued for the UNGA resolution for the elaboration of the principle of Permanent Sovereignty over Natural Resources (PSNR) adopted in 1952.[168] It was felt by the developing nations that the concept and tool of PSNR will be instrumental in remedying their historical colonial exploitation, promoting and strengthening social and economic development, building technical financial, legal and administrative capacity building in developing nations and ensuring equitable resource exploitation by all the nations and promote peoples right to self-determination.[169]

As pointed out by Simon Chesterman, the Asian states did not participate in the vast majority of negotiations forming most of the international legal instruments, and as for the original members of the United Nations only India was represented among the South Asian nations.[170] The newly independent countries have determinedly demanded their participation and to create a system of International Law based on

---

164 Edward Kwakwa, 'Emerging International Development law and Traditional International law Congruence or Cleavage?' (1987) 17 Ga. J. Int'l & Comp. L, 431–455.

165 Nico Schrijver, 'Sovereignty Over Natural Resources: Balancing Rights and Duties', (Cambridge University Press 2008)33.

166 B.V.A. Roling, *International Law in an Expanded World* (Amsterdam, 1960) 50–51.

167 *Declaration on the Granting of Independence to Colonial Countries and Peoples, Adopted by General Assembly Resolution 1514 (XV) of 14 December 1960,* A/RES/1514(XV), available at: http://www.refworld.org/docid/3b00f06e2f.html accessed 10 February 2018.

168 Kenneth J. Vandevelde, 'A Brief History of International Investment Agreements' (2005) UC Davis J. Int'l L. & Pol'y 158–194, 165, see Dominique Carreau and Patrick Juillard, *Droit International Cconomique* (Dalloz, 2003) 402, Michael P. Todaro, *Economic development in the Third World* (Longman, 1989) 598, Francis Snyder and Peter Slinn, (ed.), *International Law of Development: Comparative Perspectives*, (Abingdon, England).

169 Subrata Roy Chowdhury, 'Permanent Sovereignty over Natural Resources—Substratum of the Seoul Declaration' in P. De Waart et al. (eds) *International Law and Development'*, (Dordrecht,1988) 61, 64–65. Karol N. Gess, 'Permanent Sovereignty over Natural Resources—An Analytical Review of the United Nations Declarations and Its Genesis', (1964) 13(2) Int'l & Comp. L. Q 398–449, Antonio Cassese, *Self-determination of Peoples—A Legal Appraisal*, (Cambridge 1995), 55–56; N. Schrijver (167) 22–24.

170 Chesterman (n 158) 946.

legitimacy and contemporary realities.[171] The period also witnessed the multi-fora effectively initiated effort of the developing nations to promote and achieve economic development and the developed countries to continue their economic domination.[172] In this continuous economic struggle for maintaining the status quo and supremacy by the developed world and the catching up the struggle initiated by the developing nations, the environment was neglected and suffered terrible degradation.[173]

The initial phase of international environmental law was characterized by the concern for conservation of nature, which spurred the first concerted international environmental regulations in the early 1900s.[174] The attention of the first phase of the environmental law development was on conservation.[175] Steadily, the focus of environmental protection got widened to include pollution aspects and other concerns.[176] However, these advances were not genuinely comprehensive, wide ranging and global, it was mostly measures introduced like ad hoc and piecemeal approach focusing on control and command mechanisms.[177] Similar to the evolution and growth of international law, the initial progression of environmental law was achieved with the only peripheral participation of the developing nations including the South Asian states without analysing the economic and social needs of underdeveloped countries and their societies.[178] The 1972 U.N. Conference on Human Environment (UNCHE)[179] held at Stockholm for the first time saw international community focusing its concerted attention on environmental problems.[180] By the time of Stockholm conference, the international community in the formal sense was based on the equality and participation of states.[181] However, developing countries were sceptical about the whole process and participated in the forum reluctantly.[182] For developing coun-

[171]Surya Prakash Sinha, 'Perspective of the Newly Independent States on the Binding Quality of International Law' (1965) 14(1) Int'l & Comp. L. Q, 121–131, Georges M. Abi-Saab, 'The Newly Independent States and the Rules of International Law: An Outline', (1962) 8 HLJ 95.

[172]R.P. Anand, *Environment, Development and the Developing Countries, Changing Dimensions of International Law: An Asian Perspective* (Martinus Nijhoff Publisher, 2006) 27.

[173]Ibid.

[174]Daniel Bodansky, *Craft of International Environmental Law*(2011) 22, Alexandre Kiss and Dinah Shelton, *International Environmental Law* (Transnational Law Publishers, 2004) 40.

[175]Ibid, 40. Philippe Sands, *International Environmental Law* (Cambridge, 2003) 26.

[176]Ibid, 41; see Sands, 30.

[177]Bodansky (n 176) 24.

[178]Ulrich Beyerlin, 'Bridging the North-South Divide in International Environmental Law', (2006) 66 ZaöRV, 259–296.

[179]Declaration of the United Nations Conference on the Human Environment, adopted June 16, 1972, U.N. Doc. A/CONF.48/14, reprinted in 11 I.L.M. 1416 (1972) [hereinafter Stockholm Declaration].

[180]Edith Brown Weiss, 'The Evolution of International Environmental Law', (2011) 5(4) Japanese Y.B. Intl. L, 1–27, 5, See generally Patria Birnie, Alan Boyle, and Katherine Redgwell, *International Law and the Environment*, (Oxford University Press 2009).

[181]DA Kay and E. B. Skolnikoff (eds) *World Eco-crisis: International Organizations in Response.* (Madison, University of Wisconsin Press, 1972).

[182]Adil Najam, 'Developing Countries and Global Environmental Governance: From Contestation to Participation to Engagement', (2005) 5 *International Environmental Agreements* (Springer)

tries, the agenda of the protection of environment did not assume the same gravity as for the developed nations and were illustrative of their experience with industrialization.[183] They turned the attention of the world to the fundamental distinction between the 'pollution of affluence' and the 'pollution of poverty' a difference ascribed in the conference.[184] The difference between pollution of affluence and poverty reflects the dynamics, which are central to the global efforts to reconcile the conflicting demands of conservation and development and became a catalyst in the structural evolution and development of international environmental law as a worldwide concern.[185]

Najam while tracing the participation of developing nations in the international environmental law negotiations points that 'some developing countries distrusted the entire process, doubted the positions of developed nations and saw Stockholm conference as an attempt to focus on environmental matters, which will enhance the existing unequal economic relations and technical dependence, and miring in poverty and perpetual underdevelopment forever'.[186] They questioned the need for such a conference and viewed it not just as a diversion, but also as a threat to their development pursuits fearing that environment conservation will take precedence over poverty and underdevelopment.[187] The developing countries main concern was on addressing the metasymmetrical power structure and questions of legitimacy in environmental governance rather than on the effectiveness and normative content of environment protection.[188]

Though the participation of the developing nations was marked by reluctance and distrust, unlike in the field of general international law, the developing countries participated and contributed in shaping the structure and contents of international environmental law. Stephen Krasner points out that the Third-World countries have been able to alter the norms, policies, structure, nature and policies of international law.[189] Though the structural impediments to developing countries in using international law to their advantage remain, the present international environmental law principles and structure owe a great deal to the collective wisdom and camaraderie exhibited by the developing nations.[190] The participation of developing countries in the Stockholm negotiation brought into focus the issue of development, issues of

---

303–321, see Tim E.J. Camphell, 'The Political Meaning of Stockholm: Third World Participation in the Environment Conference Process' (1973) 8 Stan. J. Int'l Stud. 138–153.

[183]John Ntambirweki, 'The Developing Countries in the Evolution of an International Environmental Law', (1991) 14 Hastings Int'l & Comp. L. Rev., 905–928.

[184]Bjørn-Oliver Magsig, *International Water Law and the Quest for Common Security*, (Routledge 2015) 165.

[185]Ibid, See R.P. Anand, *Studies in International law and History: An Asian Perspective*, (Springer, 2013) 260.

[186]Najam (n 184), 303–321; Susanna Hecht and Alexander Cockburn, 'Rhetoric and Reality in Rio', (1992) 254 (24) The Nation, 848–854.

[187]Kiss and Shelton, (n 176) 35.

[188]Najam (n 184) 307.

[189]Stephan D Krasner (ed.) *International Regimes* (Cornell University Press, 1983) 66–94.

[190]Ntambirweki (n 185) 907–908.

political and territorial sovereignty, population control, etc.[191] In spite of the desire of the developed nations to have a legally binding Declaration,[192] the reluctance of countries mostly consisting of developing countries forced the conference to produce a seminal non-binding soft law Declaration indicating its moral and ethical character rather than the legal dimension to guide states in evolving the law governing the environment protection.[193] Stockholm Declaration prominently linked environment with human rights and development.[194] The developing countries were able to articulate that the crisis of development and environmental protection is based on equity, which requires countries to cooperate and assistance to the developing nations in reducing the economic gap.[195]

From a legal point of view, two developments are essential; Stockholm conference led to the creation of a new institutional structure in the form of UNEP to coordinate and assist in the progressive development of international law.[196] The establishment of UNEP is symbolic of the vociferous position of developing nations: UNEP is the first international organization to be headquartered in developing nation-states.[197]

Stockholm represents the first instance where a compromise between the positions of developed and developing nations and the understanding between the environmental protection and development were achieved.[198] The compromise between environmental protection and economic growth got a pivotal boost with the articulation of the idea of sustainable development[199] that was delineated as a development 'that meets the needs of the present era without negotiating the ability of future generations to meet their own needs'.[200] In the legal sense, the principle of sustainable development was enunciated as the leading concept of international environmental law with the Rio Declaration[201] and Agenda 21.[202]

---

[191]Camphell (n 184) 139.

[192]Philppe Sands, *Greening International Law,* (Routledge 2014) 2.

[193]Ibid 2.Sands, (n 177) 36, see generally Birnie, Boyle Redgwell (182).

[194]Anita M.Halvosren, 'The Origin and Development of International Environmental Law', in Shawkat Alam, Md Jahid Hossain Bhuiyan, Tareq M.R. Chowdhury and Erika J. Techera (eds), *Routledge Hand Book of International Environmental Law,* (London: Routledge) 25–42.

[195]See generally, Christopher C. Joyner, 'Stockholm in Retrospect Progress in the International Law of Environment', (1974) 136(4) World Affairs 347–363; Stockholm Declaration (n 181) Preamble.

[196]Carol Annette Petsonk, 'The Role of the United Nations Environment Programme (UNEP) in the Development of International Environmental Law', (1991) 5(2) Am. U. Int'l. L. Rev., 351–391, 354.

[197]Paul G Harris, *Routledge Handbook of Global Environmental Politics*, (Routledge, 2013) 13.

[198]Elen Stokes, *Environmental Law,* (Oxford University Press, 2017) 149.

[199]Birnie, Boyle and Redgwell (n 182) 50.

[200]Brundtland Commission, Report of the World Commission on Environmental & Development: Our Common Future, ch. 2, 1, U.N. Doc. A/42/427 (Oct. 1987) (hereinafter Brundtland Report).

[201]United Nations Conference on Environment and Development, Rio de Janeiro, Braz., June 3–14, 1992, Rio Declaration on Environment and Development, U.N. Doc. A/CONF.151/26/Rev.1 (Vol. I) (hereinafter Rio Declaration).

[202]U.N. GAOR, 46th Sess., Agenda 21, U.N. Doc. A/CONF.151/26 (June 14, 1992) (hereinafter Agenda 21), See also Stokes (n 200) 150.

Rio conference represented the collective euphoria, bargaining and resilience of the developing nations.[203] From the stable position of contestation, which has pigeon-holed the position of developing nation-states in the Stockholm, the negotiation leading to the Rio was marked by reluctant participation, collective bargaining and broader agenda of institutional and legal reformulation in the structure of north–south economic relationship.[204] For the South Asian countries, if the Stockholm conference represented the articulation of Indian Prime Minister, the Rio conference 1992 marked the opportunity for Pakistan's leadership as the Chair of Group 77 to lead the south.[205] The numerical majority of the developing nations immensely assisted them in articulating the special needs, assistance and concession for the developing nation-states.[206] The legacy of Rio Declaration continued and had a rippling effect on the climate negotiations taking place at the global level.[207]

Formal climate negotiations were initiated in 1990, and the negotiations' history reveals the collective bargaining strength of developing nations.[208] The language of equity, sustainable development, participation and environmental justice permeated early in the phase of climate negotiations, characterized by fierce and opposing positions between developed and developing countries.[209] The developing countries interpreted the notion and assignment of responsibilities based on environmental justice according to historical contributions and the overwhelming greenhouse emissions by the developed nations.[210] They also emphasized on the capacity of developed nations to undertake emissions cut and bear the burden of

---

[203] Jacqueline Noga and Gregor Wolbring, 'An Analysis of the United Nations Conference on Sustainable Development (Rio +20) Discourse Using an Ability Expectation Lens', (2013) 5 Sustainability 3615–3639.

[204] Nigam (n 184) 318 see Dennis Soden and Brent S. Steel, *Handbook of Global Environmental Policy and Administration*, (CRC Press, 1999) 227.

[205] Parvez Hassan, Human Rights and The Environment: A South Asian Perspective, A keynote address delivered at the 13th Informal ASEM Seminar on Human Rights, coorganized by the French Ministry of Foreign and European Affairs, the Raoul Wallenberg Institute the Philippine Department of Foreign Affairs and the Asia-Europe Foundation (ASEF) and hosted by the Danish Institute of Human Rights and the Danish Ministry of Foreign Affairs, on 21–23 October 2013, at Copenhagen, Denmark, 12 https://www.asef.org/images/docs/Keynote%20speech-Parvez%20Hassan_Human%20Rights%20and%20Environment%20A%20South%20Asian%20Perspective.pdf accessed 10 March 2018, 11.

[206] William L. Ascher and Natalia Mirovitskaya, *Guide to Sustainable Development and Environmental Policy*, (Duke University Press, 2002) 8.

[207] Oberthür Sebastian and Ott, Hermann E., *The Protocol International Climate Policy for the 21st Century* (Springer 1999) 43.

[208] Joydee Gupta, 'Negotiating Challenges and Climate Change', (2012) 12 Clim. Pol'y 630–644, 631.

[209] Ellen Wiegandt, *Climate Change, Equity, and International Negotiations in International Relations and Global Climate Change*, (MIT Press, 2010) 128, Paramjit S. Jaswal and Stellina Jolly, Fairness and Rule of Law in Climate Change Discourse: A Critical Analysis', (2010) Vol 52(3&4) JILI, 366–379.

[210] Jutta Brunnée, and Charlotte Streck, 'The UNFCCC as a Negotiation Forum: Towards Common But More Differentiated Responsibilities' (2013) 13(5) Clim. Pol'y 589–607.

obviating climate change.[211] As noted by Rajmani, 'developing countries also have fewer financial resources for investing in robust infrastructure that is resistant to the negative impacts of climate change'.[212] The need to differentiate between commitments between nations based on the contribution and capacity became the central issue.[213] This was resolved through the adoption of 'Common but Differentiated Responsibilities' (CBDR) engrained in the 1992 UNFCCC.[214] The principle based on the notion of equity and pragmatism became the overarching concept of the climate change legal framework and got further strengthened after its adoption under the Kyoto Protocol to the UNFCCC architecture.[215] It is believed that the formulation of CBDR corresponds to the proposal submitted by India, which underscored the collective responsibility and differential responsibility for the industrialized countries.[216] The CBDR principle marked a moral and legal victory for the collective stand of developing nations.[217]

As noted by Philippe Cullet, the broader concept of environmental protection cannot be looked at from the narrow prismatic and one-dimensional view of environmentalism; the issues are deeply entrenched in the social, economic and political concerns.[218] Unlike the development of international law, which was achieved without the support of the developing nations in the context of environmental protection, developing nations were able to influence and steer the direction of international environmental law through the emphasis on development, environmental justice, sustainable development and the principle of CBDR.[219]

The position, participation and perspectives of developing nations in climate negotiations have gone through different stages, with countries aligning differently based

[211] Yoshiro Matsui, 'Some Aspects of the Principle of Common but Differentiated Responsibilities', (2002) 2(2) *International Environmental Agreements: Politics, Law and Economics* 155, cited by Rachel Boyte, 'Common but Differentiated Responsibilities: Adjusting the Developing/Developed Dichotomy in International Environmental Law', (2010) 14 Nz J. Env. L 70–71.

[212] Lavanya Rajamani, *Differential Treatment in International Environmental Law,* (Oxford University Press 2006) 136.

[213] Justin Lee, 'Rooting the Concept of Common but Differentiated Responsibilities in Established Principles of International Environmental Law', (2015) 17 Vt. J. Envtl. L. 27–50, 47.

[214] Malgosia Fitzmaurice, 'Responsibility and Climate Change', (2010) 53 GYIL, 89–138, 106.

[215] Kyoto Protocol (n 150).

[216] Antto Vihma, 'India and the Global Climate Governance: Between Principles and Pragmatism' (2011) 20(1) J Envt & Dev. 69–94, See also, Christina Voigt and Felipe Ferreira, 'Differentiation in the Paris Agreement' (2016) 6(1 and 2) Clim. L. 58–74, 59. Tuula Honkonen, 'The Development of the Principle of Common But Differentiated Responsibilities and its Place in International Environmental Regimes', in Tuomas Kuokkanen (ed), *International Environmental Law-making and Diplomacy: Insights and Overviews.* (Routledge, 2016) 160, Beyerlin (n 180) 259–296.

[217] Philippe Cullet, 'Differential Treatment in Environmental Law: Addressing Critiques and Conceptualizing the Next Steps', (1999) 10(2) 549–582, 561 Jutta Brunnée, 'Promoting compliance with Multilateral Environmental Agreements' in Lavanya Rajamani, Jutta Brunnée and Meinhard Doelle (eds) *Promoting Compliance in an Evolving Climate Regime* (Cambridge University Press, 2012) 48–49.

[218] Philippe Cullet, *Differential Treatment in International Environmental Law*, (Ashgate, 2003) 15.

[219] Beyerlin(n 180) 259–296.

on their specificities regarding socio-economic and geographical aspects that determine their relative peculiarities and vulnerabilities.[220] Developing nations attempted to present a collective voice, identity and desire to negotiate as a collective block, even though the 'south' never constituted a homogenous unit.[221] However, the stronger, emerging and powerful nations among the developing south assert their separate position at international fora.[222] This dichotomy is more pronounced as the economic disparity among the Southern states intensified,[223] and environmental policy-making has shifted its focus from Declaration to concentrate on substantive and procedural implementation.[224] At the same time, today's multifarious environmental issues have presented an opportunity to vulnerable nations among developing nations to align themselves and present positions, demand concessions and receive financial assistance from the stronger nations.[225] However, the polarized positions of the negotiating blocks have at times seriously impeded the commitments and complicate the introduction of legal obligations.[226]

The general position of developing countries has been consistent throughout the negotiation, with the rejection of any mandatory emission cuts and obligations for developing nations.[227] According to Rajmani 'generally, developing countries conceptualizes climate change as a development issue and believe that unless the current generation generates and sustains high levels of economic growth, future generations will inherit earth that is highly vulnerable to climate change.'[228] Hence, equity demands that the current generation prioritizes development as a matter of urgency.[229]

---

[220]Hemanta Raj Poudel, 'Developing Countries in the International Climate Negotiations with the Prospective of International Relation Theories: A Case of Nepal' (BIBSYS Brage 2013), 17–23 https://brage.bibsys.no/xmlui/bitstream/handle/11250/187883/Hemanta%20Raj%20Poudel.pdf?sequence=1 accessed 25 April 2018.

[221]Tuula Honkonen, *The Common But Differentiated Responsibility Principle in Multilateral Environmental Agreements: Regulatory and Policy Aspects*' (Kluver Law International 2009) 9, Adil Najam, 'Dynamics of the Southern Collective: Developing Countries in Desertification Negotiations, Global Environmental Politics, and International Environmental Agreements' (2005) 5 *Global Envt. Politics* 303–321.

[222]Bandeira, Luiz Alberto Moniz, 'Brazil as a Regional Power and Its Relations with the United States.' (2006) 33(3) Latin American Perspectives 12–27, see Steven Radelet, *The Great Surge: The Ascent of the Developing World*, (Simon and Schuster, 2015).

[223]Fen Osler Hampson, and Paul Heinbecker, 'The New Multilateralism of the Twenty First Century', (2011) 17(3) Global Governance 299-310.

[224]Najam (n 184) 320.

[225]Joydee Gupta, *The Climate Change Convention and Developing Countries: From Conflict to Consensus*, (Springer, 2013)150–151.

[226]Wytze Van Der Gaast, *International Climate Negotiation Factors: Design, Process, Tactics*, (Springer 2016).

[227]Mark.J. Mwandosya '*Survival Emissions: A Perspective from the South on Global Climate Change Negotiations*' (DUP, 2000) 37.

[228]Lavanya Rajmani, 'India's Negotiating Position on Climate Change: Legitimate but not Sagacious' (CPR Issue Brief India 2007) 1–4 www.cprindia.org/system/tdf/policy-briefs/1253776076-PolicyBrief.pdf? accessed 26 April 2018.

[229]Ibid.

In spite of the success in steering the direction of climate discourse through the CBDR, developing countries look at the negotiation with suspicion and mistrust and feel that the transparency and information sharing are lacking at the negotiation process.[230] Developing nations historically looked at the negotiation process as creating an additional burden to them as they lack enough resources and are not part of informal decision-making.[231]

For the South Asian countries, the G-77 represented the strongest collective position to articulate their collective voices.[232] However, individual countries aligned themselves with alliances pursuing their collective often incorporating a range of conflicting interests on different occasion. During the negotiations that culminated into the adoption of the Kyoto Protocol, India sided with the Alliance of Small Island States (AOSIS) rather than G-77 and demanded a reduction of 20% that industrialized nation-states must meet by the year 2000 as measured against the levels of 1990.[233] This step led to the creation and establishment of the green group within G-77 and played a crucial role in securing the Berlin Mandate.[234] The negotiating groups of countries in South Asia exhibit divergence. Bangladesh, Bhutan and Nepal are the members of the LDC Negotiating Bloc. However, the Maldives being an island nation and faced with the catastrophic climate threat is aligned with the Association of Small Island Nation-States. India is allied with the BASIC alliance that consists of counties like Brazil, South Africa, India and China having 38.56 per cent of the world's population represented by the group. Its economic prominence and rising greenhouse emissions confers a unique relevance of this group in the negotiation.[235] The SAARC grouping has been conferred the position of a regional bloc in the capacity of having an observer status by the UNFCCC; however, SAARC is not a negotiating stakeholder.[236] Nepal aligned itself with the Group of Mountain

[230]Michael Richards, 'A Review of the Effectiveness of Developing Country Participation in the Climate Change Convention Negotiations' (Forest Policy and Environment Group Overseas Development Institute, 2011) 28 https://www.odi.org/publications/3618-review-effectiveness-developing-country-participation-climate-change-convention-negotiations accessed 26 April 2018.

[231]Joydee Gupta 'North-South Aspects of the Climate Change Issue.' (2000) 3(2) IJSD 115–135.

[232]Marc Williams, 'The Third World and Global Environmental Negotiations: Interests, Institutions, and Ideas' (2005) 5(3) *Global Environmental Politics* 48–69, Vihma, Antto, Yacob Mulugetta, and Sylvia Karlsson-Vinkhuyzen, 'Negotiating Solidarity? The G77 Through the Prism of Climate Change Negotiations' (2011) 23(3) Global Change, Peace and Security 315–334.

[233]Katharina and Axel Michaelowa, 'India in the International Climate Negotiations: From Traditional Nay-Sayer to Dynamic Broker', (Center for Comparative and International Studies (CIS) University of Zurich) 2011, 7 https://www.ethz.ch/content/dam/ethz/special-interest/gess/cis/cis-dam/Research/Working_Papers/WP_2011/2011_WP70_Michaelowa.pdf accessed 25 April 2018.

[234]Ibid.

[235]Karl Hallding and others, 'Together Alone: Brazil, South Africa, India, China (BASIC) and the Climate Change Conundrum', (Stockholm Environment Institute) 2011, 57 https://www.sei-international.org/mediamanager/documents/Publications/Climate/sei-basic-preview-jun2011.pdf, accessed 25 April 2018; Xinran Qi, 'The rise of BASIC in UN climate change negotiations' (2011) 18(3) SAJIA 295–318.

[236]Chandra Lal Pandey, 'Climate Change in South Asia: Green Bridging between India and Nepal', in Iain Watson, Chandra lal Pandey (eds), *Environmental Security in the Asia-Pacific*, 2015, 95–126.

Landlocked Developing Countries and had been demanding the recognition of the particular vulnerabilities of the Small Mountainous Developing Nations.[237] Along with aligning themselves in different negotiation groups, the South Asian Countries differ in ideologies as far as some of the core climate change strategies are concerned.

At the Copenhagen summit 2009, SAARC submitted its common position that given the overwhelming historical greenhouse emissions by the developed nations' adherence to the principle of CBDR is a reflection of equity.[238] They also recommended separate funding mechanisms for mitigation adaptation and technology transfer.[239] For India and Pakistan, the priority was on enhancing access to Clean Development Mechanism. However, the other South Asian nations preferred to highlight access to adaptation fund to fight climate change effects.[240] All the South Asian countries except India aim to limit the temperature rise to below 1.5-°C with the peak year of 2015.[241] Whereas India with its large coal reserves to generate electricity refers to the maintenance of temperature to the level of 2 °C and is opposed to the pegging the peak year at 2015.[242] The difference between Indian position and the rest of the South Asia countries is prominent as far as the fundamental question of legally binding obligation was concerned. India was opposed to any mandatorily legally binding mitigation obligation and supported the concept of 'nationally appropriate mitigation actions' for all the countries.[243] While Nepal and other LDCs had been demanding a legally binding commitment to keep the temperature rise to 1.5 °C, the demand has been included in the Paris Agreement as an aspirational goal.[244] As far as climate displacement or refugees are concerned, Bangladesh and Maldives have been active in demanding concrete legal measures, and NGOs in Bangladesh have advocated for a new Protocol to the UNFCCC and opined that such a mechanism would be instrumental 'to ensure social, cultural and economic rehabilitation of the "climate refugees" through recognizing them as "Universal Natural Persons"'.[245]

The climate change negotiating strategy followed by the developing nations has been described as defensive.[246] Experts highlight the main shortcomings of such a strategy as including reluctance to make new proposals, lack of well-articulated, comprehensive and thought-out strategies, reductionist approach of plummeting the

---

237 Hemanta Raj Poudel (n 222) 27.

238 Chandra Lal Pandey (n 238) 112.

239 Ibid.

240 Overview of the SAARC Country Positions on International Climate Negotiations, (Equitybd) 2015, 25 http://www.equitybd.net/wp-content/uploads/2015/10/Overview-of-the-SAARC-Country-Positions-on-International-Climate-Negotiations.pdf accessed on 25 April 2018.

241 Ibid.

242 Ibid.

243 Ibid.

244 Paris Agreement, (n 151) article 2 (a).

245 Equity BD, 'Climate Change Induced Forced Migrants: In Need of Dignified Recognition under a New Protocol' http://www.equitybd.org/English/Press%20040409/English%20Position%20paper.pdf accessed 10 November 2009.

246 Joydee Gupta (n 227) 179–180.

entire issue of climate change to the general north-south divide.[247] For instance, India attempted to block the discussion on Climate Change at Security Council in 2007 claiming scientific uncertainty.[248] Under the UNFCCC and Kyoto Protocol, developing nations were successful in preventing developed countries from imposing emissions reduction targets on developing nations.[249] Their collective voice was able to withstand the gradual evolving participatory approach argued by the developed nations for the participation of developing nations.[250] Specifically, the collective voice of the developing nations achieved the Warsaw International Mechanism for Loss and Damage intended to assist the countries from the adverse impacts of climate change in developing countries marked a paradigm shift in institutional mechanism driven by justice imperative.[251] The establishment of loss and damage mechanism can be crucial in the context of climate displacement in protecting the rights of the displaced population. However, the negotiations leading to the establishment of loss and damage mechanism clearly showed the conflicting positions of developed and developing nations.[252] G-77 and China wanted the establishment of loss and damage mechanism as an independent entity separate from mitigation and adaptation.[253] However, the positions of developed nations were to include the mechanism under the broader head of adaptation.[254] Nonetheless, despite the magnitude of these developments, there has been relatively scant attention paid to developing countries' perspectives on climate change at a normative level. K. Mickelson has argued that 'international environmental law as a discipline has failed to respond to Third-World concerns in a meaningful fashion'.[255]

[247]Ibid.

[248]Katharina Michaelowa (n 235) 8.

[249]Mathew Coghlan, 'Prospects and Pitfalls of the Kyoto Protocol to the United Nations Framework Convention on Climate Change', (2002) 3 MJIL, 165–184 see also Adil Najam, Saleemul Haq, YouboSokona, 'Climate Negotiations Beyond Kyoto: Developing Countries Concerns and Interests', (2003) 3(3) Clim. Pol'y 221–231.

[250]Mwandosya (n 239) 38
Lavanya Rajmani (n 230) 1–4.

[251]Lisa Vanhala and Cecilie Hestbaek, 'Framing Climate Change Loss and Damage in UNFCC negotiations', 2016 16(4) Global Env'l. Politics 111–130, 112.

[252]Ibid, 120, see Zapletal, Nathalie, 'Climate Change and Developing Countries; Examining the Motives for Participation in International Negotiations', (CUNY Academic Works, 2014) 30 http://academicworks.cuny.edu/cc_etds_theses/3, accessed 26 April 2018.

[253]Ibid, 124.

[254]Karen Elizabeth McNamara, 'Exploring Loss and Damage at the International Climate Change Talks', (2014) 5 Int J Disaster Risk Sci 242–246.

[255]Karin Mickelson, 'South, North, International Environmental Law, and International Environmental Lawyers', (2000) Yearbook of International Environmental Law, Oxford Academic 52–81.

5.4 Domestic Legal and Institutional Mechanism to Protect Climate Refugees in South Asia

The climate change has come to be perceived as the most substantial challenge of the present century that is destined to reshape our destiny on this planet. In policy discourse, climate change is more considered to be an environmental issue instead it is a challenge that is impregnated with potentially far-reaching ramifications for almost all fields of development, from health care to livelihood and displacement. Expert's prediction suggests an alarming figure of climate-induced displacement/migration for South Asia.[256] Catastrophic climate events and consequent mass displacement will threaten the very fabric of the nation and implicate constitutional rights of citizens including the right to life, and the rights held to be an integral part of this right to life.[257] As home to a fourth of the humanity, South Asian position and legal response are critical to the future of the international climate regime and for the development of a legal framework on climate-induced displacement. The occasion presents challenges as well as a unique opportunity for the region to steer South Asia's development path sustainably.[258] The essence of the domestic debate in South Asia concentrated on the implementation of the principle of CBDR, technological and financial assistance, poverty eradication, developmental priorities, mitigation and adaptation.[259] South Asian countries have taken policy measures at national level and tried to cooperate at regional as well as at international level.[260] The following section will analyse the domestic legal initiatives undertaken by South Asian jurisdictions in response to climate displacement. Since climate change issues are dealt primarily by the environmental law regimes, the discussion will involve the extent of legal attention and protection guaranteed to climate-displaced populations under the domestic environmental law frameworks and how far the legal response to climate change has accounted for climate displacement.

---

[256] IOM, 'Assessing the Climate Change Environmental Degradation and Migration Nexus in South Asia', (International Organization for Migration (IOM) Bangladesh, 2016) 28 https://publications.iom.int/system/files/pdf/environmental_degradation_nexus_in_south_asia.pdf, accessed 25 April 2018.

[257] Jane McAdam, *Climate Change, Forced Migration, and International Law*, (Oxford University Press, 2012) 63.

[258] Jayant Sathaye, P. R. Shukla and N. H. Ravindranath, 'Climate Change, Sustainable Development and India: Global and National Concerns', (2006) 90(3) Curr. Sci, 314–325.

[259] Vijay P. Ojha, 'Climate Change Mitigation Strategies in South Asia', in Raghabendra Jha (ed.), *Routledge Handbook of South Asian Economics, London and New York*, (Routledge, 2011) 309–322.

[260] Ibid.

5.5 Legal and Institutional Mechanisms to Protect Climate Refugees Under the Environmental Laws

5.5.1 Afghanistan

> Afghanistan is highly prone to natural disasters throughout its 34 provinces. As a result of climate change, it is anticipated that the incidence of extreme weather events, including heat waves, floods, and droughts will likely increase, as will climate change-linked disasters such as glacial lake outflows. The majority of Afghanistan's population relies directly or indirectly on the available natural resources for their livelihoods so with these climatic changes the foundation of the country's economy, stability, and food security are under threat.[261]

The above statement from the INDC submitted by Afghanistan captures the vulnerability profile of the country in respect of climate change. However, the legal responses and policy development are in the nascent stage as the concentration of the country is fixed on what is perceived to be more urgent problems of conflict and economic crisis.[262]

Afghanistan, a civil law country, has a mixed legal system, the formation of which goes back to nineteenth century. The modernization of Afghanistan's legal system started after the adaption of the 1930 Constitution.[263] The Constitution was based on Islamic values; in addition to the formal written positive laws and Islamic law (sharia), the customary law on informal dispute resolution is another component of the legal system of Afghanistan.[264] A new Constitution was adopted in 1964, which created a constitutional monarchy and a bicameral legislature.[265] The efficacy of constitutionalism was seriously jeopardized by the conflict and civil war, which plagued the country.[266] Following the overthrow of the Taliban, a constitutional drafting commission was charged with writing a new constitution for the country and adopted the Country's new Constitution in 2004.[267] Afghanistan envisages a Presidential form of government with a bicameral national assembly.[268] Article 15 of the Afghanistan

[261]UNFCCC, 'Intended Nationally Determined Contribution Submission to the United Nations Framework Convention on Climate Change', (UNFCCC, 2015) 2 http://www4.unfccc.int/ndcregistry/PublishedDocuments/Afghanistan%20First/INDC_AFG_20150927_FINAL.pdf, accessed 25 April 2018 (hereinafter INDC Afghanistan).

[262]Matthew Savage, Bill Dougherty, Mohammed Hamza, Ruth Butterfield and Sukaina Bharwan 'Socio-Economic Impacts of Climate Change in Afghanistan', (Oxford: Stockholm Environment Institute, 2008) 23.

[263]Abdul Satar Sirat, 'The Modern Legal System of Afghanistan An Introduction to the Study of Comparative Law' (1968) 16(4) AJCL, 566–571.

[264]Ibid.

[265]Chris Johnson, William Maley, Alexander Thier and Ali Wardak, *Afghanistan's Political and Constitutional Development*, (Overseas Development Institute, 2003) 23.

[266]J. Alexander Thier, 'The Making of a Constitution in Afghanistan', (2006) 51 N. Y. L. SCh. L. Rev., 558–579.

[267]Ibid, 566.

[268]Erik Gilbert Jensen and Larry Kramer 'An Introduction to the Law of Afghanistan, Afghanistan Legal Education Project' (Stanford University) 2009, 41.

Constitution directs the state to adopt the requisite measures to protect and improve the forests as well as the living environment.[269] Unlike other South Asian jurisdictions, there is no judicial statement, which has come from Afghanistan on the expansive interpretation of the right to life to include right to clean environment. Following the constitutional mandate, the Afghanistan's principal environmental regulatory body was conceived and established in 2005 that is known as the National Environmental Protection Agency (NEPA) along with other institutional mechanisms.[270]

As the primary institutional mechanism, NEPA primarily focuses on the sustainable management of the country's environment.[271] NEPA with the active assistance from the UNEP drafted the country's comprehensive environment law in 2005.[272] The Act is based on international standards and lays down a framework for the attainment of a healthy environment, ecological consumption and sustainable management of Afghanistan's natural resources basis, hard-nosed environmental governance and safeguarding the involvement of the stakeholders towards achieving the country's ecological development and socio-economic targets.[273] One of the prominent features of the Act is in prescribing rights for individuals and society. The Act does not explicitly mention climate change. However, the Act is based on the understanding of environmental degradation and consequent displacement as a result of desertification, and climate concerns can be interpreted in provisions about the pollution control and natural resource management.[274] Additionally, the Environmental Impact Assessment Policy of Afghanistan prepared after reviewing the best practices from other jurisdictions offers a strong platform to consider the climate concerns.[275]

The National Adaptation Programme 2009 recognizes the 'forced migration' but with a specific reference to in the situation of drought and desertification.[276] The programme refers explicitly to the deplorable situation of mobility of Kuchi Nomadic peoples.[277] Currently, the NEPA has been developing the National Climate Change

---

[269] Constitution of the Islamic Republic of Afghanistan, 200(n 44) article 15.

[270] UNEP, 'A Guide to Afghanistan's 2007 Environment Law' (United Nations Environment Programme, 2007) 7 https://postconflict.unep.ch/publications/afg_env_law.pdf accessed 25 April 2018.

[271] Ibid, 7–8.

[272] Ibid.

[273] McKenzie F. Johnson, 'Institutional Change in a Conflict Setting: Afghanistan's Environment Law', (2017) 23(1) EJIL 168–191.

[274] NEPA & UNEP. (2015) Climate Change and Governance in Afghanistan, (Kabul National Environmental Protection Agency and United Nations Environment Programme) 3–4, see also Ole Kristian Fauchald, David Hunter, Wang, Xi (2009) 20 *Yearbook of International Environmental Law* (Oxford University Press).

[275] National Environmental Impact Agency, 'National Environment Impact Assessment Policy, An Integrated Approach to Environmental Impact Assessment in Afghanistan', (Ministry of Mines and Petroleum, 2007), http://mom.gov.af/Content/files/National_EIA_Policy.pdf accessed 25 April 2018.

[276] UNFCCC, 'National Capacity Needs Self-Assessment for Global Environmental Management (NCSA) and National Adaptation Programme of Action for Climate Change (NAPA) Afghanistan' (UNFCCC, 2009) http://unfccc.int/resource/docs/napa/afg01.pdf accessed 25 April 2018.

[277] Ibid at 18.

Strategy and Action Plan for Afghanistan (ACCSAP).[278] The action Plan proposes to integrate and mainstream the climate change agenda into the national development framework and reduce the vulnerability of the Afghanistan and its people through boosting resilience and adaptive capacity.[279] Though the county has started focusing its attention on climate change, it has not adopted any specific policy or legislative mechanisms regulating climate change and climate displacement.[280] The adoption of specific climate policy remains one of the goals under INDC submitted by Afghanistan to the UNFCCC.[281] If an effective climate policy has to be enacted and implemented, country has to overcome the acute research gaps, enhance human and institutional capacity and expertise as well as basic environmental awareness, both within its government institutions including NEPA and among the general population.[282]

5.5.2 Bangladesh

Bangladesh has been portrayed as the world's most climatically vulnerable country that has been bracing the climate change-related risks and resultant climate change-induced displacement.[283] Bangladesh has been active at the international negotiations to formulate effective and strong mitigation and adaptive measures to address climate change concerns and climate-induced displacement.[284] At the domestic level, there is no explicit recognition of right to environment under the Constitution of Bangladesh. Article 18(1) of the Constitution of the Peoples Republic of Bangladesh adopted through the 15th amendment provides for environmental protection in the form of

---

[278] National Environmental Protection Agency, 'United Nations Environment Programme, Climate Change and Governance in Afghanistan' (NEPA, 2015) https://postconflict.unep.ch/publications/Afghanistan/Report_CC_Governance_Afghanistan_EN_v2.pdf accessed November 13, 2017.

[279] Ibid.

[280] USAID, 'Climate Change Risk Profiles' (USAID, 2016) https://www.climatelinks.org/sites/default/files/asset/document/2016%20CRM%20Fact%20Sheet%20-%20Afghanistan_2.pdf accessed 13 November 2017.

[281] INDC Afghanistan (n 263).

[282] John F. Shroder and others, *Transboundary Water Resources in Afghanistan: Climate Change and Land-Use Implications*, (Elsevier, 2016)147.

[283] ADB, 'Addressing Climate Change and Migration in Asia and the Pacific', (ADB, March 2012) 23 http://www.adb.org/sites/default/files/pub/2012/addressing-climate-changemigration, accessed 20 April 2018; See also Katha Kartiki, 'Climate Change and Migration: A Case Study From Rural Bangladesh', (2011) 19(1) Gender and Development 23–37.

[284] Jane Mc Adam and Ben Saul, 'Displacement with Dignity: International Law and Policy Responses to Climate Change Mitigation and Security in Bangladesh' (2010) 53 German Y.B. Int'l L 233–287, 270

Tahera Akter, *Climate Change and Flow of Environmental Displacement in Bangladesh* (Unnayan Onnesha-The Innovators, 2009) 11–12.

non-justiciable principles.[285] It exhorts the nation-states to go for protecting and enhancing the quality of environment, preserving and safeguarding the biodiversity, forests, wetlands, natural resources and wildlife for the present and posterity.[286] Judiciary has enunciated the right to have a clean environment the as part of the constitutional right to life in the landmark case of *M. Farooque v. Bangladesh and Others*.[287]

Environmental lawmaking in Bangladesh followed the Stockholm Declaration with the enactment of the Environmental Pollution Control Ordinance 1977, which superseded the Water Pollution Control Ordinance.[288] The comprehensive and umbrella legislation to protect the environment was passed in the year 1995 in the form of Bangladesh Environment Conservation Act 1995.[289] The Act does not contain any provisions on climate change. However, the provisions relating to control of environmental protection[290] environmental pollutant[291] and harm[292] can be employed to push for actions on climate change.

About climate change, the government of Bangladesh had adopted the National Adaptation Plan (NAPA) 2005, which aims to address the negative impacts of the climate change and to promote sustainable development of the country.[293] The Plan recognizes migration from a rural area to urban areas as result of climate-induced disasters.[294] The NAPA process has been provided much-needed impetus through the adoption of the Bangladesh Climate Change Strategy and Action Plan 2009

[285] Constitution of Bangladesh 1972(n 58) article 18A. 'The State shall endeavour to protect and improve the environment and to preserve and safeguard the natural resources, bio-diversity, wetlands, forests and wild life for the present and future citizens.

[286] Ibid.

[287] *M. Farooque v. Bangladesh and Others* (1997) 49 DLR (AD) 1.

[288] Alexandra Clemet, 'A Review of Environmental Policy and Challenges in Bangladesh', (Assets Publishing Services, 2015) 4 https://assets.publishing.service.gov.uk/media/57a08c2fe5274a31e0001058/R8161-section2.pdf, Accessed 30 March 2018.

[289] Bangladesh Environment Protection Act, 1995, Act No.1 of 1995, See Nazrul Islam, *Environmental Law in Developing Countries: Selected Issues*, (IUCN, 2002) 39. See Noor Mohammad, 'Development of Environmental Law and Policy in Bangladesh: An Overview', (2013) 38(8) IJAMS 96–106. See Faruque, Abdullah Al & Karim, Md Saiful, 'Environmental Law of Bangladesh', in Robinson, Nicholas A., Burleson, Elizabeth, & Lye, Lin-Heng (Eds.) *Comparative Environmental Law and Regulation.* (Thomson Reuters, 2016) 7A:1-7A:16.

[290] Bangladesh Environment Protection Act, 1995 (n 298) section 2 (f) 'protection of environment' means the qualitative and quantitative improvement of the different components of the environment and prevention of the deterioration of qualitative and quantitative standards.

[291] Ibid; 2(e) 'environmental pollutant' means any solid, liquid or gaseous substance likely to be, or tending to be, injurious to environment and shall also include heat, noise and radiation.

[292] Ibid; section 8 Information of the Director General regarding environmental pollution or deterioration. (1) Any person damaged or apprehending to be damaged on account of environmental pollution or deterioration of the environment may, in such manner as may be prescribed by rules, apply to the Director General for remedying against the damage or apprehended damage.

[293] MOEF, 'National Adaptation Programme of Action (NAPA)' (MOEF Bangladesh, 2005), http://old.moef.gov.bd/Climate%20Change%20Unit/National%20Adaptation%20Programme%20of%20Action(NAPA).pdf, Accessed 30 April 2018.

[294] Ibid.

(BCCSAP).[295] BCCSAP is a project that has to be completed within the stipulated time frame, i.e. 2009–2018, and aims at increasing the country's resilience to climate change; reduces the impacts/risks of posed by climate change and aims at low-carbon development of the country.[296]

In the early five years, the Plan directed at raising and progressing in the six areas, namely (1) food and social security, and health care; (2) comprehensive and consolidated natural disaster management; (3) building and creating infrastructure; (4) research & development (R&D), and knowledge production and management; (5) reduction and low-carbon discharge and development; and (6) capacity building measures and institutional development amplification.[297] The Plan recognizes the complicated scenario of climate displacement as an undesirable consequence[298] and describes migration along with crime as an outcome of loss of livelihood due to decreasing livelihood opportunities and lowering agricultural productivity.[299] The action Plan recognizes the impact of sea-level rise and states that it will cause displacement of millions of people and lays down three-pronged action Plans:

(a) Development of monitoring mechanism for the internal and external migration;
(b) A Protocol to support resettlement and rehabilitation of people due to climate change;
(c) Enhancing capacity through educating and training to enable their preparing in a new environment.[300]

However, in spite of the explicit, acknowledgement of the issue, the Plan does not lay down any specific provision to combat the same. The Plan does not comprehensively address the issue of climate refugees by treating them as a distant threat.[301]

In 2010, the Bangladesh Climate Change Trust Fund Act was enacted to implement the provisions of the BCCSAP and redress the adverse impact of climate change on Bangladesh.[302] The aims of the Act specifically lay down that the funds will be used to Plan and implement strategies to deal with the people or groups of people of the affected and risky areas resulting from climate change[303] (i.e. climate dis-

---

[295]Sabrina Persaud, 'Eyes on Bangladesh's Disappearing Coasts: Proposed Constitutional Protections for Coastal Communities Particularly Vulnerable to Climate Change', (2015) 6 J. Sustainable Dev. L. & Pol'y 52–82, 65; Ministry of Environment and Forest, 'Bangladesh Climate Change Strategy and Action Plan' (Ministry of Environment and Forest, Government of the People's Republic of Bangladesh, 2009) https://www.iucn.org/downloads/bangladesh_climate_change_strategy_and_action_plan_2009.pdf accessed 13 November 2017.

[296]Ibid.

[297]Ibid 28–30.

[298]Ibid.

[299]Ibid.

[300]Ibid, 22, 59.

[301]Sanjoy Biswas, Md. Akterul Alam Chowdhury, 'Climate Change Induced Displacement and Migration in Bangladesh: The Need for Rights-Based Solutions' (Refugee Watch, 2012) 171 http://www.mcrg.ac.in/rw%20files/RW39_40/13.pdf accessed 13 November 2017.

[302]The Climate Change and Trust Act, 2010, section 1 (Trust Act).

[303]Trust Act, section 5.

placement). Objectives of the fund also include the use of fund in facing the risk arising from climate change, formulate and implement suitable action Plan for climate change, sustainable development and development of human resource at the grass root level, to undertake necessary in-action research in the mechanisms to deal with climate change and dissemination of the conclusions, to assist the institutional framework in implementing its Plans and projects, to assist any emergency arising out of climate change causing a natural disaster.[304] The challenges faced by the Trust Fund includes complexity of the financing procedure, institutional incapacity, reduced capacity of the fund to perform owing to lack of funds and inadequate technology, and weak control and coordination with different climate focal points.[305]

To build a disaster-resilient society, the government of Bangladesh developed Standing Orders on Disaster, which outlines the disaster management arrangements.[306] McAdam and Saul have critiqued the standing orders as ineffective.[307] The National Plan for Disaster Management from 2010 to 2015 was adopted to provide a vision and guidance to improve the scenario and deal with the extreme disasters.[308] The Plan makes a departure from the earlier position of a reactionary method to concentrate on pre-disaster mitigation and preparedness of the people and focusing on strengthening of community institutional mechanisms; and empowering community at risk particularly women, the poor and the disadvantaged.[309] The vulnerabilities leading to displacement is acknowledged under that Plan.[310] However, the Plan does not contain concrete strategies and programmes primarily in the scenarios of slow disasters.[311] The legislative framework on climate displacement got a boost with the

---

304 Trust Act, section 6.

305 Md. Zahangir Alam, 'Capacity Building on Climate Finance ESCAP Policy Dialogue on Infrastructure Finance Strategies for Sustainable Development', (UNESCAP, 2017) 12, http://www.unescap.org/sites/default/files/Session%205%20-%20Climate%20Finance%20-%20Bangladesh.pdf, Accessed 30 April 2018.

306 Mizanur Rahman Bijoy Tapas Ranjan Chakraborty, 'An Overview on Gaps and Opportunities of Disaster Management in Bangladesh Managing Climate Induced Disasters', (Network on Climate Change, Bangladesh, 2017) 5 http://www.nccbtrust.org/wp-content/uploads/doc/An%20overview%20on%20gaps%20and%20opportunities%20of%20Disaster%20_Disaster%20booklet_Eng.pdf, Accessed 30 April 2018; See Lisa F. Schipper, 'Meeting at the Crossroads? Exploring the Linkages between Climate Change Adaptation and Disaster Risk Reduction. Climate and Development' (2009)1(1) Clim. & Dev. 16–30.

307 McAdam and Saul(n 286) 271,286.

308 National Plan for Disaster Management 2010–2015, Bangladesh.

309 Ibid, See Ahsan Uddin Ahmed, Saleemul Haq, Mahbuba Nasreen and Abu Wali Raghib Hassan, 'Sectoral Inputs Towards the Formulation of Seventh Five Year Plan (2016–2021) Climate Change and Disaster Management 2015' (Planning Commission, 2015) 44 http://www.plancomm.gov.bd/wp-content/uploads/2015/02/11a_Climate-Change-and-Disaster-Management.pdf Accessed 25 April 2018.

310 Ibid, Climate Change and Disaster Management, 20.

311 Matthew Walsham, 'Assessing the Evidence: Environment, Climate Change and Migration in Bangladesh' (IOM, 2010) 33–34 https://publications.iom.int/books/assessing-evidence-environment-climate-change-and-migration-bangladesh Accessed 25 April 2018.

adoption of the Disaster Management Act of Bangladesh, in 2012.[312] The Act links the issue of climate-induced loss and damage within the ambit of the definition of disaster. The definition of disaster explicitly includes reference to climate change. Disaster was defined as '*any incidents created by nature or human or created due to climate change and its massiveness and devastation cause such damage to cattle, birds and fisheries including life, livelihood, normal life, resources, assets of community and the environment of the damaged area or create such level of hassle to that community whose own resources, ... any unnatural incident or a misfortune causing massive life loss and damage*'.[313]

Though the Act represents significant progress in the modification of institutional framework about disaster management, the Act has not established any institutional arrangements or task force to deal with the climate displacement issue.[314] The climate change responses and the disaster framework laws in Bangladesh have analysed the issue of climate displacement from the perspective of adaptation. However, an effective adaptation strategy depends on the technical and financial resources and coordination among multiple agencies and departments. The constraints of resources and coordination gaps to regulate relocation and rehabilitation of climate-displaced population seriously impede the policy implantation process in Bangladesh.[315]

5.5.3 Bhutan

Bhutan was a monarchy and transitioned to democracy in 2007. The King of Bhutan is the head of State. Bhutan has a bicameral Parliament system with executive power is vested in the Council of Ministers, headed by the Prime Minister. The administrative arrangement in Bhutan is divided into Dzongkhags (Districts) and 205 Geogs (the smallest administrative unit within a district).[316] Bhutan has adopted a vibrant decentralized form of administrative set-up after the passing of the Local government Act 2007 formalizing local government system of governance.[317]

---

[312]Disaster Management Act, 2012, Bangladesh.

[313]Ibid, section 2(11).

[314]National Strategy on the Management of Disaster and Climate Induced Internal Displacement, Comprehensive Disaster Management Programme (CDMP II) Ministry of Disaster Management and Relief (MoDMR) 2015, 5.

[315]See generally Mostafa Mahmud Naser, 'Climate-Induced Displacement in Bangladesh: Recognition and Protection under International Law', (2013) 82 Nordic J. Int'l L. 487–527, Mizanur Rahman Bijoy and Tapas Ranjan Chakraborty, *An overview on gaps and opportunities of Disaster Management in Bangladesh managing climate induced disasters*, (NCC, 2013) 41.

[316]Maud Poissonnier-Lescuras and François Gemenne, 'Bhutan Case study in the framework of the project Climate Migration: Climate-related migration and the need for new normative and institutional frameworks', (IDDRI, 2013) 10 http://www.humanrights.at/climmig/wp-content/uploads/Bhutan-ClimMig.pdf Accessed 20 April 2018.

[317]Ibid.

About environment protection, Constitution of Bhutan has elaborate provisions. Article 5 of the Constitution reads

> Every Bhutanese is a trustee of the Kingdom of Bhutan's natural resources and the environment for the advantage and benefit of the present and the future generations, and it is the fundamental duty of every citizen to contribute to the protection and conservation of the rich biodiversity and natural environment of Bhutan and prevention of all forms of ecological degradation including noise, visual and physical pollution through the adoption and support of environment-friendly practices and policies.[318]

This provision beautifully outlines the responsibilities of the people and the government to protect and conserve the pristine environment and safeguard Bhutan's wildlife. The Constitution of Bhutan protects the environmental conservation as an inalienable part of the country's policies as no generation has the right to destroy the legacy of the agreed and inherited rich and opulent natural environment.[319]

Article 5(2) further empowers the royal government to (a) conserve, protect and improve the primeval environment and safeguard the rich biodiversity of Bhutan; (b) put a stop to pollution and ecological degradation; (c) safeguard ecologically well-adjusted sustainable development while encouraging justifiable socio-economic; and (d) ensuring a safe and healthy environment.[320] The constitution under article 5 (3) categorically authorizes the government to maintain and conserve the Bhutan's natural resources and to ward off the degeneration of the Bhutan's ecosystem; a modicum of 60% of Bhutan's total land shall be maintained under forest cover for all time.[321] It is interesting to analyse, in spite of the heavy reliance and emphasis placed on the protection of environment under the Constitution of Bhutan, the protection, and preservation of environment has not been elevated to the position of fundamental rights. One explanation could be the reliance of Bhutanese society on the duty of individuals to protection of the environment.

At the institutional level, National Environment Commission (NEC) has been established to coordinate, frame and implement policies for the protection of the environment.[322] The objective of the commission is to conserve and protect the environment through consolidated action aiming at sustainable development and regulation of all environmental impacts.[323] NEC strives to increase Bhutan's participation in regional, sub-regional and international environmental negotiations.[324] In 2012, the Disaster Management Act was passed with the purpose of establishing and augmenting the institutional capability of Bhutan to deal with the natural disas-

---

318 The Constitution of the kingdom of Bhutan, 2008(n 65).

319 Government of Bhutan, 'A Guide to the Constitution of kingdom of Bhutan' (Judiciary, 2008) 13 http://www.judiciary.gov.bt/education/high%20court%20book.pdf accessed 25 April 2018.

320 The Constitution of the kingdom of Bhutan (n 65) article 5(2).

321 Ibid, article 5(3).

322 National Environment Commission, 'Vision, Mission and Mandate' http://www.nec.gov.bt/nec1/index.php/about-nec/vision/ accessed 13 November 2017.

323 Ibid.

324 Ibid.

ters and developing its management, mainstreaming the reduction of disaster risk by guaranteeing the active public participation.[325]

Since its ratification of the UNFCCC in August 1995, Bhutan has presented its adaptation and reduction strategies roadmap through the First and Second National Communication to the UNFCCC.[326] Bhutan has committed to the UNFCCC to remain carbon neutral.[327] National Adaptation Programme adopted in 2006 does not explicitly refer to the threat and vulnerability of climate displacement. Rural to urban migration is referred to as an environmental vulnerability.[328] However, in spite of the specific emphasis on environmental protection in the legislative framework, a dedicated analysis of climate displacement is missing from the policy and legal responses governing climate change in Bhutan.

5.5.4 India

India's environmental laws flow from the constitutional and legal provisions enacted at the domestic level and international commitments made at the global level.[329] In the beginning, the Constitution of India did not contain any specific provisions on the environment and only had provisions having a bearing on the environment such as improvement of public health,[330] and organization of agricultural and animal husbandry on modern and scientific lines.[331] In the context of India, environmental consciousness as understood in the contemporary legal parlance owes its origin to the Stockholm Declaration 1972.[332] The Supreme Court of India in multiple cases has categorically acknowledged the impetus provided by the Stockholm Declaration in initiating an environmental discourse in the country.[333] The Declaration undertaken a flurry of domestic regulatory and legislative mechanisms.[334] Through a constitu-

---

325 Bhutan Disaster Management Act (2013) http://www.ddm.gov.bt/download/dm_act_2013.pdf accessed 10 November 2017.

326 Second National Communication to the UNFCCC, Kingdom of Bhutan, 2011, http://unfccc.int/resource/docs/natc/snc_bhutan.pdf accessed 10 November 2017.

327 UNFCCC, National Environment Commission, 'Communication of INDC of the Kingdom of Bhutan' (2015) http://www4.unfccc.int/submissions/INDC/Published%20Documents/Bhutan/1/Bhutan-INDC-20150930.pdf accessed 13 November 2017.

328 Bhutan, 'National Adaptation Programme of Action' (2006).

329 Kailash Thakur, *Environmental Protection Law and Policy in India* (Deep and Deep Publication 1997) 185.

330 Constitution of India, 1950 (n 92) article 47.

331 Ibid, article 48.

332 Stellina Jolly, 'Application of Solar Energy in South Asia: Promoting Intergenerational Equity in Climate Law and Policy' (2014) 7(1) Int. J. Private Law 20, 23.

333 See *Rural Litigation and Entitlement Kendra v State of U.P* (1986) Supp SCC 517; *Vellore Citizens Welfare Forum v Union of India* (1996) 5 SCC 647; *M. C. Mehta v Union of India*, (1992) 1 SCC 358; *M.C. Mehta v Kamal Nath*, (2000) 6 SCC 213.

334 Water (Prevention and Control of Pollution) Act (1974) Forest Conservation Act (1981), The Air Prevention and Control of Pollution Act (1986), the Environmental Protection Act (1986), the

tional amendment in 1976, articles 48-A and 51-A (g) were incorporated as directive principles of state policy and as a fundamental duty in the Constitution of India, respectively.[335] However, unlike fundamental rights, which are direct, enforced directive principles of state policy and Fundamental duties are not justiciable rights.[336] However, the proactive role of Indian judiciary steered the country's commitments towards environmentalism through the harmonious interpretation of directive principles[337]and under the expanded notion of the right to life[338] guaranteed under the constitution of India.[339] The judicial trends from other jurisdictions reveal that the fundamental rights provision under the constitution can be utilized to address the impact of climate change on the basic human rights of citizens. The discussion on the role of judicial response to climate change will be undertaken in the next chapter.

In addition to the constitutional authorization, India has a multitude of national policies and legislation governing environmental issues, which can be invoked in the case of climate change concerns. Relevant legislation includes the Environment (Protection) Act (1986), the Indian Forest Act (1927), the Forest Conservation Act (1980), National Green Tribunal Act 2010, The Energy Conservation Act of 2001 and other legislations related to air, water and energy. The legislative provisions can be interpreted to initiate mitigate climate actions and also to take actions against entities contributing to climate change through liberal access to litigants and princi-

National Environment Tribunal Act (1995), the National Environment Appellate Act (1997) and the Biodiversity Protection Act (2002). See also Stellina Jolly and Sandeep Mahajana, 'Climate Change and Security: Forging a Cooperative Mechanism in South Asia' (2014) 10(6) Int. J. Public Policy 315, 332.

[335]Constitution of India (n 92) 42nd amendment 1976, Constitution of India, 1950 article 48A. The State shall endeavour to protect and improve the environment and to safeguard the forests and wildlife of the country, article 51 A (g) (g) to protect and improve the natural environment including forests, lakes, rivers and wildlife, and to have compassion for living creatures; Both these provisions are non-justiciable, see Mahendra Singh (ed.) *V. N. Shukla's Constitution of India* (Eastern book company 2010) 936; Stellina Jolly, (n 334) 23 see generally, Stellina Jolly, 'Role of Judiciary in Bio Diversity Enforcement and Compliance: Comparison between International Court of Justice and Indian Judiciary' in Usha Tandon, Mohan Parasaran, Sidharth Luthra (eds) Biodiversity: Law, Policy and Governance, (Taylor and Francis 2017) 231–255.

[336]Bhaskar Kumar Chakravarty, 'Environmentalism: Indian Constitution and Judiciary' (2006) 48 JILI 99–102,103.

[337]Md Zafar Mahfooz Nomani, 'Enviro-Constitutional Ethos in Right Duty Discourse: Towards the Creation of an Equitable and Sustainable Socio-Legal Order' (2000) 42 (2–4) JILI 432–442.

[338]*Subhash Kumar v. State of Bihar*, A.I.R 1991 S.C 420, 423; *M.C. Mehta v. Union of India* (Delhi Crushing Case) (1992) 3 S.C.C 256,256; and *Virendar Gaur v. State of Haryana* (1995)2, S.C.C577,581, *Parmanand Katara v. Union of India.* AIR 1989 SC 2039, *Kharak Singh v. State of U.P.* AIR 1963 SC 1295; *Pashchim Bang KhetMazdoor Samiti v. State of West Bengal* AIR 1996 SC 2426;*Consumer Education and Research Centre v. Union of India* (1995) 3 SCC 42; *Kirloskar Brothers v.Employees State Insurance Corporation* (1996) 2 SCC 682; *State of Punjab v. Mohinder Singh Chawla* AIR 1997 SC 1225; *N.D. Jayal v. Union of India*, (2004) 9 SCC 362.

[339]Chakravarty (n 338) 102 99; Stellina Jolly and Amit Jain, *Climate Change, Changing Dimensions of law and Policy*, (MD Publications, 2008) 51, 52.

pled and benevolent legal framework.[340] The Environment Protection Act 1987 is the primary legislation for the protection of the environment, preventing and decreasing the pollution. The directions issued under the Act especially about Environmental Impact Assessment can prove useful in the context of climate change.[341] The Air Act defines Air pollutant as 'any liquid or gaseous substance or solid material including the existing noise in the atmosphere in such concentration as may be or tend to be detrimental humans or other living beings or plant varieties or property or the environment'.[342] The issue of GHGs can be regulated and controlled under this Act.[343] However, these provisions are insufficient to deal with issues of climate displacement and the rights of the displaced population. With specific to climate change, country's obligation flows from the International commitments undertaken at the global level.[344] India signed and ratified the UNFCCC,[345] Kyoto Protocol[346] and Paris Agreement.[347] India has submitted its regular National Communication on Climate Change (NATCOMs)[348] and also the INDC under the Paris Agreement.[349] India's Initial Communication to the Convention has recognized climate displacement and vulnerability in the case of sea-level rise, land loss and health hazards.[350] Vulnerability assessments were undertaken as part of the country's national communication in the areas of water resources, agriculture, natural ecosystem, health, coastal zone management and climate modelling.[351] However, discussion on climate displacement as a significant consequence of climate change was absent from these assessments. With specific to climate change, the country has not drafted dedicated legislation and has adopted National Action Plan on Climate Change 2008.[352]

The National Action Plan (NAP) on Climate Change 2008 formulated to guide the future direction of climate action in India aims at adopting a low-carbon energy path for achieving sustainable development without compromising on economic growth

---

[340]Lavanya Rajmani and Shibani Ghosh, 'India' in Richard Lord and others (eds) *Climate Change Liability: Transnational Law and Practice* (Cambridge University Press, 2011) 13–241, 158.

[341]Ibid, 157.

[342]Article 2(1) Air (Prevention and Control of Pollution) Act, 1981, India.

[343]Rajmani (n 342) 159.

[344]M K. Ramesh, 'Environmental Justice Delivery in India: In Context', (2002) 2 IJEL 12.

[345]UNFCCC (n 149).

[346]Kyoto Protocol (n 150).

[347]Paris Agreement (n 151).

[348]Initial National Communication to the United Nations Framework Convention on Climate Change, Government of India, 2004, India submitted its second communication on 2008 and Biennial update report on 2016.

[349]UNFCCC, 'India's Intended Nationally Determined Contributions: Working Towards Climate Justice' http://www4.unfccc.int/ndcregistry/PublishedDocuments/India%20First/INDIA%20INDC%20TO%20UNFCCC.pdf accessed 12 November 2017.

[350]Initial National Communication to the United Nations Framework Convention on Climate Change, Government of India (2004), 109 https://unfccc.int/resource/docs/natc/indnc1.pdf accessed 10 February 2018.

[351]Press Release, Ministry of Environment and Forests, Wednesday, October 22, 2008.

[352]National Action Plan On Climate Change, India (2008).

and quality of life.[353] The Plan is based on multi-pronged, time-honoured and unified strategies for realizing different targets in the context of climate change.[354] NAPCC aimed at directed to strike an equilibrium between the requirement to maintain the huge growth rate for advancing the living standards of the people and the requirement to alleviate the vulnerability of the population to the negative impacts of climate change.[355] The operational thrust of NAPCC is in the shape of eight missions.

The implementation of the Plan has been left to the lead ministries.[356] They are directed to develop objectives, implementation strategies, timelines, and monitoring and evaluation criteria, to be submitted to the Prime Minister's Council on Climate Change.[357] The NAPCC has laid down a national strategy on climate change about adaptation and mitigation of greenhouse gas emissions.[358] The Plan has been criticized by many as non-progressive and lacking a social vision.[359] NAPCC reiterates India's adherence to the Kyoto principle of CBDR and does not specify well-defined goals and targets to be achieved under the missions in a time-bound manner.[360] The Plan primarily focuses on technological and infrastructural developments without drawing evident, interrelations between political, economic and social consequences.[361] Byravan and Rajan consider lack of attention to developmental concerns and the situation of the poor and vulnerable sections as the main weakness of the Plan.[362] As a result, climate displacement/migration does not figure in the focus of the Plan and has only briefly been mentioned under the Mission on Strategic Knowledge.[363] The individual mission reports released subsequently had attempted to address the human and social impacts.[364] The National Mission on Sustainable Habitat and National Mission on Sustainable Agriculture refer to climate migration. However, the mission discusses migration only in the context of possible rural–urban

---

[353] Ibid, Principles, NAPCC.

[354] *India Energy Policy, Laws and Regulations Handbook Volume 1 Strategic Information and Basic Laws* (International Business Publications 2015) 216.

[355] Wayne Ganpat and Wendy-Ann Isaac, *Environmental Sustainability and Climate Change Adaptation Strategies* (IGI Global 2016) 61.

[356] NAPCC (n 154) 6.

[357] Ibid 6.

[358] Dewald Van Niekerk, 'Climate Change Adaptation and Disaster Law' in Jonathan Verschuuren (ed), *Research Handbook on Climate Change Adaptation Law* (Edward Elgar Publishing, 2013) 148.

[359] Editorials, 'Climate Change: Not Vision, Not Plan' (2008) 43 EPW.

[360] NAPCC (n 354).

[361] Ingrid Boas, *Climate Migration and Security: Securitization as a Strategy in Climate Change Politics* (Routledge 2010) 129.

[362] Sujatha Byravan and Sudhir Chella Rajan, 'An Evaluation of India's National Action Plan on Climate Change' Project conducted by Centre for Development Finance (2012) 10 http://ifmrlead.org/wp-content/uploads/2016/05/NAPCC%20Evaluation.pdf accessed 10 November 2017.

[363] Boas (n 363) 32.

[364] Ibid.

migration as a result of crop failure and food security rather than on cross-border mass movements and conflict.[365]

The NAPCC has been translated into state-specific Plans, called State Action Plans on Climate Change (SAPCCs) as an extension aligned with eight national missions.[366] The development of these State-level Plans presents an opportunity for the country to incorporate considerations of displacement in its approach to tackling climate change. Kerala State Climate Action Plan identifies permanent displacement leading to water and food insecurity, as a consequence of climate change impact on the food system.[367] Assam State Action Plan on Climate Change makes special reference to displacement as a result of flash floods, droughts and heatwaves.[368] Though the State Climate Action Plans have identified climate displacement in certain situations, they have failed to visualize clear strategies intended to understand the resilience of people and increase the local participation. Most of the programmes are executed at the external level based on the belief that local people are ignorant about climate change and they could be victims, and mitigation and adaptation mechanism should come from the experts. However, this is a wrong assumption as communities have specialized knowledge about the local surroundings. The experience over generations gives communities special knowledge of nature and climate.[369] It makes practical sense to understand the perception of local people in situ about their interaction with environment and climate[370] in a dynamic manner as the nature of local knowledge keeps evolving.[371]

The legislative framework, which can be considered to be the closest regarding dealing with climate displacement, can be found under the disaster management framework. Though intervened in complexity, the scientific assessment of the IPCC reports highlight the inter-linkage between climate change and considerably increased frequency and magnitude of disaster.[372] International community through

---

365 National Mission for Sustainable Agriculture, India, (2010) 18 http://www.nicra-icar.in/nicrarevised/images/Mission%20Documents/National%20Mission%20For%20Sustainable%20Agriculture-DRAFT-Sept-2010.pdf accessed 19 November 2017.

366 As per the direction of the Government of India State action Plans on Climate Change is being prepared. So far 25 states have developed the state action Plan on climate change.

367 Kerala State Action Plan on Climate Change, 'Department of Environment and Climate Change, Government of Kerala' (2014) 51 http://www.moef.gov.in/sites/default/files/KERALA%20STat.%20ACTION%20PLAN%20ON%20CLIMat.%20CHANGE.pdf accessed 23 July 2017.

368 Assam State Action Plan on Climate Change, (2015–2020) 63, http://www.moef.gov.in/sites/default/files/Final%20draft%20ASAPCC%20document.pdf accessed 10 November 2017.

369 Julie Dekens, *Local Knowledge for Disaster Preparedness: A Literature Review, International Centre for Integrated Mountain Development* (ICIMOD 2007) 11.

370 Steve Langill, *Indigenous Knowledge: A Resource Kit for Sustainable Development Researchers in Dryland Africa* (People, Land and Water Program Initiative, IDRC, 1999) 9 http://idrinfo.idrc.ca/archive/corpdocs/114518/ikkit.pdf accessed 2 August 2017.

371 Lori Ann Thrupp, 'Legitimizing Local Knowledge: From Displacement to Empowerment for Third World People' (1989) 6(3) Agriculture and Human Values 13, 20.

372 Robert R.M. Verchick, 'Disaster Law and Climate Change', in Daniel A. Farber and Marjan Peeters (eds), *Climate Change Law* (Edward Elgar Publishing, 2016). See Intergovernmental Panel on Climate Change (IPCC), *Managing the Risks of Extreme and disasters to Advance Climate*

United Nations Sustainable Development Goals (SDGs), Sendai Framework for Disaster Risk Reduction 2015–2030[373] and the Paris Climate Agreement adopted in 2015 has recognized the correlation between disasters and climate change and displacement. The inter-linkage of climate change and disaster cannot be comprehended from the narrow prismatic view of environmentalism. The consequences of climate change and disasters will shake the very foundation of human existence and are multifarious.[374] The most affected will be those nations; people and communities who have the least adaptive capacity and added vulnerability are leaving the most marginalized and disadvantaged to deal with the consequences of climate change leading to mass displacement.[375]

It is appropriate to mention that the international climate change law and international disaster law developed independently, without adopting any positive synergy mechanisms. The threat of climate change as a critical factor in increasing the disaster risk has brought climate change law and the disaster risk reduction in close coordination at both the domestic and international levels.[376] The global climate regime incorporated the disaster risk through adaptation and has envisaged fund mechanism, financial and technological assistance, and loss and damage mechanism to provide for relief.[377] Both climate change and disaster management programmes aimed at reducing the vulnerability of people and communities through the prism of adaptation.

Change Adaption (Cambridge University Press, 2012) https://www.ipcc.ch/pdf/specialreports/srex/SREX_Full_Report.pd accessed 10 August 2017.

373 United Nations, Sendai Framework for Disaster Risk Reduction 2015–2030, http://www.unisdr.org/files/43291_sendaiframeworkfordrren.pdf accessed 20 October 2017, see generally for a discussion on the evolution of disaster law and responsibility, Stellina Jolly,' Interrogating the Pedagogy of state Responsibility and Individual Rights in Disaster Law, in Amita Singh (ed) *Disaster Law Emerging Thresholds,* (Taylor and Francis 2017).

374 M. V. Ogra and R. Badola, 'Gender and Climate Change in the Indian Himalayas: Global Threats, Local Vulnerabilities, and Livelihood Diversification at the Nanda Devi Biosphere Reserve' (2015) 6 Earth Syst. Dynam, 505–523, 507.

375 See generally Leith Dunn (ed), *Gender Climate Change and Disaster Risk Management*', Working Paper Series, 7, (Institute of Gender and Development Studies Mona Unit, The University of the West Indies, 2013); Andrea Warnecke, Dennis Tänzler and Ruth Vollmer, 'Climate Change, Migration and Conflict: Receiving Communities Under Pressure', The German Marshall Fund of the United States (2010) 1–12, 4 http://www.ehs.unu.edu/file/get/7105 accessed 10 October 2017.

376 Geoff O'Brien, Phil O'Keefe, Joanne Rose, Ben Wisner, 'Climate Change and Disaster Management' (2006) 30(1) Disasters 64–80, 72; Tom Mitchell, Maarten Van Aalst and Paula Silva Villanueva, 'Assessing Progress on Integrating Disaster Risk Reduction and Climate Change Adaptation in Development Processes, Strengthening Climate Resilience', Discussion Paper No. 2 (Institute of Development Studies 2010) 1–31, 10.

377 Decision 3/CP.18, 'Approaches to address loss and damage associated with climate change impacts in developing countries that are particularly vulnerable to the adverse effects of climate change to enhance adaptive capacity', FCCC/CP/2012/8/Add.1 (28 February 2013), para 26, Non-economic Losses in the Context of the Work Programme on Loss and Damage, Technical paper, FCCC/TP/2013/2 (2013).

The international community through the Nansen principles has highlighted the complex interlinkage between climate change disasters and displacement.[378] Comprehensive principles were adopted to guide the international community frame policy response while dealing with climate displacement.[379]

Principle-I emphasized that basic principle of legal responses to be guided by the human rights principles and international cooperation.
Principle-II highlighted the need to protect the needs of the most vulnerable population affected by climate change.
Principle-III emphasizes the engagement with local governments and communities, civil society for an inclusive approach.
Principle-IV, V and VI focus on regional and international cooperation to build resilience and protecting the rights of displaced.
Principle-VII, VIII and IX underscore the normative value of existing international law and look to fill the protective gaps.
Principle-X underlines the foundational norms to be followed in developing and implementing legal norms to manage climate displacement, based on non-discrimination, consent, empowerment and participation.[380]

In October 2012, Norway and Switzerland launched the 'Nansen Initiative' conceived as an informal state-owned consultative process, operating through a bottom-up approach aiming to address cross-border displacement in the context of sudden- and slow-onset disasters.[381]The consultation aims to enhance the common understanding on the magnitude and dimension of the climate displacement, identify and facilitate the good practices for the protection of persons displaced in climate change, develop basic principles on cooperation, protection standards of displace people, and make recommendation for furthering and identifying the protection gaps and operational goals including funding mechanisms and responsibilities at national regional and international levels.[382] The government of Bangladesh hosted the South Asian Intergovernmental Consultation by the Nansen Initiative.[383] The consultations resulted in nuanced understanding from diverse stakeholders, community perspectives and knowledge.[384] The international community in October 2015 adopted the Agenda for Protection of the Cross-Border Displaced Persons in the Context of Envi-

[378]The Nansen Conference Climate Change and Displacement in the 21st Century, Oslo, Norway, June 5–7 (2011) http://www.unhcr.org/4ea969729.pdf accessed on 21 August 2017.

[379]Walter Kalin, 'The Nansen Initiative: Building Consensus on Displacement in Disaster Contexts' (2015) 49 FMR 5, 7.

[380]Ibid.

[381]Ibid.

[382]Ibid.

[383]Climate Change, Disasters, and Human Mobility in South Asia and Indian Ocean Khulna, Bangladesh, 3–5 April 2015, Background Paper Nansen Initiative South Asian Regional Consultation, http://www.nanseninitiative.org/wp-content/uploads/2015/04/Nansen-Initative-South-Asia-Regional-Consultation-Background-Paper.pdf accessed 2 October 2017.

[384]Climate Change, Disasters and Human Mobility in South Asia and the Indian Ocean Khulna, Bangladesh, 3–5 April 2015, Summary Of Conclusions, Nansen Initiative Regional Consul-

ronmental Disasters and Climate Change, which suggest hybrid policy steps integrating human rights, environment, and humanitarian principles that the nation-states can take at the national, regional and international levels.[385] The Nansen initiative is relevant to climate change and displacement though not legally binding can form the basis on which the cooperation between states can be founded.[386]

At the domestic level, Disaster Management Act 2005 is the fundamental legislation coordinating the implementation of Plans for disaster preparedness and management in the country.[387] The Act defines disaster as a calamity mishap, or grave occurrence from natural or human-made causes, which is beyond the coping capacity of the affected community.[388] The definition does not account for the slow on the set of disasters brought about by climate change. The Act envisages a three-tier institutional structure of governance at national, state and the district levels.[389] At the apex level, the National Disaster Management Authority (NDMA) is constituted to formulate policies and approve the national Plan for disaster management apart from coordinating 'the enforcement and implementation of the policy and Plan for disaster management'.[390] The Act focuses on both the pre- and post-disaster phases including prevention, mitigation, preparedness, response, rehabilitation, reconstruction and recovery.[391] The National Disaster Management Plan (NDMP) has been largely associated broadly with the objectives and priorities stipulated in the Sendai Framework for Disaster Risk Reduction.[392] The vision of the Plan is to 'make India disaster resistant accomplish significant disaster risk reduction, and substantially decreasing the losses of life, livelihoods, and assets—economic, physical, sociocultural and environmental—by maximizing the capability to cope with disasters at all levels of administrative hierarchy as well as among the communities'.[393] Though the Plan highlights climate change as a disaster risk multiplier, the Plan lacks a clear vision, risk reduction strategies, goals and timelines for building up resilience and capacity building for managing and coping up with climate-induced disasters and consequent displacement.

---

tation in South Asia, https://disasterdisplacement.org/wp-content/uploads/2015/04/Summary-of-Conclusions-FINAL-5-April-2015.pdf accessed 1 November 2017.

[385] Jane McAdam, 'From the Nansen Initiative to the Platform on Disaster Displacement: Shaping International Approaches to Climate Change, Disasters and Displacement' (2016) 39 U.N.S.W.L.J. 1518–1546.

[386] Francesca Mussi, 'Climate Induced Migration: What Legal Protection for Climate Migrants at the International level' in Laura Westra and others (eds), *The Common Good and Ecological Integrity: Human Rights and the Support of Life* (Routledge 2016) 177–191.

[387] The Disaster management Act, 2005 India.

[388] Ibid, sections 2(d) and (e).

[389] Ibid, Rahjendra Kumar Pandey, 'Legal Framework of Disaster Management in India' 2016 ILI Law Review, 172, 180.

[390] Disaster Management Act, 2005, (n 389) section 6 (2) (f).

[391] Ibid.

[392] National Disaster Management Plan, India 2016, Vision.

[393] Ibid.

The principal reason for the lack of clear strategies and specificity in disaster management is the centralized and top-down model of disaster management. The top-down model of disaster management does not account for the internal differences between regions and communities about the extent, category and magnitude of disasters. In India, the involvement of Panchayats or Gram Sabha, the basic unit of administration, is in a way very restricted. The 73rd constitutional amendment has created Panchayats and given them constitutional status.[394] Panchayat Raj institutions aim to increase the inclusive governance based on the understanding that administration could be handled best at the basic level of governance.[395] Though the Act has acknowledged the role of local authorities in disaster preparation, mitigation relief and response operations,[396] the lack of technical and financial aspects restricts the involvement of Panchayat Raj institutions in disaster relief and rehabilitation operations.[397] For instance, the governance of the Chief Ministers relief fund is entrusted with the revenue officials thus restricting the involvement of the local bodies.[398] Regarding disaster preparedness, Panchayat bodies could be involved and trained in early warning, information about population, safe location and arrangements for evacuation, etc.[399] The active involvement of the Panchayat Raj institutions will be the first step in the direction of envisaging and institutionalizing a locally driven disaster management Plan. When local stakeholders and resources lead the disaster management, new local capacities and resources will emerge. Further, a locally driven programme will have better receptivity among the people as based on socio-economic practices.[400] The policy framework should concentrate on disaster reduction and preparedness, including reducing vulnerabilities, building upon the resilience, early warning signs, strengthening humanitarian response and participa-

---

[394]Constitution of India (n 92) 73rd amendment Act 1992, India The amendment recognize Gram Sabha as the basic unit of Administration.

[395]Vijantra Singh, *Panchayat Raj and Village Development* (Sarup & Sons, 2003).

[396]Disaster Management Act, 2005 (n 389) section 41. 'Functions of the local authority. (1) Subject to the directions of the District Authority, a local authority shall—

(a) ensure that its officers and employees are trained for disaster management;

(b) ensure that resources relating to disaster management are so maintained as to be readily available for use in the event of any threatening disaster situation or disaster;

(c) ensure all construction projects under it or within its jurisdiction conform to the standards and specifications laid down for prevention of disasters and mitigation by the National Authority, State Authority and the District Authority;

(d) carry out relief, rehabilitation and reconstruction activities in the affected area in accordance with the State Plan and the District Plan.

(2) The local authority may take such other measures as may be necessary for the disaster management'.

[397]Kamal Thori, *Disaster Management Through the Panchayat Raj* (Concept Company 2005) 221.

[398]Ganapathy Palanithurai, *Panchayats in Disaster: Preparedness and Management, Gujarat, Orissa, Andhra Pradesh, and Tamil Nadu with Successful Case Studies* (Concept Publishing Company 2009) 106.

[399]Ibid,108.

[400]Asmita Tiwari, *The Capacity Crisis in Disaster Risk Management: Why Disaster Management Capacity Remains Low in Developing Countries and What can be Done* (Springer 2016) 196.

tory mechanisms and specificity.[401] Policy measures need to be enacted to protect the rights of displaced people based on non-discrimination and easy access.[402]

A review of the existing legal framework for addressing the climate displacement in India under the environmental laws indicates that the legal system has not given specific attention to the issue of climate displacement. The current legal framework regulating climate change is in a state of fragmentation. There is an urgent requirement to enhance the inter-linkage between climate change laws and disaster laws on a common platform, with an emphasis on information dissemination, capacity building and democratic participation to adequately protect the people in climate displacement scenario.[403]

5.5.5 Nepal

Nepal is one of the most susceptible and helpless countries to climate change negative impacts due to its ragged mountain topography, abject poverty, and tragic national and local institutional probity and operational capacity.[404] Nepal is the only South Asian jurisdiction, which has incorporated the specific right to environment as part of the constitutional provisions. The article 30 of the Constitution of Nepal 2015 assures the right of every citizen to live in a clean environment. The article also provides that the victim of environmental contamination, pollution and degradation shall have the right to be compensated and reimbursed by the pollutant as provided under the law.[405] The provision can be utilized for the protection of rights of the population displaced by climate change. Article 35(5) requires the state to keep the environment clean and to render priority to the security and advancement of the environment and the avoidance of further destruction to the environment due to the developmental initiatives and to take extraordinary measures for the protection of wildlife, vegetation and forests.[406]

---

[401] 'Climate Change and Internal Displacement', Brookings LSE-Project on Internal Displacement (2014) 10, https://www.brookings.edu/wp-content/uploads/2016/06/Climate-Change-and-Internal-Displacement-October-10-2014.pdf accessed on 25 November 2017.

[402] Ibid.

[403] Sunil Chauhan, 'Climate Change, Disasters and Security Issues, Concerns and Implications for India' 16 http://www.umdcipe.org/conferences/epckdi/9.PDF accessed 1 October 2017.

[404] Douglas S Massey, William G. Axinn and Dirgha J. Ghimire, 'Environmental Change and out-Migration: Evidence from Nepal' (2016) 32 Population and Environment, 109–136; Institutional Responses to Local-Level Climate Change Adaptation in Nepal, Policy Research Brief No 4, 2012, Regional Climate Change Adaptation Knowledge Platform for Asia, 1–9, 1 http://resourceshimalaya.org/wp-content/uploads/2017/07/AKP-PolicyBrief4-DAC-V1-1.pdf accessed on 15 November 2017.

[405] The Constitution of Nepal, 2015 (n 125) See also article 16, The Interim Constitution of Nepal, 2008.

[406] Ibid.

Besides the constitutional mandate, the country has enacted various legislations for the protection of the environment along with authorizing the Ministry of Population and Environment with the administrative and executive power to implement the international conventions and treaties.[407] The Environment Protection Act, 1996 is an the umbrella legislation for the entire range of cross-sectoral environmental issues directed to sustain the clean and healthy environment and contribute to sustainable development.[408] The Environment Protection Act makes the impact assessment mandatory for the major development projects which can be alluded to in the context of climate change impacts.

Nepal has been an active member of International environmental negotiations. The country has ratified the UNFCCC in 1994[409] Kyoto Protocol in September 2005[410] and Paris Agreement on 2016.[411] The country submitted its national communications to the UNFCCC and INDC to Paris Agreement. The second national communication to UNFCCC elaborately points to the impact of climate change on human settlements. It states, 'most vulnerabilities and impacts of human settlement and infrastructure to climate change are cross-thematic in nature and are largely related to climate-induced disasters. Human settlement planning process has increasingly become more challenging due to climate-induced rural-urban migration'.[412] The INDC underscores the interconnection, between earthquake and climate change-induced displacement, disasters and migration that has enhanced degree of susceptibilities, vulnerabilities and dangers to water availability, hygiene and food security causing further exacerbation of Nepal's defencelessness to the climatic change hazard.[413]

The recurrent and high degree of environmental disasters faced by the country has prompted the embracing of Disaster Risk Mitigation and Management Bill 2017

---

[407]Handbook of National Environmental Legislations and Institutions in Nepal, South Asia Cooperation for Environmental Program (SACEP) 2–3 http://www.sacep.org/pdf/Reports-Technical/2001-UNEP-SACEP-Law-Handbook%20-Nepal.pdf accessed 12 November 2017; Colombo, Sri Lanka, 2001 http://www.sacep.org/pdf/Reports-Technical/2001-UNEP-SACEP-Law-Handbook%20-Nepal.pdf accessed 12 November 2017.

[408]The Environment Protection Act 1996 Nepal.

[409]Nepal UNFCCC ratification status http://unfccc.int/tools_xml/country_NP.html accessed on 4 October, 2017.

[410]Status of Ratification of the Kyoto Protocol, http://unfccc.int/kyoto_protocol/status_of_ratification/items/2613.php accessed 1 November 2017.

[411]Paris Agreement—Status of Ratification, http://unfccc.int/paris_agreement/items/9444.php accessed on 10 November 2017. See generally Sumit Vij and others, 'Climate Adaptation Approaches and Key Policy Characteristics: Cases from South Asia' (2017) 78 Environ. Sci. Policy 58–65.

[412]Second National Communication to United Nations Conference on Climate Change, Nepal 124–125 http://unfccc.int/resource/docs/natc/nplnc2.pdf accessed 2 November 2017.

[413]Government of Nepal Ministry of Population and Environment, Intended Nationally Determined Contributions (INDC), Communicated to the UNFCCC Secretariat in February 2016, 2 http://www4.unfccc.int/submissions/INDC/Published%20Documents/Nepal/1/Nepal_INDC_08Feb_2016.pdf accessed 20 November 2017.

that replaces the Natural Calamity Relief Act, 1982.[414] The new Act under the scope covers human-made disasters and envisages institutional mechanism in the form of National Disaster Risk Mitigation and Management Centre with a decentralized structure at a provincial and local level for a participatory approach.[415] National Adaptation of Programme of Action 2010 prepared, as a mandatory requirement under the UNFCCC for the Least Developed Countries[416] has described 'displacement and migration' as a climate-induced disaster.[417] Nepal has not enacted a dedicated climate legislation and has promulgated the Climate Change Policy 2011, which aims to promote environmental conservation, human development and sustainable development.[418] The thrust of the policy is on adaptation.[419] The policy has emphasized on climate acclimatization and natural disaster risk reduction; climate resilience, access to financial resources and utilization and capacity building form public's participation, association and empowerment.[420]

The issue of climate displacement has not been explicitly mentioned in the policy document, though the issue of displacement can be covered under the broader policy goals of adaptation and disaster risk reduction, in the absence of mainstreaming climate displacement in developmental and disaster policies, the adequacy of programmes are questionable. Further, as a stand-alone policy, there are no legal mechanisms to give a legal backing or framework to the policies.[421] There are several hurdles that the implementation of the policy is facing. Primarily, there is no credible institution delivery information transcending the general socio-economic space.[422] There is a severe lack of coordination between different institutions and statutory bodies for the implementation of policies and assimilation and management of financial resources.[423] One of the laudable features of most of the recent environmental legislation in the country has been its emphasis on participation and

[414]Tika R. Pradhan, House Endorses Disaster Risk Management Bill http://kathmandupost.ekantipur.com/news/2017-09-26/house-endorses-disaster-risk-management-bill.html accessed 19 November 2017.

[415]Ibid.

[416]Hemant R. Ojha and others, 'Policy without Politics: Technocratic Control of Climate Change Adaptation Policy Making in Nepal' (2016) 16 Clim. Poli'y 415, 420.

[417]Nepal, National Adaptation of Programme of Action to Climate Change (2010) http://unfccc.int/resource/docs/napa/npl01.pdf accessed 5 November 2017.

[418]Climate Change Policy 2011, Nepal.

[419]Helvates Swiss Inter-cooperation Nepal, 'Nepal's Climate Change Policies and Plans: Local Communities Perspective' (2011) 1–32, 4 http://www.adaptation-undp.org/sites/default/files/downloads/nepals_climate_change_policies_and_plans_2011.pdf accessed 19 November 2017.

[420]Krishna Lamsal and others 'Climate Change Policy in Nepal: Challenges, Opportunities and Imperative' 1–4 http://libird.org/app/publication/view.aspx?record_id=142 accessed 12 October 2017.

[421]Ibid.

[422]Ibid.

[423]Ibid.

involvement in the developmental process.[424] Though the climate policy emphasis on the participation of the local people and local governments, the overall thrust of the policy is not representative of an inclusive approach and has been adopted without a noticeable public contestation in Nepal.[425] Since it is the poor and the backward are the most affected by the climate change displacement, it is incumbent upon the government to take concrete measures to build up mechanism of dialogue, and a participatory mechanism to accommodate the concerns of the vulnerable community groups affected by climate change.[426]

5.5.6 *Pakistan*

Lately, the report on 'Global Climate Change Risk Index 2017' published by the Green Watch, a Bonn-based NGO, assessed that Pakistan had been among the ten nation-states who was most adversely affected owing to climate change between 1995 and 2015.[427] The country suffered an economic loss of 3.8 million dollars as a result of disasters and climate change.[428] The INDC submitted by the Pakistan to the UNFCCC states that 'repetitive exposure to life-threatening climate change-induced calamities like droughts, earthquakes, landslides, famines, floods, cyclonic activities, recession of glaciers, global warming, glacial lake outburst flooding (GLOF) and heatwaves that have headed the Pakistan to the rank among top ten most climate change-affected countries on the Global Climate Risk Index'.[429]

There is no clear provision that provides 'right to environment' in the Constitution of Pakistan. The subject matter of 'environmental protection and ecology' was placed under the concurrent legislative list. However, after the 18th constitutional amendment, the subject matter has been shifted and put under the provincial list of jurisdiction.[430] The environmental laws of Pakistan were shaped and influenced by the international legal developments. After the independence of Pakistan until the late 1950s, the inception of environmental legislations in the country was almost absent

---

[424]Surya Subedi, 'Access to Environmental Justice in a Politically Unstable Environment: A Case Study of Nepal' in Andrew Harding, *Access to Environmental Justice, A Comparative Study*, (Martinus Nijhoff Publishers, 2007) 175.

[425]Climate Change Policy 2011, Nepal (n 420) See Ojha (n 418) 415–433.

[426]Ibid.

[427]Sönke Kreft, David Eckstein and Inga Melchior, 'Who Suffers Most From Extreme Weather Events? Weather-related Loss Events in 2015 and 1996 to 2015.' Briefing Paper, Global Climate Risk Index 2017, Briefing Paper, (German Watch 2017) 1–32, 9 https://germanwatch.org/en/download/16411.pdf accessed 13 November 2011.

[428]Ibid.

[429]Pakistan's Intended Nationally Determined Contribution (Pak-Indc) 1–31, 2 http://www4.unfccc.int/ndcregistry/PublishedDocuments/Pakistan%20First/Pak-INDC.pdf accessed 12 November 2017.

[430]Ibid.

or negligible.[431] The Stockholm Declaration of 1972, provided the momentum and the Environment Ministry, was created in 1975.[432] A National Conservation Strategy (NCS) was established in response to the World Conservation Strategy (WCS) introduced by the International Union for the Conservation of Nature and Natural Resources (IUCN) in 1980 and devised in 1992.[433] The Pakistan Environmental Protection Act was passed in 1987 to address the Environmental Protection in Pakistan comprehensively.[434] The legislation also contemplated and emplaced the Pakistan Environmental Protection Council (PEPC) as the supreme policy-making body at the federal level and Environmental Protection Agencies (EPAs) at the provincial level.[435] The Environment Protection Act specifies the basic framework for environmental regulation and enforcement in Pakistan.[436] The Act aims to offers conservation, protection, rehabilitation and the improvement of the environmental standards, for the prevention and control of pollution levels, and promotion and preservation of sustainable development.[437] Its provisions can be invoked to provide for methods dealing with mitigation of climate change. However, the issue of climate change displacement and the rights of the displaced people cannot be addressed under the Act. The judicial establishment of Pakistan has positively interpreted the article 9,[438] article 14[439] whereunder the right to life and the right to dignity of man, respectively, protected and promoted the environmental right jurisprudence.[440]

The response of Pakistan to the challenges of climate change has been closely tuned with its international environmental obligations and the principle of sustainable

[431]Ibid.

[432]Irum Ahsan and Saima Amin Khawaja, '*Development of Environmental Laws and Jurisprudence in Pakistan*' (Asian Development Bank, 2013, 1, Abdul Matin Khan, 'Brief on Environmental Legislation in Pakistan, Sustainable Development Policy Institute, (1999) 1, https://www.sdpi.org/publications/files/Pb7Brief%20on%20Environmental%20Legislation%20in%20Pakistan.pdf accessed 10 November 2017.

[433]Mujahida Naureen, 'Development of Environmental Institutions and Laws in Pakistan' (2009) 30 (1) Pakistan Journal of History and Culture, 93–112, 96.

[434]Ahsan and Amin (n 435) 1.

[435]Ibid, 5, Naureen (n 436) 96, see also government of Pakistan, Gazette of Pakistan 1983 (Islamabad: Ministry of Law and Parliamentary Affairs, 1983) 57.

[436]Ibid, 4–5.

[437]Pakistan Environmental Protection Act (1997), Parvez Hassan, 'Handbook On National Environmental Legislation And Institutions In Pakistan, (*Law-Handbook-Pakistan-* UNEP-SACEP 2001) http://www.sacep.org/pdf/Reports-Technical/2001-UNEP-SACEP-Law-Handbook-Pakistan.pdf/ accessed 3 November 2017.

[438]The Constitution of the Islamic Republic of Pakistan (130) article 9 Security of person. No person shall be deprived of life or liberty save in accordance with law.

[439]Ibid article 14 of the constitution of Pakistan, Inviolability of dignity of man, etc. The dignity of man and, subject to law, the privacy of home, shall be inviolable, No person shall be subjected to torture for the purpose of extracting evidence.

[440]Ahsan and Amin (n 435) 15, see also Muhammad Khurshid, 'Environmental Initiatives: Global and National Perspective' (SN 2004) Parvez Hassan (n 207).

development. Pakistan has ratified the UNFCCC, Kyoto Protocol[441] and the Paris Agreement.[442] The government of Pakistan has successfully submitted its national communications to the UNFCCC along with its INDC under the Paris Agreement.[443] In the year of 2012, the government of Pakistan has formulated and adopted the National Climate Change Policy designed to achieve the development goals incorporated in the Planning Commission's Vision Document 2030 and to ensuring the mainstreaming of the climate change in the socio-economic sectors of Pakistan for taking its economy towards climate change resilient development models.[444] The policy adopts a social vision and emphasizes on participatory approach for climate adaptation and mitigation.[445] The policy acknowledges explicitly the issue of climate displacement and points that the super floods of 2010 singularly uprooted the 20 million people from their homelands in just one stroke, making it by far the biggest human displacement caused by any climate-induced single event.[446] Climate displacement is identified as a significant climate change threat and proposes to regulate rural to urban migration as a mainstay of urban planning.[447]

To guarantee the execution and mainstreaming the climate change apprehensions into local and national and developmental agenda, the framework for the Implementation of the Climate Change Policy (FICCP) 2014–2030 was commenced.[448] The framework adumbrated the susceptibilities and exposures of the different fields to climate change and acknowledges suitable adaptation and mitigation measures. The FICCP has identified increased health risks and climate change-induced migration and rural to urban migration as a vulnerable factor.[449] However, the framework is silent on the cross-border climate-induced displacement. In March 2017, Pakistan enacted the Climate Change Act 2016[450] becoming one of the few nations to have specific legislation dealing with the global climate change.

The purpose and objective of the Act states '*An Act to meet Pakistan's obligations under international Conventions on climate change and address the effects of climate change.*'[451] As a signatory to the Paris Agreement and other international climate change treaties, Pakistan is obliged to respect the provisions and streamline domestic

---

[441]Kyoto Protocol-Status of Ratification https://treaties.un.org/Pages/ViewDetails.aspx?src=TREATY&mtdsg_no=XXVII-7-a&chapter=27&clang=_en accessed 10 March 2018.

[442]Paris Agreement - Status of Ratification http://unfccc.int/paris_agreement/items/9444.php accessed 9 November 2017.

[443]Pakistan's Intended Nationally Determined Contribution (PAK-INDC) http://www4.unfccc.int/ndcregistry/PublishedDocuments/Pakistan%20First/Pak-INDC.pdf accessed 11 November 2017.

[444]National Climate Change Policy, 2012, Pakistan, Goals.

[445]Ibid.

[446]Ibid.

[447]Ibid.

[448]Framework for Implementation of Climate Change Policy (2014–2030) government of Pakistan http://www.gcisc.org.pk/Framework%20for%20Implementation%20of%20CC%20Policy.pdf accessed 17 November 2017.

[449]Ibid.

[450]The Pakistan Climate Change Act, 2017 (Climate Change Act).

[451]Climate Change Act, pt 1.

policy with its international obligations. For example, under the Paris Agreement, all member states have to direct their efforts towards cutting down the global temperature rise to 2C and eventually 1.5C.[452] The Act defines climate change *as 'a change in the climate system which is caused by the significant changes in the concentration of the greenhouse gases as a direct or indirect consequence of human activities and which is in addition to the natural climate change that has been observed during a considerable period'*.[453] The Act covers mitigation and adaptation. However, it did not refer to climate-induced displacement and failed to underline adaptive measures concerning disasters and climate displacement.

The Act contemplates the structural reforms by envisaging the two statutory bodies, and one is domestic fund such as Pakistan Climate Change Council (PCCC),[454] Pakistan Climate Change Authority (PCCA)[455] and the Pakistan Climate Change Fund (PCCF).[456] The Climate Change Authority is a body corporate consisting of a Chairperson, Member (Adaptation), Member (Mitigation), Member (Climate Finance), Member (Coordination) and one member from each Province.[457] These persons shall be academicians, scientists, retired or serving government servants, agriculturalists, or other technocrats with at least 15 years of practice in the area of climate change and a distinguished service record.[458]

The Act has carefully divided the functions between the council and the Authority. The primary role of the Authority is administrative in nature. It is responsible for administering the implementation of the Act, the international treaties and the sustainable development goals.[459] The function of the Authority is more instrumental than tangential because it is responsible for the formulating policies and projects for realizing the climate change goals outlined in the international obligations.[460] It also plays a crucial role in acting as an advisor to the government regarding appropriate legislative, policy and implementation measures.[461] The role of authority could be critical in the coming years as far as the issue of adoption of polices aimed at adaptation to reduce the impact of climate displacement. By the 18th amendment Act to the Constitution of Pakistan, the environment was a subject matter of jurisdiction for the provincial governments, however, with the Climate Change Act coming into the picture; it has centralized the issue and created a federal mechanism with

[452]Daniel Bodansky, 'The Paris Climate Change Agreement: A New Hope?' (2016) 110 AJIL 288–319.

[453]Climate Change Act (n 453) s. 2.

[454]Climate Change Act, s. 3.

[455]Climate Change Act, s. 5.

[456]Climate Change Act, s. 12.

[457]Ibid s. 6(1).

[458]Ibid s. 6(2).

[459]Ibid s. 4.

[460]Ibid s. 8.

[461]Ibid s. 8(j)-(k).

provincial representation to combat climate change.[462] The situation demands carefully threading and the centralization brought in by the Act could wipe the authority of the provinces granted under the 18th amendment.[463] The council and Authority should equip the provincial governments to take measures to mitigate and adapt to climate change-induced displacement and formulate policies to deal with climate displacement and their rehabilitation and relocation.

Regarding disaster management, a three-tier regime was established by the National Disaster Management Act 2010, with National Disaster Management Authority (NDMA) at the federal level, the Provincial Disaster Management Authorities (PDMAs) at the provincial level and the District Disaster Management Authorities (DDMA) on the district levels.[464] The Commission is mandated to lay down policies on disaster management; issue guidelines for all levels of governance regarding disaster management; and arrange funds for mitigation measures, preparedness and response efforts.[465] The Act defines disaster as means a serious disruption of the functioning of society, which poses a significant, widespread threat to human life, health, property or the environment, whether arising from accident, nature or human activity, whether developing suddenly or as the result of 'long-term processes'. The broad definition of disaster covers both natural and human-made disasters, and the term 'long process' provides enough hooks to incorporate climate disasters under the ambit of the Act.[466] Adoption of the and National Disaster Risk Reduction Policy in 2013 further provided a fill up for the mainstreaming the climate change, particularly in socio-economically vulnerable sectors of the national economy.[467]

In spite of having a specific Climate Act enacted, and a comprehensive disaster management legal framework in place, Pakistan faces the following challenges. National Disaster Reduction Policy while acknowledging the inter-linkage between climate change disasters has pointed to low levels of risk awareness and knowledge, lack of community participation, ineffective standard methodology or institutionalized capacity to conduct a multi-hazard risk analysis, inefficient integration of disaster management into developmental policies and lack of institutional capacity.[468] This is in spite of the decentralization of core responsibilities brought in by the disaster

[462]Pakistan Passes Climate Change Act, Experts Remain Skeptical https://www.thethirdpole.net/2017/03/28/pakistans-passes-climate-change-act-experts-remain-sceptical/ accessed 15 November 2017.

[463]Ibid.

[464]National Disaster Management Act 2010, Pakistan; Atta-Ur-Rahman, 'Introduction and Approaches of Disaster Risk Reduction in Pakistan' in Atta-Ur-Rahman and others (eds), *Disaster Risk Reduction Approaches in Pakistan* (Springer 2015) 15.

[465]Climate Change Act (n 453) s. 6.

[466]IFRC, 'International Disaster Response Law (IDRL) in Pakistan: A Desk Review of the Legal Framework for Facilitating and Regulating International Disaster Assistance' (IFRC 2011) 70.

[467]Ibid, 29 Rahman (n 467) 16.

[468]National Disaster Risk Reduction Policy (NDRRP), 2013, Pakistan, 4, 29 http://www.ndma.gov.pk/plans/drrpolicy2013.pdf accessed 20 November 2017, Rahman (n 467) 23.

legislation.[469] In 2015, Pew Research Centre conducted a survey of twenty nations of which it concluded that Pakistan was the one country that is minimally likely to settle with the fact that environment should be safeguarded irrespective of the possible cost.[470] Therefore, as long as the stakeholders are not made aware of the threats that climate change is posing to the world, no policy or project at the federal level will create a difference. Since climate change is a collective activity, there is a need for more cooperation with regional and international bodies like SAARC and the UN.[471]

5.5.7 *Maldives*

'For the Maldives, where 80% of the land area is less than 1 m above sea level, even the most benign climate change scenarios point to deep vulnerabilities',[472] and the country has engaged and shaped positively, resolutely and constructively with the international climate change regime.[473]

The Constitution of Maldives provides for the protection of the environment. Article 22 of the Constitution states '*The State has a fundamental duty to protect and preserve the natural environment, biodiversity, resources, and the beauty of the country for the benefit of the present and future generations. The State shall undertake and promote the desirable socio-economic goals through the ecologically balanced sustainable development and shall take measures that are necessary to foster the conservation, prevent pollution, the extinction of any species and the ecological degradation from any such goals*'.[474]

Further article 23 imposes an obligation on the state to ensure progressive realization of various rights including a healthy and ecologically balanced environment[475]; like Nepal, the environment protection has been positioned as a fundamental right, which will significantly influence the implementation and enforcement of the protec-

---

469 Ibid NDRRP 4, Zubair Ahmad, 'Disaster Risks and Disaster Management Policies and Practices in Pakistan: A Critical Analysis of Disaster Management Act 2010 of Pakistan' (2015) 4 IJDRR, 15–20.

470 Global Attitude Project, 2010, Chapter 8 Environmental Issues, (Pew Research Centre, 2010) 69 http://www.pewglobal.org/2010/06/17/chapter-8-environmental-issues-2/ accessed 13 November 2017.

471 Radoslov S Dimitrov, 'International Negotiations', in Robert Falkner (ed) *The Handbook of Global Climate and Environment Policy* (John Wiley & Sons, 2016) 348.

472 UNDP—United Nations Development Programme, 'Human Development Report 2007/2008. Fighting Climate Change: Human Solidarity in a Divided World' (2007) 384 http://hdr.undp.org/sites/default/files/hdr_20072008_summary_english.pdf accessed 13 November 2017.

473 Maldives Climate Change Policy Framework, Maldives, 2014, 18 http://www.file://Users/sau/Downloads/Maldives_Climate_Change_Policy_Framework_draft_3(7).pdf accessed 17 November 2017.

474 Constitution of the Republic of Maldives, 2008.

475 Ibid article 23(d).

tion.[476] Environmental Protection and Preservation Act 993 of the Maldives provides for the institutional and normative mechanism for the protection of the environment.[477] Though the Act does not specify climate change, it does give the government the power and jurisdiction to terminate any projects that are deemed to have an undesirable impact on the environment.

With specific to climate change adaptation, National Adaptation Programme of Action adopted in 2007 refers to displacement and relocation as an active adaptation strategy.[478] Maldives Climate Change Policy Framework 2014 (MCCPF)[479] considers climate change as the biggest developmental and security threat.[480] The policy Framework acknowledges the longer-term impacts of sea-level rise and the resultant migration including cross-border migration.[481]

The five strategic goals include sustainable financing; low emission development; adaptation and opportunities; capacity building and fostering sustainable development.[482] The strategic goals of adaptation, capacity building and sustainable development are crucial to climate displacement. Most importantly, the policy highlights the incorporation of local and traditional knowledge for mitigating the effects of climate change and combating it.[483] However, as a stand-alone document, the policy framework lacks legal backing and has to be backed by legislation for the effective implementation of the strategies and Plans envisaged under the framework. Besides for the Maldives, the concern is not the mere threat of internal or cross-border migration; the question is about the inundation of territory and loss of the land, which raises questions on security, statehood, nationality and cultural identity.[484] The subject of climate displacement cannot be tackled at domestic level, a global and regional action and cooperation is a prerequisite to sustaining habitable earth.[485] However, among the South Asian countries, the Maldives has paid the maximum attention to climate change-induced human displacement due to its vulnerability to littoral land mass erosion and its archipelagic settings in the region. Unfortunately, global community particularly Global North countries avoiding their global climatic obligations and countries like the Maldives have been suffering for no faults of their own.

---

[476]David R Boyd, *The Environmental Rights Revolution: A Global Study of Constitutions, Human Rights, and the Environment*, (UBC Press, 2011) 65.

[477]Environment Protection and Preservation of the Maldives, http://www.unesco.org/culture/natlaws/media/pdf/maldives/maldives_act_11_08_1998_engl_orof.pdf accessed 10 March 2018.

[478]National Adaptation of Program of Action, Republic of Maldives, 2007 http://unfccc.int/resource/docs/napa/mdv01.pdf accessed 14 November 2017.

[479]Maldives Climate Change Policy Framework 2014.

[480]Ibid.

[481]Ibid.

[482]Ibid.

[483]Ibid.

[484]Michael Gagain, 'Climate change, Sea Level Rise and Artificial Islands: Saving the Maldives' Statehood and Maritime Claims through the Constitution of the Oceans' (2012) 23 Colo. J. Int'l Envtl. L. & Pol'y 77–120, 88.

[485]Anton and Shelton (n 115) 296.

5.5.8 *Sri Lanka*

Like other island nations, Sri Lanka is susceptible to climate change hazards ranging from extreme weather conditions to heavy rainfall and flash floods to dry spells causing draughts.[486]

In Sri Lanka, the 1978 Constitution neither endorses explicitly a right to a clean environment nor does it embody a right to life.[487] However, in the chapter on directive principles, there is a specific reference to environmental protection.[488] The Constitution declares the protection of nature and conserving its riches a fundamental duty of every person in Sri Lanka.[489] In 1987, the 13th amendment in the Constitution of Sri Lanka established an amount of decentralization and delegated the subject matter of environment to the concurrent list and enhanced the powers to Provincial councils.[490] The Draft Constitution 2000 has proposed to incorporate the right to life but fell short of integrating the right to clean environment. In *Wattegedera Wijebanda v. Conservator General of Forest and Eight Others*, the court noted that while environmental rights are not specifically alluded under the category fundamental rights, the right to clean environment is inherent in a meaningful reading of constitution by the equal protection provisions contained in article 12(1) of the Constitution.[491] Contrasting the other South Asian jurisdictions where the right to clean environment has been sculpted out of the right to life. However, in Sri Lanka, the same has been accomplished through the solicitation of the right to equality. In the case of Sri Lanka, the absence of environmental rights in either the 1978 Constitution or the 2000 Draft Constitution occurred as a result of historical factors, which focused on stability and modernization.[492]

---

[486]Sri Lanka: State of the Nation on Climate Change, Civil Society Position Paper on Policies and Strategies—People's Secretariat on Climate Change The Green Movement of Sri Lanka, 2010, 2 https://www.preventionweb.net/files/22113_15417stateofthenationlowfilezise1.pdf accessed 15 November 2017. See also Yasuaki Hijioka and others, 'Asia', in C.B. Field and others (eds), *Climate Change 2014: Impacts, Adaptation, and Vulnerability. Part B: Regional Aspects. Contribution of Working Group II to the Fifth Assessment Report of the Intergovernmental Panel on Climate Change* (Cambridge University Press, 2014) 1327–1370.

[487]Sumudu Attapattu, Sustainable Development Myth or reality, Sustainable Development under International Law and Srilankan Law, A Survey' (2001) 14 Geo. Int'l Envtl. L. Rev 265-300.

[488]Ibid.

[489]Ibid See The Constitution of the Democratic Socialist Republic of Sri Lanka 1978 (n 136) article 28(f).

[490]Ibid, See Shyami Puvimanasinghe, 'Public Interest Litigation, Human Rights and the Environment in the Experience of SriLanka' in Nico J. Schrijver and Fried Weiss (eds) *International Law And Sustainable Development: Principles And Practice*, (MartinusNijoff, 2004) 665.

[491]Ibid., article 12, (1): All persons are equal before the law and are entitled to the equal protection of the law. (2) No citizen shall be discriminated against on the grounds of race, religion, language, caste, sex, political opinion, place of birth or any such grounds.

[492]Joshua C. Gellers, 'Environmental Constitutionalism in South Asia: Analyzing the Experiences of Nepal and Sri Lanka' (2015) 4 Transnat'l. Envtl. L, 395–423.

Sri Lanka as a party to the UNFCCC submitted its national communications[493] and had established a clean development mechanism under Kyoto Protocol.[494] The National Climate Change Policy 2012 of Sri Lanka is the primary policy mechanism to deal with climate change in Sri Lanka.[495] The objective of the policy is to take measure to increase the resilience of community to the effect of climate change and minimize the GHG emissions in line with the mandate of Kyoto Protocol.[496] It is aimed at guiding and capacity building of all stakeholders to address the adverse impacts of climate change.[497]

The policy Plans are categorized under the following broad headings—vulnerability, adaptation, mitigation, sustainable consumption and production, knowledge management and general statements.[498] At the institutional level, National Advisory Committee on Climate Change (NACCC) has been formed to bring together the principle actors and coordinate all activities related to climate change.[499] The policy does not explicitly refer to climate displacement; however, the provisions on assessing the vulnerability and disaster management can be invoked in the context of climate displacement.

In this context, a reference to the Peninsula Principle developed in the context of climate displacement can be beneficial for the countries in the region exploring legal avenues of action and cooperation. The Peninsula Principles on Climate Displacement within states were emerged out of the activities of the NGO displacement solutions.[500] The principles were adopted in 2013 by an international group of climate change experts.[501]

The Peninsula principles focus more on the internal displacement.[502] They have adopted a broader definition of the term 'climate-displaced persons' as including 'individuals, households or communities who face or experience climate displacement'.[503] The definition of climate displacement covers 'movement of people caused

---

[493]Initial National Communication under the United Nations Framework Convention on Climate Change, Sri Lanka 2000 http://unfccc.int/resource/docs/natc/srinc1.pdf accessed 10 November 2017.

[494]Ibid.

[495]The National Climate Change Policy of Sri Lanka, 2012 http://www.climatechange.lk/CCS%20Policy/Climate_Change_Policy_English.pdf, accessed 13 November 2017.

[496]Ibid, goal and objective.

[497]Ibid.

[498]Ibid, policy statements.

[499]Climate Change Secretariat Sri Lanka, 'National Advisory Committee on Climate Change (NACCC)' http://www.climatechange.lk/About_us.html accessed 13 November 2017. See also Sagarika Dutt and Alok Bansal, *South Asian Security: 21st Century Discourses* (Routledge 2013).

[500]Scott Leckie and Chris Huggins, 'Repairing Domestic Climate Displacement: The Peninsula Principles' (Routledge 2015).

[501]Ibid.

[502]The Peninsula Principles on Climate Displacement Within states, 2013 http://displacementsolutions.org/wp-content/uploads/FINAL-Peninsula-Principles-FINAL.pdf accessed 14 November 2017.

[503]Ibid, Principle 2: Definitions (b).

by sudden- and slow-onset environmental events transpiring either alone or in amalgamation along with other ecological reasons'.[504] The Peninsula principles offer a normative soft law framework for applied action based on principles of international law, international human rights obligations and good practices.[505] The basic premise of Peninsula principles emphasizes non-discrimination as a core value.[506] The primary responsibility for prevention and adaptation assistance has been assigned to the national states.[507] States are to initiate transparent, accountable domestic law and policies, prioritizing the prevention of displacement.[508] The principle also emphasizes the importance of international cooperation to prevent and respond to climate displacement adequately.[509] The principle lists out consolidated rights including land, housing, property and livelihood.[510] The principle also focuses and places the victim on the centre stage.[511] The Principle prescribes that states should consult with climate-displaced persons and obtains their consent prior to relocating them, except when there is an impending threat to life or limb.[512] Protection and promotion activities undertaken by the states should be executed in a manner that respects both the cultural sensitivities and understandings prevailing in the affected area and the principles of sustaining family and community harmony and cohesion.[513] The principle prescribes the following core features, which need to be addressed before the relocation of population.

- The acquisition of land;
- The preferences of the community;
- The transitional refuge and permanent housing;
- The preservation of prevailing sociocultural institutions and places of climate change-induced displaced persons;
- The accessibility to public services;
- The support required during the transitional period;
- The family and community integration and cohesion;
- The concerns of the country of reception and host community;
- The monitoring or supervisory mechanisms; and
- The grievance procedures and pragmatic remedies.

---

[504] Ibid, Principle 2: Definitions (c).

[505] Ibid, Principle 1: Scope and purpose (b).

[506] Ibid, Principle 3: non discrimination, rights and freedoms (a).

[507] Ibid, Principle 6: Provision of adaptation assistance, protection and other measures.

[508] Ibid, Principle 7: National implementation measures.

[509] Ibid, Principle 8: International cooperation and assistance.

[510] Ibid, Principle 15 Housing and livelihood.

[511] Ibid, Principle 10: Participation and consent; See Bonnie Dochert, 'Climate Change Migration And Social Innovation'(2014) Harvard College Review of Environment and Society, 22–24 http://www.hcs.harvard.edu/~res/2014/05/climate-change-migration-and-social-innovation/ accessed 3 November 2017.

[512] Ibid, Principle 10: Participation and consent.

[513] Ibid, Principle 14: State assistance to those climate displaced persons experiencing displacement but who have not been relocated (b).

Resorting to an interdisciplinary and hybrid approach, the Peninsula principle stipulates that the nation-states have to ensure the legal and practical rights of climate change-induced displaced population based on international legal standards, customary international law, and best practices and experiences available from across the globe.[514] Though lacking legally binding obligations, the peninsula principles articulate the legal and practical steps required to be resorted by the states. South Asian jurisdictions have signed and ratified multiple international human rights, humanitarian law and environmental law instruments. Further Constitutions in rights guaranteed under South Asian jurisdictions prescribe fundamental rights in the form of equality, non-discrimination, and right to life including livelihood, housing shelter, etc. The peninsula principle prescribes state action by the country's domestic and international obligation for the protection of climate-displaced population.

5.6 Protection Gaps in National Climate Laws and Policies in South Asia

South Asia with its low-lying coastal areas, diverse geography, complex social stratification and mammoth vulnerable population under poverty has been identified as a hot spot for climate displacement.[515] The IDMC has placed India to be one of the leading countries, which will face displacement on account of disasters.[516] Of the people internally displaced in 2015 due to natural disasters, 41 percent were in South Asia.[517] The South Asian countries have already been witnessing complex patterns of displacement/migration ranging from simple rural to urban migration,[518] to temporary to permanent migration and displacement/migration as a strategy of gradual adaptation and planned resettlement.[519] In spite of the clear acknowledgement of climate displacement and the threat it poses to society and security in South Asia, the policy and legal responses are in the nascent stage and are inadequately embodied.[520]

---

[514]Scott Leckie (n 503) See also Sumudu Atapattu, *Human Rights Approaches to Climate Change: Challenges and Opportunities* (Routledge 2015).

[515]'Addressing Climate Change and Migration in Asia and the Pacific' (n 285).

[516]IDMC Gribid, 'Global Report on Internal Displacement', 2017, 10 http://www.internal-displacement.org/global-report/gribid2017/ accessed 10 November 2017.

[517]IDMC, 'Global Estimates 2014: People displaced by disasters,' (2014) 40 http://www.internal-displacement.org/publications/global-estimates-2014-people-displaced-by-disasters accessed 19 November 2017.

[518]McAdam and Saul (n 286) 233.

[519]Sheila C. McAnaney, 'Sinking Islands—Formulating a Realistic Solution to Climate Change Displacement' (2012) 87 N.Y.U. L. Rev. 1172–1209, 1182.

[520]Tariq Waseem Ghazi, A.N.M. Muniruzzaman and A.K. Singh, *Climate Change and Security in South Asia, Cooperating for Peace*, (Global Military Advisory Council on Climate Change, 2016) 18.

South Asian nation-states have developed national adaptation Plans at national level. Climate change adaptation policy adopted by the countries[521] recognizes human mobility/displacement. However, South Asian countries lack a comprehensive explicit and dedicated legislative mechanisms dealing with a climate-induced displacement that specifically targets climate displacement and provide protective measures for people who have been displaced as a result of climate change.[522]

Several reasons can be attributed to the legal vacuum. The core international legal mechanisms under international human rights law, international refugee law and international environmental law are silent about the protection of climate-induced displacement.[523] Additionally, the legal and policy responses are in a state of denial as far as the climate displacement is concerned. The Oslo principles promulgated in 2015 recognize the complexities of climate change and the fragmented legal systems and the obligations about climate change resulting in ineffective implementation and coordination.[524] The Oslo principles while acknowledging the grave and imminent threat posed by climate change underscore the role of international environmental law and international law to the progression of the legal tenets.[525] However, the principles are conspicuous by its absence to have any reference to international human rights or refugee law as far as it is dealing with climate displacement and mobility.[526] The omission is not an accident; principles specifically state that the primary focus of climate change regime is on prevention and not on issues of adaptation and refugees.[527] However, it is noticeable that while the Paris Agreement refers to mobility in the context of climate change, the Oslo principle employs the term climate refugees.[528] Nagra has highlighted the limitation of this approach as the Refugee Convention has yet to be extended to climate change displacement.[529] The legal vacuum existing at the international level about climate displacement has a spilling effect at the domestic level.

Secondly, the essential requirement for a robust and sustainable legal and policy framework mandates understanding of the causes and dynamics of climate-induced displacements and climate vulnerability etched in the sociopolitical framework.[530]

---

[521] Uma Kothari (n 12).

[522] Tahera Akter (n 286) 11–12.

[523] Susan Martin, 'Climate Change, Migration, and Governance' (2010) 16 Global Governance 397; Daniel Bodansky, Jutta Brunnée, and Lavanya Rajamani, *International Climate Change Law* (Oxford University Press, 2017).

[524] Expert Group on Global Climate Change Obligations, Oslo principles on Global Climate Change Obligations (1 March 2015) http://globaljustice.macmillan.yale.edu/sites/default/files/files/OsloPrinciples.pdf accessed 12 June 2018 (Oslo principles).

[525] Ibid.

[526] Satvindar Nagra, 'The Oslo Principles and Climate Change Displacement' (2017) 11 CCLR 120–135.

[527] Ibid.

[528] Ibid.

[529] Ibid.

[530] Bonnie Docherty and Tyler Giannini, 'Confronting a Rising Tide: A Proposal for a Convention on Climate Change Refugees' (2009) 33 Harv. Envtl. L. Rev. 349, 355.

The dynamics and vulnerability involved in climate displacement vary greatly among individuals, communities and countries based on the social and economic conditions.[531] In the context of South Asia, the issue of climate displacement has not figured a place of prominence in the climate change discourse and narrative and has not been mainstreamed into environmental developmental and climate change polices.[532] The approach of government has been to consider the displacement as part of an adaptation strategy or simply as a disability to adapt to climate change.[533] The complex human and social impacts of climate change and the vulnerability associated with it required the government to look into the issue of climate displacement with a human-centric approach taking into account the local community perspective.

One of the main loopholes of the current environmental governance structures in South Asian jurisdictions is the top-down approach that fails to account for the local perspectives and ensure active participation.[534] The relevance of public participation is a well-accepted principle of international environmental law, and[535] some scholars even argue that public participation has become a general principle of international law.[536] The contours of public participation have been acknowledged and elaborated under the international environmental law through principle 10 of the Rio Declaration. It states that

'Environmental issues are best tackled with the involvement of all concerned citizens, at the relevant level. At the national level, each person shall have reasonable access to information relating to the environment that is kept by the governmental authorities, including dissemination of information on hazardous substances and activities in their communities, and the opportunity to participate in decision-making process. States shall facilitate and encourage the public awareness and the participation by making information widely available. The effective access to the judicial and administrative proceedings, including redress and remedy, shall be provided'.[537]

Public participation is also emphasized in Agenda 21, which states that:

> One of the fundamental prerequisites for the achieving the sustainable development goals is broad public participation in the decision-making. Furthermore, in the more specific context of the environment and development, the need for new forms of participation has also emerged. This includes the need of the individuals, groups, and organizations to participate in the environmental impact assessment procedures and to know about and involve in the

---

[531] McAnaney (n 522) 1177; See generally Susanna Price and Jane Singer (eds) *Global Implications of Development, Disasters and Climate Change: Responses to Displacement from Asia Pacific* (Routledge 2015).

[532] Addressing Climate Change and Migration in Asia and the Pacific' (n 285) 45.

[533] Ibid.

[534] See generally Jonas Ebbesson and Phoebe Okowa, *Environmental Law and Justice in Context* (Cambridge University Press, 2009).

[535] Kiss and Shelton (n 176) 375; John Scanlon and Francoise Burhenne-Guilmin (eds.), *International Environmental Governance: An International Regime for Protected Areas* (IUCN 2004) 2.

[536] Jeroen Van Bekhoven, 'Public Participation as a General Principle in International Environmental Law: Its Current Status and Real Impact' (2016) 11 NTLUR 220.

[537] Ibid.

> decision-making, particularly those which potentially affect the societies and communities in which they live and work. Individuals, groups, and organizations should have access to information germane to environment and the development held by national authorities, including the information on products and activities that have or are likely to have a significant impact on the environment, and information on the measures for the environmental protection.[538]

The Agenda 21 also illustrates the categories including women, indigenous youth communities and farmers whose participation is extremely necessary for the promotion and protection of sustainable development.[539] Article 12 of the Paris Agreement also says that '*parties shall cooperate in taking measures, as appropriate, to enhance climate change education, training, public awareness, public participation and public access to information*'.[540]

South Asian countries are party to these international instruments and have incorporated constitutional, legal and statutory provisions about public participation and access to information. For instance the article 50(1) of the Constitution of Nepal states that '*It shall be the political objective of the Nepali State to strengthen a federal democratic republican system for ensuring an atmosphere where the democratic rights are exercised by acknowledging the sovereignty, independence and integrity of the Nepal to be of utmost importance; by protecting the freedom, equality, property and all citizens through the rule of law; by embracing the norms and the values of fundamental rights and human rights, gender equality, proportional inclusion, participation and social justice; and by maintaining a just system in all spheres of the national life in order to establish a government system aimed at the public welfare, while maintaining the relations between the federal units on the basis of collaboration between them, and internalizing the norm of inclusion in the governance system on the basis of local autonomy and the decentralization*'.[541] The Constitution of Nepal defines Nepal as a three-tier structured federal democratic republic organized around federal, state and local governments.[542] 18th amendment to the Constitution of Pakistan through article 140(A) has made it mandatory for the establishment of the local government system.[543] By the amendment, the provinces have passed the Local government Act. In India, the 73rd constitutional amendment has created Panchayats as the basic unit of administration and given them constitutional status.[544] Panchayat Raj institutions aim to increase the inclusive governance-based democratic participation and decentralization on the understanding that administration could be handled best at the basic level of governance.[545]

---

538 Agenda 21 (n 204) Chapter 23, preamble.

539 Kiss and Shelton (n 176) 376.

540 Paris Agreement (n 151) article 12.

541 Nepalese Constitution, 2015,(125) article 50(1).

542 Ibid.

543 The Constitution of The Islamic Republic of Pakistan, 1973 (n 130) article 140(1).

544 The Constitution of India, (n 92) 73rd amendment Act 1992, India The amendment recognize Gram Sabha as the basic unit of Administration.

545 Ibid; P.D. Sharma, 'Women leaders in Panchayat Raj: A Case Study' in M. P. Dubey and Munni Padalia (eds), *Democratic Decentralisation and Panchayati Raj in India* (Ananmika Publishers, 2002) 147–170.

In the case of *Orissa Mining Corporation v. Union of India,*[546] Supreme Court of India highlighted the significance of public participation and promotion of environmental democracy in decision-making. The Court directed the 'Gram Sabha' the smallest units of local governance to decide whether mining could go on in the mining project in Odisha's Niyamgiri Hills. The Supreme Court decision legitimized the role of the gram sabha[547] and recognized its significance in the constitutional governance structure.[548] The decision recognized consent and consultation as part of attainment of sustainable development.[549] The decision depicts the importance attached by the Supreme Court to the recognized human rights to information and participation of affected stakeholders recognized at international level, a principle of significance proclaimed at Rio Declaration.[550] Inspite of these legal provisions, South Asian nation-states will have to enhance the existing threshold of their public participation in the climate governance and make it a mass movement. National governments in the region are not sensitive enough to attend climate concerns due to their political priorities, bureaucratic hassles, less emphasis on information access in the urban and rural South Asia, lack of public awareness and lopsided development models in which ordinary stakeholders are denied their due.

Right to information was brought into the legal sphere only in the last two decades in South Asia. Among the South Asian Jurisdictions, Pakistan was the first country to adopt the Freedom of Information Ordinance 2002.[551] This was followed by the legislative initiatives in other South Asian jurisdictions.[552] In Nepal and Pakistan, the right to information is unequivocally safeguarded under the constitution.[553] In the context of India and Bangladesh, right to information was held to be part of

---

[546]*Orissa Mining Corporation v. Union of India (1998) 9 SCC 93, Orissa Mining Corporation v* Ministry of Environment and Forests &others, [2013] 6 S.C.R. 881, para 3 (Herein after Vedanta); See Balraj K Siddhu, 'The Niyamgiri Hills Bauxite Project- Balancing Resource Extraction and Environment Protection' (2011) 41(3) Envi'l Poli'y & L. (IOS Press, Amsterdam) 166–171.

[547]The Constitution of India,1950 (n 92) (73rd) amendment Act 1992 provided for local self-government and established Gram Sabha in a village or a group of villages as a system of governance.

[548]Ipshita Chaturvedi, 'A Critical Study of Free and Prior Informed Consent in the Development of the Right to Development –Can Consent be withheld?' (2014) 5 JILS, 37, 51.

[549]Ibid, 51.

[550]Rio Declaration (n 203) Principle 10, See generally David A. Wirth, 'The Rio Declaration on Environment and Development: Two Steps Forward and One Back, or Vice Versa' (1995) 29 Ga. L. Rev 599–653.

[551]Challenges of Right to Information in South Asia. Institute of Informatics and Development (IID), Dhaka, Bangladesh (2012) 8.

[552]Ibid; Access to Information Law, Afghanistan 2014 http://transparencyadvisorygroup.org/uploads/Afghanistan_ATI_Law_draft_-English.pdf accessed 12 October 2017; Right to Information Act No. 12 of 2016; Parliament Of The Democratic Socialist Republic Of Sri Lanka https://www.rti.gov.lk/images/resources/RTI_Act_Sri_Lanka_E.pdf accessed 13 November 2017; Maldives Right to Information Act 2014 http://www.file://Users/sau/Downloads/Maldives_Right_to_Information_Act-_2014.pdf accessed 10 November 2017.

[553]The Constitution of The Islamic Republic of Pakistan, 1973 (n 130) article 19A through 18th amendment, Citizens Access to Information in South Asia, Regional Synthesis Report, The Asia Foundation, 2014 https://asiafoundation.org/resources/pdfs/CitizensAccesstoInformationinSouthAsia.pdf accessed 11 November 2017.

the fundamental rights as the right to freedom of speech and expression guaranteed under the Constitution.[554] In India, Right to Information Act was passed in 2005. Apart from making information available on demand, Act also provides that public authority should provide information *suomotu* available to the public at regular intervals through various means of communications, including the internet, so that the common man has minimum access to the use of this Act to obtain and utilize the information.[555]

In spite of the legislative development of decentralization and access to information, actual levels of public awareness, implementation, and confidence in the use of apparatus vary across the region. Even as people and civil society are increasingly becoming sensitive to the need of accountability and inclusive participation, there are malaises, which hamper the effective functioning of local-level governance in South Asia. Mainstreaming of public participation has not been integrated into the general administrative set-ups. Taking the case of disaster management in India, studies reveal that the centralized and top-down model of disaster management does not account for the internal differences between regions and communities about the extent, category and reason for vulnerabilities.[556] The lack of technical and financial aspects restricts the involvement of Panchayat Raj institutions in disaster relief and rehabilitation operations.[557] For instance, the governance of the Chief Ministers relief fund is entrusted with the revenue officials thus restricting the involvement of the local bodies.[558] From a policy perspective, the involvement of Panchayat institutions in disaster management will promote the much-needed equity as it ensures the participation of local communities.[559] Regarding disaster preparedness, Panchayat bodies could be involved and trained in early warning, information about population, safe location, arrangements for evacuation, etc.[560] The active involvement of the Panchayat Raj Institutions will be the first step in the direction of envisaging and institutionalizing a locally driven disaster management Plan. When local stakeholders and resources lead the disaster management, new local capacities and resources

---

[554]Constitution of India, 1950 (n 92) article 19; See *Raj Narayan v State of Uttar Pradesh (1975)*AIR 865; The Constitution of the People's Republic of Bangladesh, 1972 (n 58) article 39.

[555]Right to Information Act 2005- India section 4.

[556]Susanta Kumar Jena, 'Disaster management and Local Institutions in India', in T.M. Joseph (ed) *Local Governance in India: ideas, Challenges, and Strategies,* (Concept Publishing Company, 2007) 536.

[557]Kamal Thori (n 399) 221.

[558]Ganapathy Palanithurai, *Panchayats in Disaster: Preparedness and Management, Gujarat, Orissa, Andhra Pradesh, and Tamil Nadu with Successful Case Studies* (Concept Publishing Company 2009) 106.

[559]73rd amendment to the Constitution of India had created 33 per cent reservation for women in Panchayat Raj institutes. On 27th August, 2017 The Cabinet of India has approved the amendment of article 243 (d) of the Constitution to reserve 50 per cent of the total number of seats in Panchayats filled by direct election for women.

[560]Palanithurai (n 561) 108–112.

will emerge. Further, a locally driven programme will have better receptivity among the people as based on socio-economic practices.[561]

The socio-economic factors, lack of awareness and absence of mainstreaming of public participation restrict the public participation in South Asian jurisdiction, and there is a dire need to attune its institutions and political culture to the spirit of public participation. Junaid Chowdhury assessing the Bangladesh legal contexts highlights that the concept of public participation is neglected under the Bangladesh Climate Change Strategy and Action Plan 2009, and Climate Change Trust Fund Act 2010.[562] He also points out that no laws and policies including the comprehensive Bangladesh Environment Conservation Act 1995 provide the public with an opportunity to ask for a review or enactment of legal mechanisms.[563] He cites the Environmental Bill of Rights 1993 enacted by Ontario which grants powers to the public to ask for repeal or amendment to the law for the protection of the environment.[564] Public participation can assume multiple forms ranging from an active role in legal reform and policy formations, part of government delivery services of schemes, and implementation of programmes. Public participation is significant in the context of acknowledging the rights of the displaced population and South Asian countires require a shift from the bureaucratic closed process in climate actions to a public deliberation process.[565] The benefits of public participation include local empowerment, better receptivity to government schemes and initiatives leading to effective implantation adaptability, sustainability and give a fillip to the implementation of right to information.[566]

Whilst analysing the protection gaps in the national climate change displacement frameworks in South Asia, it is pertinent to note that the region comprises of developing and least developed countries and their primary goal is economic development and poverty eradication. Paris Agreement requires its member states to prepare Nationally Determined Contributions to ascertain the goals of the Agreement. For South Asian countries, pledging ambitious goals and realising them is beset with much difficulty. One of the crucial challenges is being the lack of financial resources in these countries. It is estimated that only India itself shall require approximately USD 2.5 trillion (at 2014–2015 prices) to implement its climate actions by 2030. Apart from

---

561 Umma Habiba,Rajib Shaw,and Md Anwarul Abedin, 'Community Based Disaster Reduction Approaches in Bangladesh' in Rajib Shaw, Fuad Mallick, Aminul Islam (eds) *Disaster Risk Reduction Approaches in Bangladesh*, 259–280, 277.

562 Junayed Ahmed Chowdhury, *Public Participation in Bangladesh's Response To Climate Change Issues* (Advocates for International Development 2012) http://workspace.unpan.org/sites/internet/documents/UNPAN90580.pdf accessed 10 March 2018.

563 Ibid, 12.

564 Ibid, 13.

565 Scott Leckie and Chris Huggins, *Repairing Domestic Climate Displacement: The Peninsula Principles*, (Routledge, 2015) 77.

566 Bulkeley Hareit and Arthur P.J.Mol 'Participation and Environmental Governance: Consensus, Ambivalence and Debate' (2003) 12 Environmental Values, 143–54 http://www.erica.demon.co.uk/EV/EV1207.html accessed 5 October 2017. Peter J.Taylor, and Frederick H. Buttel, 'How Do We Know We Have Global Environmental Problems? Science and the Globalization of Environmental Discourse' (1992) 23 Geo forum, 405–416. http://www.sciencedirect.com/science/article/pii/001671859290051S accessed 12 October 2017.

the financial aid, countries in this region cannot envisage steps without technological assistance and capacity building from the global community. The scenario points to the clear fact that for countries relying on purely domestic approach in adaptation and mitigation will not be the viable option. Climate change and climate-induced displacement know no boundaries.[567] In the initial stage, the trend of displacement may be more conventional following a rural to urban and internal displacement. However, the situation will eventually reach alarming proportions leading to cross-border mass displacement.[568] The situation requires deepening and strengthening the mutual dialogues and negotiation between the countries of the region.[569] The regional and international cooperation can play a catalytic role in promoting sustainable development, the rule of law, governance and cooperation.[570] Hence, the South Asian jurisdictions should address the legal vacuum and formulate regionally coordinated mechanisms and policies, which aim to reduce the causes of climate-induced displacement and protect the rights of displaced people.[571] Such a step would be vital in forging cooperative mechanisms significant for addressing the climate change-security linkage.

5.7 Migration Governance in South Asia: National Security Narratives

The traditional concept of security in international relationships can be understood as something concentrating on territorial sovereignty and its government, which is apprehensive with maintaining political independence and defending through military power and other *de rigueur* means, the people and assets within its territory from aggression by outside enemies engrained in the collective consciousness of the UN system.[572] However, in the wake of growing dependency among nations and the multitude of global concerns in the environment and social issues, the conventional notion of security came under significant scrutiny and expanded to issues

---

567 Jane McAdam, 'Climate Change and Displacement: Multi Disciplinary Perspectives' (Bloomsbury, 2010) 86.

568 Ibid, 9.

569 Ibid., See generally for regional cooperation, Ilona Miller and Kylie Wilson, Towards a Climate Change Displacement Facility in Benoît Mayer and François Crépeau, (eds) *Research Handbook on Climate Change, Migration and the Law* (Edward Elgar Publishing, 2017) 420–448.

570 Christel Cournil, 'The Protection of Environmental Refugees in International law' in Etienne Piguet and others (eds), *Migration and Climate Change* (Cambridge University Press, 2011) 359–386, 372.

571 Mc Adam and Saul (n 286) 233.

572 Arnold Wolfers, 'National Security as an Ambiguous Signal' in Arnold Wolfers (ed), *Discord and Collaboration: Essays on International Politics* (The Johns Hopkins Press, 1962).

of social and environmental concerns with a human-centric approach.[573] The Security Council recognized that social and ecological conflicts could constitute threats to international peace and security in a summit Declaration in 1992.[574] The newly identified and perceived social, economic and environmental issues which could be labelled as a security threat and brought under the purview of securitization[575] are buttressed with a mixture of normative arguments, which can be readily discernible from the conventional military defence conception.[576] The new and increased security threats are not limited to one specific state, and its impacts are perceived across borders.[577] These issues are not the premeditated creation of a state against other states. This gives an opportunity for the states to collaborate with each other to resolve the issues.[578] Among the expanded notions of security, the question of whether climate change and the consequent displacement/migration constitute a security issue is a highly contested issue among scholars[579] and empirical evidence is fraught with the methodological problem.[580] The general argument in support of the securitization of climate displacement includes the possible resource conflict between locals and displaced population, loss of territory, economic damage and destruction,[581] instability in the region and climate wars highlighting the human dimension.[582]

---

[573]Rita Floyd, 'The Environmental Security Debate and its Significance for Climate Change' (2008) 43 It. J. Int'l. Affairs 51–65; David A Baldwin, 'The concept of Security' (1997) 23 Review of International Studies (1997) 5–26.

[574]See generally The Secretary-General, An Agenda for Peace: Preventive Diplomacy, Peacemaking and Peace-keeping, U.N. Doc. S/24111, A/47/277 (June 17, 1992) (identifying environmental damage as new stability risk); U.N. Sec. Council, Statement by the President, U.N. Doc. A/47/253 (Jan. 31, 1992).

[575]O. Wæver, 'Aberystwyth, Paris, Copenhagen: New Schools in Security Theory and the Origins between Core and Periphery', Paper presented at the International Studies Association Conference, Montreal, March 2004, cited in Rita. Taureck, Securitization Theory and Securitization Studies', (2006) 9 Journal of International Relations and Development 53, 54.

[576]Jolly and Mahajana (n 336) 318.

[577]Ibid.

[578]Ibid.

[579]Jon Barnett 'Security and Climate Change' (2003) 13 Glob. Environ. Challenges 7–17; Wallace Broecker 'Unpleasant Surprises in the Greenhouse?' (1987) Nature, 328(6126), 123–126; Thomas Diez, Franziskus von Lucke and Zehra Wellmann, The Securitisation of Climate Change: Actors, Processes and Consequences (Routledge 2016); Floyd (n 576)18.

[580]Ben Saul, 'The Security Risks of Climate Change Displacement in Bangladesh' (2012) Journal of Human Security, Sydney Law School Research Paper No. 12/58. Available at https://ssrn.com/abstract=2138006.

[581]High Representative for the Common Foreign and Security Policy and the European Commission on Climate Change and International Security, *Climate Change and International Security (S113/08, 14 March* 2008*)*, http://www.consilium.europa.eu/uedocs/cms_data/docs/pressdata/en/reports/99387.pdf accessed 13 November 2017.

[582]German Advisory Council on Global Change (WGBU), Climate Change as a Security Risk 2007,23, See Christina Voigt, 'Security in a "Warming World": Competences of the UN Security Council for Preventing Dangerous Climate Change', in C. Bailliet (ed), *Security: A Multi disciplinary Normative Approach* (Leiden: Brill Publishers) 291–312.

The experts who argued against the securitization have often pointed to the unreliability of the speculative and exaggerated data about climate displacement.[583] They are of the view that in the initial stage of climate displacement, the pattern of displacement will be localized and internal than having a cross-border effect.[584] However, it needs to be emphasized that in the case of climate change the speculation and threat is futuristic.[585] Ben Saul argues in the context of Bangladesh that even if the climate displacement results in cross-border displacement, the security implications will depend on context-specific situations based on historical patterns of movements.[586] He specifically cites the migration from Bangladesh to India and presents the difficulty in distinguishing between Indian Bengalis and Bangladeshi Bengalis sharing the same cultural milieu.[587]

The securitization theory of Barry Buzan and Ole Waever justifies the securitization of climate change by the belief that problems can be transformed into a security issue if a political community agrees on that through speech Act.[588] There are various examples of speech Act delivered by leaders from South Asia reflective of the acknowledgement of security concerns of climate change in general. Jolly and Mahajana quote India's former Defence Minister A. K. Anthony, observation in his Presidential address in the '14th Asian security conference' that 'climate change too has an impact on security'.[589] They also cites the speech by Bangladesh Prime Minister Ms. Sheikh Hasina in her opening address in the International Meeting of Parliamentarians on climate change: Durban and beyond wherein she stated 'analysts in Bangladesh indicate that a mere rise by a metre in sea level, would submerge a fifth of the land mass in the country uprooting about 20 million people. It would lead to a mass movement to the cities causing developmental and security problems'.[590]

In 2007, the UN Security Council concluded its first debate on climate change-induced displacement and its ramifications for international security environment.[591] The fundamental focus of the meeting was on the possible role of Security Coun-

---

[583] Cord Jakobeit and Chris Methmann, 'Climate Refugees' as Dawning Catastrophe? A Critique of the Dominant Quest for Numbers' in Jurgen Scheffran and others (eds), *Climate Change, Human Security and Violent Conflict* (Springer 2012) 301–314.

[584] Se Katie Peters and Leigh Mayhew, 'The Securitization of Climate Change: A Developmental Perspective', in Stephen Brown and Jörn Grävingholt (eds), *The Securitization of Foreign Aid* (Palgrave Macmillan 2016) 212–236.

[585] Andrew Baldwin, 'Orientalising Environmental Citizenship: Climate Change, Migration and the Potentiality of Race' (2012) 16 *Citizenship Studies* 625–640; Maria Julia Trombetta, 'Linking Climate-Induced Migration and Security Within the EU: Insights from the Securitization Debate' (2012) 2 Crit. Stud 131–147.

[586] Saul (n 583); See also Ali Rias, 'Bangladesh' in Daniel Moran (ed), *Climate Change and National Security: A Country-Level Analysis* (Georgetown University Press, 2011) 103–114.

[587] Ibid.

[588] Ole Wæver and others, *Identity, Migration and the New Security Agenda in Europe* (Palgrave Macmillan, 1993).

[589] Jolly and Mahajana (n 336) 322.

[590] Ibid.

[591] European Commission (2008) Climate Change and International Security http://www.consilium.europa.eu/uedocs/cms_data/docs/pressdata/en/reports/99387.pdf accessed 12 July 2017.

cil to address the security implications of climate change.[592] In 2011, UN security council issued a Presidential statement on climate change where it *reiterates that the UNFCCC is the principal instrument for addressing the climate change, reminds the provisions of the UNFCCC, including an acknowledgement that the global nature of the climate change demands the broadest possible collaboration by all countries and their participation in an adequate and appropriate international response, in accordance with their CBDR obligations and respective capabilities and their social and economic conditions, and invites the relevant organs of the UNO, as appropriate and within their individual mandates to strengthen their efforts in considering and addressing climate change, including its possible security implications*.[593] The expanded narrative is evidenced by the recent IPCC report,[594] which has included a chapter on security. The IPCC reports highlight-range security threats associated with climate change, including ill-health, food shortages and natural disasters; and increased conflict, displacement and migration.[595]

Regarding South Asia, the global military council on climate change report highlights that climate change-induced disasters have resulted in unprecedented large-scale migration/displacement within and from the region and could undermine the safety and stability of the region.[596] The report also points out that food insecurity and water scarcity can result in violent conflict and instability in the area, however, climate-conflict linkages are complex and require more empirical research and evidence for the region while touching upon concerns like strained India-Pakistan relations over water sharing in the wake of climate change or military tensions over the melting of Siachen Glacier.[597] The position of South Asian countries differs as far as the issue of climate change as a threat to their national security. For the Maldives, fighting climate change is a fight for existence and considers climate change as a threat to the rights and protection of everyone on earth.[598] National Action Plan on Climate Change of India is conspicuous by its absence to even mention the term security[599]; the reason could be the fear of the politically motivated use of the Security

---

[592]Dane Warren, 'Climate Change And International Peace And Security: Possible Roles for the U.N. Security Council in Addressing Climate Change' (2015) Sabin Center for Climate Change Law, Columbia Law School 2015, 1–3.

[593]Security Council (2011) In Statement, Say 'Contextual Information' on Possible Security Implications of Climate Change Important when Climate Impacts Drive Conflict, Department of Public Information, UN, New York.

[594]W.N Adger and others, 'Human security' in C.B. Field and others (eds), *Climate Change 2014: Impacts, Adaptation, and Vulnerability. Part A: Global and Sectoral Aspects. Contribution of Working Group II to the Fifth Assessment Report of the Intergovernmental Panel on Climate Change* (Cambridge University Press, 2014) 755–791.

[595]Ibid.

[596]Tariq Wasim Khazi and others (523) 4.

[597]Ibid, 14–16.

[598]Zahid Shahab Ahmed, *Regionalism and Regional Security in South Asia: The Role of SAARC* (Routledge 2016) 98.

[599]NAPCC (n 354).

Council as a device to control the developing nation-states,[600] whereas Bangladesh has advocated a supporting approach to considering the security implications of global warming.[601] Climate policy of Pakistan regards climate change as affecting the security especially in the context of water, food and energy and has emphasized international and regional cooperation.[602] The changed position of Pakistan became visible in 2013 when it cosponsored an Arria-formula meeting to discuss the security implications of climate change.[603] Generally, the position of developing nations led by G-77 has been insisting on analysing the issue of climate change-induced displacement and migration from the perspective of sustainable development.[604] The reluctance is based on the fear that once labelled as a security threat; Security Council will seize the matter and developing nations will lose the collective strength of decision-making.[605] In this scenario of contestation, though there is a consensus on the inter-linkage between climate change and security threats, the issue has not been wholly elevated to full-fledged securitized agenda, though the recently in 2017 UNSC resolution is recognized the role of climate change as a threat multiplier in exacerbating human security.[606]

The primary point of contention in elevating climate change to a security threat lies in its anticipated consequences. There is a consensus if the result of securitization is limited to highlight the human security implications and proactive preventive measures against climate change through the negotiation strategy adopted at the international level. In other words, If the objective of securitization is to prompt governments to elevate climate concerns and responses proactively, such narratives need to be harnessed. However, the possible involvement of Security Council in addressing climate change and security link brings forth sharp contestation.[607] The main line of the opposition stems from the question of legitimacy[608] compliance issues, and mode of enforcement mechanisms mandated by the UN Charter[609] Article 39 of the UN Charter provides absolute discretion to the Security Council to decide

[600] Voigt (n 585) 296.

[601] See Sarah Dalrymple, 'Climate Change and Security in Bangladesh: A Case Study' (2009) Bangladesh Institute of International and Strategic Studies.

[602] National Climate Policy 2012, Pakistan(454).

[603] Warren (n 595) 4.

[604] Ibid, 2.

[605] Ibid.

[606] Security Council Resolution 2349 (2017) Recognizes the adverse effects of climate change and ecological changes among other factors on the stability of the Region, including through water scarcity, drought, desertification, land degradation, and food insecurity, and emphasizes the need for adequate risk assessments and risk management strategies by governments and the United Nations relating to these factors.

[607] Shirley Scott, 'Climate Change and Peak Oil as Threats to International Peace and Security: Is it Time for the Security Council to Legislate?' (2008) 9 MJIL 495, 503.

[608] Ibid, 510.

[609] Trina Ng, 'Safeguarding Peace and Security in Our Warming World: A Role for the Security Council' (2010) 15 J. Conflict Security L. 275–300.

whether a subject matter constitutes a threat to international peace and security.[610] However, the identification of a subject matter as security does not automatically open up the role of the Security Council and the traditional military enforcement measures under the United Nations Charter.[611]

The human security implications of climate change can be addressed through cooperative mechanisms as opposed to state-centric military approach. Developing cooperation in legal, administrative and policy spheres will help address some of the causes and support frameworks for managing and resolving climate change consequences before they escalate into security threats.[612] It is also crucially relevant that any initiative to bring the regional agencies together should not be aimed at establishing a military alliance, but at bringing in transparency and confidence-building measures aimed at increasing social and community resilience. The cooperative mechanisms should be people oriented responsive to the vulnerabilities and security needs of local communities.[613] The negotiations, transparency, information dissemination and sharing, capacity building, confidence-building measures (CBMs) and asymmetric reciprocity are central to the cooperative security concept[614] and cooperation for action, underscored by principles of justice.[615]

5.8 Conclusion

Climate Change affects social, economic and existential framework of individual, society, and nations. It touches upon issues ranging from health, food security, access to water, livelihood and migration/displacement patterns. The international community has responded to the heightened threat of climate change and has constructed an edifice of the legal regime in the form of UNFCCC, Kyoto Protocol and the Paris Agreement. The legal structure was built on the presumption that climate change is an environmental disaster and threat should be addressed through placing obligations on countries to reduce greenhouse gas emissions. Gradually, the discourse on climate change expanded and the legal regimes recognized the inter-linkage between climate

[610]Charter of the United Nations, 1945 (164) article 39.

[611]Voigt (n 585) 297.

[612]Zhongqin Zhao, 'Non Traditional Security and the New Concept of Security in China' in H.G. Brauch and others, *Coping with Global Environmental Change, Disasters and Security: Threats, Challenges, Vulnerabilities and Risks* (Springer 2011) 307–313.

[613]Loraine Elliot, Human Securitizing the Climate Debate, Asia Security Initiative Policy Series, Working paper No 2, 2010, 9 https://www.rsis.edu.sg/wp-content/uploads/rsis-pubs/NTS/resources/research_papers/MacArthur_working_paper_Lorraine%20Elliott.pdf accessed 1 November 2017.

[614]Arving Gupta, 'Will Cooperative Security Work in South Asia?' (2011) Institute for Defence Studies and Analyses https://idsa.in/idsacomment/WillCooperativeSecurityWorkinSouthAsia_agupta_131111 accessed 12 October 2017.

[615]Steve Vanderheiden, *Atmospheric Justice: A Political Theory of Climate Change* (Oxford Publications 2008); Edward A Page, 'Climate Change, Justice and Future Generations' (2006) 6 James Garvey, *The Ethics of Climate Change: Right and Wrong in a Warming World* (Bloomsburry 2008).

change and displacement/mobility. The issue of climate change moved to spheres including security and its co-relation with migration/displacement. The expanded narrative is evidenced by the recent IPCC report,[616] which has included a chapter on security. The IPCC reports highlight range of security threats associated with climate change, including ill-health, food shortages and natural disasters; and increased conflict, displacement and migration.[617] However, the attention of international climate regime to the heightened threat of climate displacement has been narrow and limited in analysing from the perspective of climate adaptation.

On the national level, as far as the responses to climate change are concerned, South Asian countries have developed national adaptation Plans. Climate change adaptation policy adopted by the countries recognizes human mobility/displacement. However, South Asian Countries lacked a comprehensive explicit inclusive and dedicated policy mechanisms dealing with a climate-induced displacement that targets explicitly climate displacement and provides protective measures for people who have been uprooted due to the climate change. In this scenario of national and international legal vacuums, the climate-induced displaced populations in South Asia who are internally displaced or are driven to cross borders are in a state of uncertainty as far as their legal status and protection are concerned. The cross-border migration expected out of climate change in South Asian countries in the years to come are bound to increase the vulnerability and provide an acid test to the resilience of the people. The vulnerability assumes serious proportion when we consider the refugee protection regime developed at international level and domestic level in South Asian jurisdictions. As elaborated in chapter three, under the Refugee Convention, the definition of refugees has a limited connotation and addresses only individual political refugees fleeing their countries of origin due to the state-sponsored persecution owing to race, religion, political opinion, or ethnicity and do not cover climate-induced displacement. At the South Asian level, the SAARC countries except Afghanistan are not signatory to UNCSR.

The South Asian countries at the domestic level have been granting varying degrees of protection to the political and civil refugees by their international human rights obligations. In the absence of national refugee laws in SAARC countries, there is a gulf between normative protection and processual protection; there is an interstice between legal protection and practical protection, and there is policy volatility in refugee protection. In South Asian region, refugee problem has been gaining new crisis proportions in the absence of refugee-specific national law and refugees have been suffering all kinds of abuse in our part of the world. The refugee protection challenges in South Asia require being addressed on the anvil of practical and ingenious solutions. It is distressing to note the magnitude of the refugee crisis in South Asia. However, realistic and pragmatic solutions can make the discernible distinction in their conditions. Therefore, these refugee challenges pose optimism about the potential changes that might be carried out in the lives of the refugees for a better world for them. Instead of being pointed to like the country that deported thousands

---

[616] Adger (n 697) 755–791.

[617] Ibid.

of helpless people such as the Rohingyas, SAARC could become a model[618] for other regions of the world about the treatment of refugees.

In this conspectus, the looming climate change migration[619] or environmental[620] displacement crisis, the existing refugee protection regime of the UN appears not adequately prepared. It is, indeed, questionable, without radical reforms, UNHCR can protect and support refugee exoduses that scaling all previous records. Therefore, the previous chapter makes a case for an Additional Global Protocol on Climate Refugees to the 1951 UNCSR. Whatever might be the nomenclature employed to express climate displacement, the situation demands a legally binding global protocol to attend and address the climate-displaced population additional to the existing international refugee law and international climate change law.

In the meanwhile, since the issue of climate displacement will pose serious cross-border concerns having significant regional consequences, it would be a utopian narrative if the South Asian countries address the refugee/displacement problems in a united manner. Envisaging a regional Convention to address the issue of climate refugees will be too much to hope at present, given the political and legal constraints under which the SAARC functions and operates. However, cooperative mechanisms and consultations and good practices involving judicial cooperation, exploring roadmap, promoting technology and capacity building could be resorted to. South Asian Declaration on Migratory Movements in South Asia must also be considered, but it should also include climate change refugees while making it the basis of any present and future understanding on the resolution of the refugee problem. Countries can consider the establishment of a South Asian Regional Refugee Database accessible to all refugee rights defenders, advocates, policy-makers, lawyers, academics, individuals, NGOs, CSOs and other organizations to enhance the refugee protection thresholds in the SAARC jurisdictions and beyond; the refugee protection research must be promoted and funded in South Asia that would produce findings and results. Such results should be utilized to evolve and develop best protection practices and to create potential spaces for refugee protection in different jurisdictions in the region and their reception beyond South Asia. NGOs, CSOs, academics and other stakeholders in tandem with institutions of higher learning must undertake advanced research to strategize novel practices for 'all classes of refugees' including the protection of climate change refugee protection specific to the South Asian region. Proactive advocacy for the protection of all classes of refugees must be pursued under a formalized framework of consultation, and its working must also be annually evaluated to document the protection practice for the guidance of civil society sector.

---

[618]Nafees Ahmad, 'As Rohingyas flee Myanmar, India Needs to Drop Religious Criteria in its Refugee Law', September 05, 2017, https://theconversation.com/as-rohingyas-flee-myanmar-india-needs-to-drop-religious-criteria-in-its-refugee-law-79648 accessed 20 January 2018.

[619]Joann. McGregor, 'Climate Change and Involuntary Migration: Implications for Food Security' (1994) 19 Food Policy 120–132,126.

[620]David Keane, The Environmental Causes and Consequences of Migration: A Search for the Meaning of 'Environmental Refugees' (2004) 16 Geo. Int'l. Envtl. L Rev 214–15.

In the meanwhile, the individual nation-states in South Asia can consider the feasibility of formulating the policy and enacting the legislative mechanisms to deal with the problem of climate change-induced displacement, migration and ‘climate refugees’ in a comprehensive and consolidated manner. Legislation by the respective national Parliaments would be a positive step in protecting the rights of the refugees/displacement. The need is to understand the complex scenario of climate displacement and envisage multifarious and coherent policy mechanisms for managing the climate-induced displacement in a human-centric way. The issue of climate change-induced displacement needs to be mainstreamed into environmental, developmental, human rights and refugee regimes through the cross-fertilization of hybrid legal principles.

Chapter 6
Climate Refugees: The Role of South Asian Judiciaries in Protecting the Climate Refugees

Abstract The people are forced to move, migrate or leave their homelands due to climate change in South Asia. Justice necessitates that climate refugees must be provided and extended the same protection as is provided to the political refugees who have an array of rights and protections under the international refugee law superintended by the UNHCR that was created following the World War II. However, the national governments in South Asia have failed to address the climate change-induced migration in the region. In this conspectus, the present chapter explores, examines and assesses the role of the regional judiciary in protecting the climate refugees in the absence of a 'climate refugee-specific law' in the SAARC jurisdictions. The role of the judiciary in environmental governance and sustainable development has been recognized as one of most important features of South Asian jurisprudence. Thus, the instant chapter evaluates the judicial reception and responses to the refugee law and international climate change law in South Asia while making a case for Intra-South Asian Judicial Interactions on 'climate refugees'. The chapter explains how the judiciary has played a proactive role in identifying the international environmental law principles to environmental protections. It also tries to explore the evolution of climate change regional constitutionalism to provide the SAARC jurisdictions a compendium of law and policies where under people crossing international borders in the region due to environmental calamities recognized as refugees by concluding and adopting a regional mechanism to protect the climate refugees.

6.1 Introduction

The modern world has graduated itself from savagery to civility and sets for itself a trajectory of high standards of justice, the rule of law and human rights. These values and vision emplaced an institutional framework of a well-crafted and well-calibrated independent legal structure called judiciary to arbitrate and adjudicate on matters ranging from socio-matrix to economics, avionics to genomics and human justice to climate justice in all democratic, plural and liberal geopolitical entities across the globe. Thus, the SAARC jurisdictions are not exceptions to this modern

S. Jolly and N. Ahmad, *Climate Refugees in South Asia*, International Law and the Global South, https://doi.org/10.1007/978-981-13-3137-4_6

embellishment of democracy with an independent judiciary. In South Asia, the quest for climate justice has been emerging with all its manifestations on an unprecedented scale that has threatened the human existence. The biggest predicament of the CDPs in South Asia is the denial of justice. Therefore, the role of South Asian judiciaries has become most prominent, pivotal and promising in addressing the plight of humanitarian refugees and refugees made by the climate change. In South Asia, the judicial establishments have adjudicated in refugee cases, but these judgments were not pronounced under the ambit of refugee legal framework.

In the field of refugee protection, the SAARC region governments are deeply mired in contradictions, and discrepancies resulting in the unequal treatment of refugees, the absence of legal principles, oscillating municipal political opinion and volatile diplomatic pomposity.[1] However, the human rights are the *lingua franca* of peace, stability and public order in protecting the refugees.[2] The genesis of the refugee protection lies in human rights perspective, and UN General Assembly has recognized the international protection of refugees as the principal mandate of the UNHCR.[3] The legal protection framework for refugees is embedded in the humanitarian principles of general international law and the interpretation of the purposes and principles of the UN Charter.[4] But, unfortunately, such an interpretation has not been expanded to incorporate the climate refugees.

Therefore, this chapter critically analyses the position of the judiciary in South Asian jurisdictions concerning the reception of international refugee law and international climate law in South Asian jurisdictions. The problems of climate change displacement and climate concerns have began to make constitutional moments before the South Asian judiciaries which have pragmatically contributed in evolving a cornucopia of judicially created law on climate concerns in their respective municipal legal systems. Thus, Pakistan witnessed the judicial activism as far as the climate change scenario is concerned with Judiciary recommending and overseeing the drafting of climate legislation for the country. The contents of these principles are drawn from the human rights law principles. The judiciary in India has also highlighted the need to account for climate concern in its disaster management laws and explain the protection of climate-induced displacement. The courts in the South Asian jurisdiction have suggested recommendations on the protection of internally displaced population including climate displacement.

[1] H Knox Thames, 'India's Failure to Adequately Protect Refugees' (1999) 7(1) HRBRIEF 20–21.

[2] G S Goodwin-Gill, 'The Language of Protection' (1989) 1(1) IJRL 6, 15.

[3] UNGA Elaboration of a Draft Convention on Territorial Asylum (adopted 9 December 1975) A/RES/3456; UNGA report of the United Nations High Commissioner for Refugees UN GAOR 30th Session Supp No 12 UN DOC A/10012 (1975).

[4] G Goodwin-Gill, *Refugee in International Law* (1st edn, OUP: Toronto 1983) 215.

6.2 Judicial Responses to International Refugee Law in South Asia

The fundamental norms of refugee protection in South Asia have not been developed and promoted by the intra-regional institutions. The issue of the refugee protection and now the protection of climate refugees remain a vertically convoluted and fragmented terrain in South Asia. Regional solutions to refugees would be of lasting significance provided legal norms of refugee protection are formally internalized and engrafted into the legal systems of the region. The history of South Asia has been replete with the instances of hosting refugees from all parts of the world, and the role of the judiciary has been positive and commendable. In the South Asian jurisdictions, judicial decisions have tried to provide humane solutions to the problems of refugees, primarily referring to the principles on *non-refoulement*,[5] right to seek asylum[6] and voluntary repatriation[7] that is categorized as one of the most preferred durable solutions.

The government of Bangladesh has hosted refugees from Urdu-speaking Bihar (India), since 1947, which had migrated to Bangladesh (then East Pakistan) in the post-partition period.[8] These people believed themselves to be Pakistani nationals and hoped to settle in Pakistan after the 1971 Bangladesh War of Liberation.[9] Though many of these settlers were repatriated to Pakistan, many others stayed in camps set up by Bangladeshi government and had to live as stateless people, with no constitutional rights.[10] In the landmark *Abid Khan case*,[11] the Supreme Court (Higher Court Division) of Bangladesh held that merely because these migrants were living in camps and had filed an application for reparation to Pakistan, it did not take away their right to citizenship in Bangladesh and thereby directed the government to register them as voters. However, the effect of this judgment was limited to the persons who had approached the court. In a subsequent case,[12] the High Court division extended this judgment to all members of the Urdu-speaking Bihari community and observed that treating them as stateless persons had not only caused misery and

---

[5]Convention Relating to the Status of Refugees (adopted 28 July 1951, entered into force 22 April 1954) 189 UNTS 137 (Refugee Convention) art 33 (1).

[6]Universal Declaration of Human Rights (adopted 10 December 1948 UNGA Res 217 A(III) (UDHR) art 14 (1).

[7]Nafees Ahmad, 'The Constitution-Based Approach of Indian Judiciary to The Refugee Rights and Global Standards of the UN Convention' (2017) 8(1) KSLR, 30–55, 47.

[8]Eric Paulsen, 'The Citizenship Status of the Urdu-speakers/Biharis in Bangladesh' (2006) 25 Refugee Survey Quarterly 54, 54–68.

[9]Sumit Sen, 'Stateless Refugees and the Right to Return: The Bihari Refugees of South Asia—Part 1' (1999) 11 IJRL 625–645, 635.

[10]Loraine Mirza, *Internment Camps of Bangladesh* (Crescent International, Ontario 1998) 66.

[11]*Abid Khan and others v Govt of Bangladesh and others* (2003) 55 DLR (HCD).

[12]*Md Sadaqat Khan (Fakku) and others v Chief Election Commissioner Bangladesh Election Commission* Writ Petition No 10129 of 2007, 18 May 2008 (Bangladesh).

suffering to those people, somewhat had also deprived the nation of any valuable contribution they could have made.

Indian courts have had the opportunity to deal with refugee cases at various levels.[13] The trial courts in India have not held a very lenient view towards refugees, and most of those who were tried under the Foreigner Act had to face imprisonment for a specified period, because having refugee status did not entitle a person to free movement within India.[14] However, in *State v. Chandra Kumar & Others,*[15] the trial court not only stopped the deportation of the refugee but also directed the government to place the Refugee and Asylum Seekers (Protection) Bill, 2006 before the Parliament. It can be concluded that the trial courts in India have been unable to develop any standard practice in cases against refugees, and have operated mostly in the hangover of colonial laws.[16]

A much more sympathetic attitude towards refugees can be seen when a study is made regarding the judgments of higher courts in India. In matters of compensation for resettlement,[17] opportunity to seek refugee status,[18] repatriation[19] and the right to Indian citizenship,[20] the High Courts have shown courage in deciding cases in favour of refugees, both on humanitarian grounds and based on law.[21] However, it is the Supreme Court of India which has given certain commendable judgments protecting the rights of the refugees. The cases (regarding refugee protection) which

---

[13]SP Sarker, *Refugee Law in India: The Road from Ambiguity to Protection* (Palgrave Macmillan 2017) 27.

[14]See *State v Montasir M Gubara* Criminal Case No 427/P/1994 (Court of Additional Chief Metropolitan Magistrate Mumbai); *State v Huson Vilvaraj* Case No 443/3 of 1997 (Court Metropolitan Magistrate New Delhi); *State v Mohd Ehsan* FIR No 435/1993 (Court of Metropolitan Magistrate New Delhi); *State v Kishan Chand and Habib Iranpur* Criminal Case No 66/96 (Court of Metropolitan Magistrate New Delhi) (in all these cases, the refugee was either not registered with the UNHRC, or failed to provide registration documents, and hence was awarded rigorous imprisonment from 1 to 6 months and fine).

[15]FIR No 78/10 Court of Metropolitan Magistrate (Dwarka) New Delhi.

[16]S.P. Sarker, *Refugee Law in India: The Road from Ambiguity to Protection* (Palgrave Macmillan 2017) 29.

[17]*Khudiram Chakma v Union Territory of Arunachal Pradesh and Ors* AIR 1992 Gau 105 (High Court of Gauhati).

[18]*Ms Zothansangpuii v The State of Manipur* Civil Rule No 981 of 1989 (High Court of Gauhati (Imphal Bench)); *Khy-Htoon and Ors v The State of Manipur* Civil Rule No 515 of 1990 (High Court of Gauhati); *Mr Bogyi v Union of India* Civil Rule No 1847/89 (High Court of Gauhati).

[19]*Gurunathan and Others v The Government of India and Others* WP No 6708 and 7916 of 1992 (High Court of Madras); *P Nedumaran and Dr S Ramadoss v Union of India and Another* WP No 12298 and 12343 of 1992 (High Court of Madras); *Aung Thant Min v Union of India* WP (CRL) 110 of 1998 (High Court of Delhi).

[20]*Namgyal Dolkar v Govt of India Ministry of External Affairs* WP (C) 12179/2009 (High Court of Delhi); *Sasikumar v State of Tamil Nadu* WP (MD) No 10080 of 2008 and MP (MD) No 2 of 2008 (High Court of Madras, Madurai Bench); *Tenzin Choephag Ling Rinpoche v Union of India* WP No 15437 of 2013 (High Court of Karnataka at Bangalore); *Nityananda Malik and Ors v State of Meghalaya and Ors* WP(C) No 235 of 2010 (High Court of Meghalaya) (Courts granted citizenship status to these refugees).

[21]S.P. Sarker (n 13) 45.

reached the Supreme Court of India are few, yet the way the Supreme Court of India has interpreted the Constitution and human rights instruments in its decisions to highlight the duty of the state to accord refugee protection is phenomenal.[22] Majority of the cases in the Supreme Court have regarded and recognized the status of Chakma refugees from Bangladesh.

The case of State of *Arunachal Pradesh v. Khudiram Chakma*[23] was put before the Supreme Court as a Special Leave Petition against the decision of the Gauhati High Court, wherein it had been held that the Chakmas were not citizens of India, or entitled to acquire land in India and the state government's order to move the community was lawful.[24] The High Court had also ordered the state government to provide compensation to the Chakmas so displaced, on humanitarian grounds. Both parties to the case then preferred a Special Leave Petition (SLP) before the Supreme Court.[25]

The Chakmas contended that since they had moved to Assam before the cut-off date under the Citizenship Act of 1955,[26] they should be recognized as citizens of India under Section 6-A of the Act. The state government argued that the Chakmas did not fulfil the requisite conditions[27] laid down under the said section and thus were not citizens of India. The conditions specified are that the person must be of Indian origin, and be an ordinary resident of Assam, as it existed at the time of signing the Assam Accord (1985). Since the Chakmas were not residents of Assam in 1985, as they had moved to Arunachal Pradesh, the state government contended that Section 6-A was inapplicable to them. The Apex Court accepted this contention by the state government. The Court also affirmed that the state had the right to restrict movement of the Chakma population and that Chakmas could not legally hold land in India as per Bengal Eastern Frontier Regulation of 1873 and the Foreigners Order of 1948. The Court also deliberated on the issue of compensation and held that this case did not require compensation to be paid. The Supreme Court, referring to the UDHR, observed that though an asylum seeker does not have a right to admission to a foreign State, any State which has granted such asylum must not later return the asylum seeker to the country where he feared persecution.

This judgment could not satisfy the Chakma population, and the matter came before the National Commission of Human Rights (NHRC), which moved the Supreme Court, seeking the issuance of a writ of *mandamus* against the state and

[22] Ibid 46; Nafees Ahmad (n 7) 45.

[23] AIR 1994 SC 1461 (Supreme Court of India).

[24] Ibid.

[25] Ibid.

[26] 1 January 1966 has been provided as cut-off date under the Citizenship Act.

[27] 'Assam' means the territories included in the State of Assam immediately before the commencement of the Citizenship (amendment) Act, 1985 (Section 6A(1)(a)) and all persons of Indian origin who came before the first day of January 1966 to Assam from the specified territory (including such of those whose names were included in the electoral rolls used for the purposes of the General Election to the House of the People held in 1967) and who have been ordinarily resident in Assam since the dates of their entry into Assam shall be deemed to be citizens of India as from the first day of January 1966. (Section 6A(2)).

union government to safeguard the life and liberty of the Chakmas. The case *National Human Rights Commission v. State of Arunachal Pradesh and Anr*[28] was filed under Section 18 of the Protection of Human Rights Act. The NHRC petitioned that the Chakmas should be provided Indian citizenship under Section 5(1) (a) of the Citizenship Act. The State of Arunachal Pradesh contended that the Chakmas faced no threat of life or liberty, and since they were not citizens of India, it was justified to move them out of the country. The Union of India, however, stated its willingness to grant citizenship to Chakma refugees but submitted that Arunachal Pradesh had not forwarded any application of such nature.

After evaluating all submissions, the Supreme Court rejected the argument by the State of Arunachal Pradesh and held that the lives and personal liberty of the Chakmas were in danger, entitling them to protection under article 21 of the Constitution of India. It also held that rejection of citizenship argument under Section 6-A could not act as a blockade against a fresh application under Section 5(1)(a) of the Act. Taking a pro-refugee stand, the Supreme Court observed that the State is bound to protect the lives and civil liberties of all and every human being, citizen or otherwise. In *Committee for Citizenship Rights of the Chakmas of Arunachal Pradesh and Ors v. State of Arunachal Pradesh and Ors*,[29] it was contended that the application made by Chakma refugees for granting citizenship was not being forwarded to the Ministry of Home Affairs, Union of India. The Court ordered the government of Arunachal Pradesh to process documents expeditiously, and any delay in granting citizenship was a violation of human rights obligations. The Supreme Court has also passed orders in favour of refugees from Myanmar,[30] Afghanistan and Pakistan.[31]

The judgments in *Maiwand's Trust of Afghan Human Freedom v. State of Punjab*,[32] *Malavika Karlekar v. Union of India*,[33] *Gramophone Company of India Ltd v. Birendra Bahadur Pandey*,[34] *Civil Rights Vigilance Committee, SLRC College of Law, Bangalore v. Union of India AIR*,[35] *Jolly George Verghese v. Bank of Cochin*[36] delivered on the non-deportation, reception and protection of refugees have upheld the global norms of human rights while appreciating the importance of national law on refugees. In *Louis De Raedt v. Union of India*,[37] the SC observed that despite the fact Chakmas are foreigners, they are entitled to have protection under article 21 of the Constitution of India. In a case[38] of Bangladeshi national who was gang-raped

---

[28](1996) 1 SCC 742 (Supreme Court of India).

[29]WP (Civil) No 510 of 2007 (Supreme Court of India).

[30]*Dr. Malavika Karlekar v Union of India and Anr* WP (CRL) No 583 of 1992 (Supreme Court of India).

[31]*ND Pancholi v State of Punjab and Others* WP (CRL) No 243 of 1988 (Supreme Court of India).

[32]CRL WP No 125 & 126 of 1986.

[33]CRL WP No 243 of 1988.

[34]AIR 1984 SC 667.

[35]1983 Kant 85.

[36]AIR 1980 SC 470.

[37](1991) 3 SCC 554.

[38]*Chairman Railway Board v Chandrimadas & Ors* 2000(2) SCC 465.

by the Railways' employees in the *Rail Yatri Niwas* maintained by the Railways, Calcutta High Court awarded Rs. 10 lakh compensation to the Bangladeshi victim. However, it was challenged before the Supreme Court of India on the ground that a foreigner cannot claim relief under article 226[39] of the Constitution of India, but the Supreme Court relied upon its earlier judgment in *Bodhisattwa v. Ms. Subdhara Chakraoborty*[40] wherein it was held that rape is a crime against humanity and violates fundamental right to human dignity enunciated in article 21 of the Constitution of India and referred to the jurisprudence developed under the UDHR[41] and the 1965 UN Convention on the Elimination of All Forms of the Racial Discrimination (CERD)[42] provisions in India and elsewhere.

The Madras High Court protected the rights of Sri Lankan refugees in the cases of *Gurunathan and others v. Government of India*[43] and *A. C. Mohd. Siddique v. Government of India and others*[44] wherein the core principle of voluntary repatriation was respected and forced deportation of refugees was rejected. In the case of *P. Nedumaran v. Union of India,*[45] the Madras High Court recognized the credibility and impartiality of the UNHCR and its functioning as an international body that has been ascertaining the voluntariness of the Sri Lankan refugees who did not wish to go to Sri Lanka and wanted to remain in refugee camps in India and they were not deported. Further, in the case of *Syed Ata Mohammadi v. Union of India,*[46] the Bombay High Court did not allow the deportation of Iranian refugee to his country of origin while preserving the principle of *non-refoulement*[47] engrafted in the international human rights law and respected by the community of civilized nations and allowed him to visit any country of his choice and alternative destination.

When faced with the issue of refugee protection, the Supreme Court of Nepal has also ruled in favour of upholding the fundamental rights of the refugees.[48] In a petition filed by HURON, Human Rights Organization of Nepal, the detention of 23 Tibetan refugees was challenged before the Court, requesting it to issue a writ of *habeas corpus*.[49] The Court ordered that the Nepalese government could not deport

[39] The Constitution of India 1950, art 226.

[40] (1996) 1 SCC 490.

[41] Universal Declaration of Human Rights (n 6).

[42] International Convention on the Elimination of All Forms of Racial Discrimination (adopted 21 December 1965, entered into force 4 January 1979) 660 UNTS 195 (CERD).

[43] WP No S 6708 and 7916 of 1992.

[44] 1998(47)DRJ(DB) 74.

[45] 1993 (2) ALT 291.

[46] CRL WP No7504/1994 (Bombay High Court).

[47] Refugee Convention (n 5) art 33 (1); Convention against Torture and Other Cruel, Inhuman or Degrading Treatment or Punishment (adopted 10 December 1984, entered into force 26 June 1987) 1465 UNTS 85 (CAT 1984) art 3.

[48] Editor, 'Nepalese Supreme Court rules against forcible return of Tibetan refugees to Tibet' *International Campaign for Tibet* (23 September 2011) www.savetibet.org/nepalese-supreme-court-rules-against-forcible-return-of-tibetan-refugees-to-tibet/ accessed 17 December 2017.

[49] Ibid.

the refugees to Nepal until a ruling is passed.[50] The Nepalese Supreme Court has on previous occasions also taken a favourable stand regarding refugee protection and has ordered the release of various Tibetan refugees from detention.[51]

As on date, the South Asian region is facing a troubling refugee problem about Rohingya refugees who are fleeing persecution from the Rakhine region of Myanmar. Bangladesh is home to approximately 8 lakh Rohingya refugees,[52] though the courts have not had an opportunity to comment on the status of these refugees. The presence of Rohingyas can also be seen in India, where around 40,000 are living as illegal immigrants.[53] The government of India citing security reasons and has ordered the deportation of these people, and a case on the issue is pending before the Indian Supreme Court.[54] The conflict has seen the Union and the NHRC taking opposite views regarding deportation, and the NHRC feels that forcing the stateless to move out would violate humanitarian principles and also India's pledge to uphold human rights.[55] Though the matter is pending and no order has been passed, the Court has observed that the problem is of great magnitude and the human rights of Rohingya refugees cannot be overlooked.[56]

The climate change has affected the sociocultural livelihoods and ecopolitical dimensions of human existence across the world. In South Asia, the climate change is producing and will produce climate refugees. However, the concept of climate justice is missing from the legal lexicon of South Asian jurisdictions.[57] There are no legal protections available to the climate refugees, and there is no legal classification of their status as a climate refugee that has created a legal gap for their protection. In this scenario, the question arises, who are climate refugees and where do they come from? Are climate refugees entitled to be granted asylum? Who will take the responsibility for the climate refugees? Would climate refugees ever be recognized

---

[50] Ibid.

[51] Sophie Richardson (ed), 'Under China's Shadow: Mistreatment of Tibetans in Nepal' (*Human Rights Watch*, 2014) 64 www.hrw.org/report/2014/04/01/under-chinas-shadow/mistreatment-tibetans-nepal#page accessed 24 December 2017.

[52] Abul Hasnat Milton et al., 'Trapped in Statelessness: Rohingya Refugees in Bangladesh' (2017) 14(8) Int J Environ Res Public Health 942–950.

[53] Krishna N. Das and Sanjeev Miglani, 'India Aims to Deport all Rohingya Muslims, Even those with UN Registration: Govt' *Hindustan Times* (14 August 2017) www.hindustantimes.com/india-news/india-aims-to-deport-all-rohingya-muslims-even-those-with-un-registration-govt/story-UhzebiCZ3zBt6qu8ujl9iN.html accessed 17 December 2017.

[54] *Mohamm'd Salimullah v Union of India* WP(C) No 793 of 2017 (Supreme Court of India).

[55] Editor, 'NHRC Issues Notice on Rohingyas' *The Hindu* (New Delhi, 18 August 2017) www.thehindu.com/news/national/nhrc-issues-notice-on-rohingyas/article19519828.ece accessed 18 December 2017.

[56] Editor, 'Supreme Court says Human Rights of Rohingya refugees cannot be Ignored' *Hindustan Times* (India 13 October 2017) www.hindustantimes.com/india-news/human-rights-of-rohingya-refugees-cannot-be-ignored-says-supreme-court/story-P1QnPPLgdQdzsQPOWkwH9M.html accessed 18 December 2017.

[57] Vernon Rive, 'Safe Harbours, Closed Borders? NewZealand Legal and Policy Responses to Climate Displacement in the South Pacific', in Paul Martin et al. (eds), *The Search for Environmental Justice* (Edward Elgar Publishing 2015) 221–238.

as refugees before the municipal and international law? Are the SAARC regional jurisdictions or international community ready to face climate refugee exoduses and floods in near future if contemporary trends of climate change displacements are realistically assessed and measured?

There are many more such legal questions that are inconvenient, intricate and intractable and not easy to answer for the national governments of South Asia in the flagrant absence of fundamental principles and substantive obligations in the international human rights law, international refugee law, international humanitarian law and international comparative constitutional law and elsewhere in international law. These legal questions are relating to the intersectionality of climate justice and climate change human mobility that requires an emplacement of the South Asian Climate Refugee Legal Regime. Therefore, the intersectionality of climate justice and climate displacement could only be addressed under a comprehensive and consolidated legal structure.[58] However, the normative legal structure envisaged by international organizations like International Bar Association (IBA)[59] and UNHCR has provided only limited norms for the South Asian judiciaries for attending climate change migration.[60] Therefore, the judiciaries in South Asia must resort to having an element of intersectional complementarity in their approach by integrating human rights norms[61] provided by these international organizations developing an 'institution of assorted asylum' in the climate refugee jurisprudence. But, unfortunately, there have not been any cases before the South Asian judiciaries regarding the protection of rights of the climate refugees. The climate justice is the bedrock of human existence that would guide the livelihoods and ecosystems in years ahead, and any derogation therefrom would decimate the humanity from its contemporary visage. Thus, an approach based on legal intersectionality on the part of South Asian judiciaries in climate refugee cases would invigorate a paradigmatic shift in eliminating the political paradoxes, legal repugnancies and sociocultural divergences among various legal frameworks and would materialize climate justice.

---

[58] Jon Barnett, and Michael Webber, 'Migration as Adaptation: Opportunities and Limits' in Jane McAdam (ed), *Climate Change and Displacement: Multidisciplinary perspectives* (Hart, Oxford 2010) 37–56.

[59] The report of the International Bar Association (IBA) and Task Force on Climate Change Justice aims to: '…ensure communities, individuals and governments have substantive legal and procedural rights relating to the enjoyment of a safe, clean, healthy and sustainable environment and the means to take or cause measures to be taken within their national legislative and judicial systems and, where necessary, at regional and international levels, to mitigate sources of climate change and provide for adaptation to its effects in a manner that respects human rights'.

[60] UNHCR 'Guidance on Protecting People from Disasters and Environmental Change through Planned Relocation' (7 October 2015) https://www.brookings.edu/wp-content/uploads/2016/06/GUIDANCE_PLANNED-RELOCATION_14-OCT-2015.pdf accessed on 15 April 2018; see also Philippe Sands, Jacqueline Peel et al., *Principles of International Environmental Law* (3rd edn, CUP 2012) 274–98, Bayes Ahmed, 'Who Takes Responsibility for the Climate Refugees?' (2018)10(1) Int.l. J. Clim. Chang. Str, 5–26.

[61] Alan Boyle, 'Human Rights and the Environment: Where Next?' (2012) 23 EJIL 613, 617–23.

6.3 Judicial Responses to International Climate Change Legal Regime in South Asia

Multilateral efforts have been made at the international level to adopt regulatory and legislative measures targeted at countering the multitude of climate hazards and mitigate and adapt to climate change.[62] The results of these initiatives have been a host of robust and dynamic legislative and institutional mechanisms at international and national levels.[63] However, even after decades of lego-institutional initiatives aimed at addressing climate change, studies and reports paint a grim picture of massive and unrelenting climate consequences including climate change-induced migration and displacement, which points to significant critical gaps in climate governance.[64]

The critical gaps in climate governance arise out of the manifold challenges presented to lawmaking by climate change. The challenges include complexities associated with science, the congregation of actors contributing to the climate change, the differential impact of climate both spatially and temporarily.[65] The matter becomes further convoluted given the fact that the climate legal framework at the international level has been constructed on climate negotiations manifested by strong north and south conflicting positions over distinctive and often diverging underlying interests concerning the distribution of the burden of climate actions.[66] Based on the consensus evolved at the negotiation by the United Nations, the climate legal framework containing the UNFCCC, the Kyoto Protocol and the Paris Agreement form the core and backbone of the international climate change regime.[67]

The Convention lays down the framework and broad objectives for the nations to cooperate and encourage dialogue and decisions to address climate change. The Convention attempts to balance the interest of developed and developing nation-states keeping in mind that the developmental process of developing country would

---

[62]See generally Philippe Sands et al. (n 60) 22–49 Philippe Sands *Greening International Law* (Earth Scan Publications, London 2011).

[63]Cinnamon P Carlarne, Kevin R Gray and Richard G.Tarasofsky. 'International Climate Change Law: Mapping the Field' in Kevin R Gray et al., *The Oxford Handbook of International Climate Change Law* (OUP 2016) 4–23.

[64]Ibid.

[65]David B Hunter, 'The Implications of Climate Change Litigation for International Environmental Law-Making' (2007) American University WCL Research Paper No 2008-14, 1–21, 8 http://ssrn.com/abstract=1005345 accessed 28 December 2017.

[66]K A Baumert, 'Participation of Developing Countries in the International Climate Change Regime: Lessons for the Future' (2006) 38 Geo Wash Int'l L. Rev 365–407; Jutta Brunnée, 'Promoting Compliance with Multilateral Environmental Agreements' in Lavanya Rajamani, Jutta Brunnée and Meinhard Doelle (eds) *Promoting Compliance in an Evolving Climate Regime*(Cambridge University Press 2012) 48–49; Sumudu Atapattu, 'Climate Change, Differentiated Responsibilities and State Responsibility: Devising Novel Legal Strategies for Damage Caused by Climate Change' in Benjamin J Richardson et al., (eds) *Climate Law and Developing Countries* (Edward Elgar 2009) 37–62.

[67]David Free Stone, 'United nations Framework Convention on Climate Change-The Baiss for the Climate Regime', Kevin R Gray et al., *The Oxford Handbook of International Climate Change Law* (OUP 2016) 97–119, 101.

not be threatened.[68] The long-standing objective of the Convention is 'to achieve the equilibrium of greenhouse gas collection and concentrations in the atmosphere at a level that would stop the dangerous anthropogenic meddling with the climate system'.[69] The convention has been complemented by the Protocol laying down quantitative binding obligations on the Annex parties through incorporating flexible market-based mechanisms.[70] The Paris Agreement represents a paradigmatic shift in the attitude of the international community in enunciating a bottom-up architecture in which countries voluntarily determined their contributions.[71] Through the evolving phases of climate governance, the focus has slowly shifted from envisaging mere preventive mechanisms to reliable and effective compliance and implementation mechanisms.[72]

The focus of the climate legal framework based on negotiation has been on climate change's global impacts: sea-level rise, glacial melts, implications for food and water security etc.[73] The emphasis on specific incidents of individual and community effects like coastal erosion in Bangladesh and the impact of the devastating floods in Pakistan affecting the justice framework has been missing at the global negotiation.[74] The legal framework has underscored the climate mitigation and adaptation and has emphasized on administrative and legal policy measures to tackle the climate change.[75] However, climate change as reality has the innate capacity to affect the very fabric of the society afflicting individual and community life.[76] Looking at the climate change from the perspective of victims has prompted the jurists and academics to look at the possibility of climate litigation as an effective tool to promote climate justice.[77] It was felt that given the complex, unstructured and multifaceted nature of climate change, existing international climate law is not designed as it stands to

[68] Cinnamon P Carlarne (n 63) 4–23; see Stellina Jolly and Amit Jain, *Climate Change: Changing Dimensions of Law and Policy* (MD Publishers 2009) 16.

[69] Daniel Bodanksy, 'The United Nations Framework Convention on Climate Change: A Commentary' (1993) 18 Yale J. Int'l L. 451–558, 497.

[70] Jane Bulmer, 'Compliance Regimes in Multilateral Environmental Agreements' in Lavanya Rajamani, Jutta Brunnée and Meinhard Doelle, *Promoting Compliance in an Evolving Climate Regime* (Cambridge University Press 2012) 55–56; Daniel Bodansky, 'A Tale of Two Architectures: The Once and Future U.N. Climate Change Regime' (2011) 43 Ariz. St. L. J 697, 698–701.

[71] Ibid.

[72] Daniel Bodansky, *The Art and Craft of International Environmental Law* (Harvard University Press 2010) 205.

[73] David B Hunter (n 65), 2–3.

[74] Ibid.

[75] Daniel Bodansky, Jutta Brunnée and Lavanya Rajamani, *International Climate Change Law* (Oxford Univers ity Press 2017) 11–12.

[76] Hannah Brock, 'Climate Change: Drivers of Insecurity and the Global South' (2012) Oxford Research Group. 1–20, 4–5 www.files.ethz.ch/isn/146109/Climate%20Change%20and%20Insecurity%20in%20the%20Global%20South.pdf accessed 28 December 2017.

[77] Daniel Bodansky, 'The Role of the International Court of Justice in Addressing Climate Change: Some Preliminary Reflections' (2017) 49 Ariz. St. L. J 689–712, 694; Philippe Sands, 'Climate Change and The Rule of Law: Adjudicating the Future in International Law' (2016) 28 JEL 19–35.

provide climate change justice to the victims.[78] They cite the slow process of climate negotiations and its meagre success in reaching a consensus on the assignment of responsibility of climate actions from states and climate victims as the major push factor for the increased preference for climate litigation.[79]

At the international level, the idea of climate litigation was advocated by the Small Island State Palau[80] for seeking an advisory opinion of the ICJ concerning the duties of states especially USA for its failure to apply due diligence to ensure that greenhouse emissions from their territory do not harm other states.[81] In 2005, an Inuit community group filed a petition in the Inter-American Commission on Human Rights, arguing that the USA has perpetuated global warming by failing to reduce its emissions thus violating the Inuit's human rights to culture, life, health and shelter.[82] Though, these cases were not successful, they were seen especially to social norm changing in the critical linkage between climate change and human rights rather than with an intention to legally proclaim the victory for the victims.[83] The idea of climate litigation was renewed later by a resolution adopted at the International Union for Conservation of Nature (IUCN) 2016 World Conservation Congress in Hawaii.[84]

The global communities have been witnessing the improved role of climate litigation across the domestic jurisdictions including USA,[85] Nigeria,[86] Pakistan,[87]

---

[78] Ibid, 19.

[79] Jacqueline Peel and Hari M Osofsky, *Climate Change Litigation: Regulatory Pathways to Cleaner Energy* (CUP 2015) 10–13; see generally William CG Bums and Hari M Osofsky (eds), *Adjudicating Climate Change: State, National, and International Approaches* (CUP 2009); Michael Faure and Marjan Peeters (eds), *Climate Change Liability* (Edward Elgar 2011); Michael G Faure and Andrew Nollkaemper, 'International Liability as an Instrument to Prevent and Compensate for Climate Change' (2007) 43 Stan. J. Int'l. 123–179.

[80] 'Palau seeks UN World Court opinion on damage caused by Greenhouse Gases' *UN News Centre* (22 September 2011) www.un.org/apps/news/story.asp?NewsID=39710#.WdCPMXFPnIU accessed 30 September 2017.

[81] Daniel Bodansky (n 77) 689; Zona Razaque, 'Access to Justice in Environmental Matters and the North–South Divide' in Shawkat Alam, Sumudu Atapattu, Carmen G. Gonzalez and Jona Razzaque, *International Environmental Law and the Global South* (Cambridge University Press 2015) 588–607, 591; see generally Eric A Posner, 'Climate Change and International Human Rights Litigation: A Critical Appraisal' (2007) 155 U.Pa.L. Rev 1925–1945.

[82] Ibid, Bodansky, 700; Sumudu Atapattu, *Human Rights Approaches to Climate Change: Challenges and Opportunities* (Routledge 2015) 85.

[83] Jacqueline Peel and Hari M Osofsky, *Climate Change Litigation* (Cambridge University Press 2015) 50.

[84] International Union for Conservation of Nature (IUCN), 'Request for an Advisory Opinion of the International Court of Justice on the Principles of Sustainable Development in View of the Needs of Future Generation' WCC-2016-Res-079-EN (10 September 2016).

[85] *Massachusetts v EPA* 549 US 497, 2007; *Am Elec Power Co v Connecticut* 564 U.S. 410, 2011; *Native Village of Kivalina v Exxon Mobil Corp* 696 F.3d 849, 9th Cir 2012; *Juliana v United States* No 6:15-cv-1517-TC 2016.

[86] *Gbemre v Shell Petroleum Dev Co Nigeria* AFR HUM RTs L REP 151 (F.H.C. Nigeria) 2005.

[87] *Leghari v Federation of Pakistan* WP No 25501/2015 (Lahore High Court) 2015 (this case was based on principles of sustainable development, precaution, and inter-generational equity).

Australia and New Zealand.[88] The cases reveal that the litigation has been used with varying degrees of success as a tool to force the states to initiate legal and policy measures to mitigate and adapt to climate change.[89] Secondly, climate change litigations have focused on the human and social impact of climate change and brought to focus the individuals and communities to the fore of climate change structure. Further, among the implementing institutions, the judiciary is strongly positioned to play an increasing role in creating and shaping a strong compliance mechanism.[90] The role of the judiciary in environmental governance and sustainable development has been recognized at the regional level.[91]

Much of the climate change litigation to date has come at the national level. In 2007, an NGO filed a lawsuit seeking a Declaration that the government of Canada has failed to comply with the UNFCCC and Kyoto Protocol Provisions.[92] The petition relied on Section 166 of the Canadian Environmental Protection Act, which requires the government to act to control pollution.[93] The petition relied upon the reports of the Canadian government indicating that the emissions are almost 40% higher than what is allowed under the Kyoto Protocol.[94] Similarly, in a landmark successful precedent founded in tort law in 2016 in the *Urgenda* case in the Netherlands, a Dutch district Court found that the government's 20% reduction target is not enough to address and prevent the dangerous climate change, and instructed that the Netherlands must do adequately to avert the prowling menace caused by the climate change in the wake of its obligation to protect, improve and care the living environment and ordered a 25% reduction target by 2020 from 1990 levels.[95] The Court observed, 'It has been proved that any anthropogenic greenhouse gas emission, no matter how minor, causes to an increase in CO2 levels in the atmosphere and therefore to hazardous climate change'.[96] The Court pointed out that the 'state has a duty to care and to take mitigation measures.' The circumstance that the Dutch contribution to the existing global greenhouse gas emissions is currently small that does not affect this'.[97] Moreover, the Court found a sufficient causal link that 'can be presumed to exist' among the Dutch emissions and radiations, global climate change and the

---

[88] *Greenpeace New Zealand v Northland Regional Council* NZHC CIV 2006404-004617 at 2006; *Genesis Power Ltd v Franklin Dist Council* NZRMA 541 (NZ) 2005 (these cases were based on domestic environmental legislation).

[89] *Friends of the earth v Canada* FC 1183 2008 (Can Fed Ct) (this case came before the Court for alleged violations of the UNFCCC and the Kyoto Protocol).

[90] Kenneth J Markowitz and Jo JA Gerardu, 'The Importance of the Judiciary in Environmental Compliance and Enforcement' (2012) 29 Pace Envtl. L. Rev 538–554, 540.

[91] Brian Preston, 'The Role of the Judiciary in Promoting Sustainable Development: The Experience of Asia and the Pacific' (2005) 9 Asia. Pac. J. of Env. L 109–211, 113–14.

[92] David B Hunter (n 65).

[93] Ibid.

[94] Ibid.

[95] *Urgenda Foundation v The State of the Netherlands* C/09/456689/HA ZA 13-1396 (24 June 2015).

[96] Ibid.

[97] Ibid.

impacts.[98] The Court specifically pointed out the stipulations included in the soft law mechanisms of UNFCCC and the Kyoto Protocol, relied on multiple international environmental principles including the 'no harm' principle; the doctrine of hazardous negligence; the principle of fairness; the precautionary principle and the norms of sustainable development to be considered when determining the state's duty to care relating to climate change.[99] The case marked the first successful attempt to elucidate state responsibility to reduce greenhouse emissions outside its statutory provisions[100] and marked the occasion where the court passed an order to limit GHG against a governmental branch.[101]

In 2010, 'the Global Legal Action on Climate Change (GLACC)' filed a petition in the Supreme Court of the Philippines requesting relief from the adverse effects of climate change and forcing the government to initiate steps under the statutory provisions about flood control and rainwater harvesting under the Republic Act 6716.[102] The case has not witnessed any progressive evolution and awaiting the decision.[103] If the emphasis of the litigation in the Netherlands was on tort law, the pending litigation in the Philippines is climate litigation based on human rights violation.[104] Greenpeace Southeast Asia and numerous other organizations and individuals filed a petition before the Commission on Human Rights of the Philippines to investigate the human rights consequences of climate change and also to assess the liabilities of carbon majors for violating the rights of people of Philippines bearing in mind the country's high vulnerability to the negative impacts of climate change.[105] They contend that the lack of statutory provisions cannot be a justification for the violation of human rights including the right to life to the highest attainable thresholds of physical and mental health, the right to adequate housing, the right to sanitation and water, the

[98]Ibid.

[99]Roger Cox, 'A Climate Change Litigation Precedent Urgenda Foundation v The State Of The Netherlands' (2015) CIGI Papers No 79, 9–10 www.cigionline.org/sites/default/files/cigi_paper_79.pdf accessed 1 January 2018.

[100]K J de Graaf and J H Jans, 'The Urgenda Decision: Netherlands Liable for Role in Causing Dangerous Global Climate Change' (2015) 27(3) JEL 517–527.

[101]Colombo, Esmeralda, 'Enforcing International Climate Change Law in Domestic Courts: A New Trend of Cases for Boosting Principle 10 of the Rio Declaration?' (2017) 35(1) UCLA Journal of Environmental Law and Policy 98–144, 131.

[102]Jolene Lin, 'Climate Change Litigation in Asia and the Pacific' in Geert Van Calster, Wim Vandenberghe and Leonie Reins, *Research Handbook on Climate Change Mitigation Law* (Edward Elgar Publishing 2015) 578–603, 596 (Global Legal Action on Climate Change v The Philippine Government) http://www.lse.ac.uk/GranthamInstitute/litigation/global-legal-action-on-climate-change-v-the-philippine-government/ accessed 10 April 2018.

[103]Ibid.

[104]Greenpeace Southeast Asia and Philippine Rural Reconstruction Movement, 'To the Commission on Human Rights of the Philippines Requesting for Investigation of the Responsibility of the Carbon Majors for Human Rights Violations or Threats of Violations Resulting from the Impacts of Climate Change' (Greenpeace) www.greenpeace.org/seasia/ph/PageFiles/105904/Climate-Change-and-Human-Rights-Complaint.pdf accessed 10 March 2018.

[105]Ibid, 7.

right to food and the right to self-determination.[106] The basis and origin of human rights can be traced back to natural law since the inception of humankind and have been in existence among the independent states irrespective of the states' capabilities and/or willingness to fulfil their international human rights obligations.[107] The case is directly significant as the petition explicitly points to the climate change-induced displacement and the inadequacy of international and national legal framework to address the rights violations. The report filed by the amicus curiae on 19 March 2018 has corroborated the contention of the petitioners on the critical inter-linkage between climate change and human rights and the extent of commission mandate to investigate the claims under the petition under national law.[108] The report recommends the Human Rights Commission to explore the various legal sources including tort laws and human rights laws in evaluating the responsibility of the carbon majors for the harms caused to the people of the Philippines resulting from the adverse impacts of the climate change.[109]

The case has inspired an era of climate litigation across the world. In 2015, some of the minor kids brought a climate case against the US government.[110] The petitioners argued that the government in spite of being fully aware of the human rights implications of climate change has failed to take appropriate steps to mitigate and reduce greenhouse emissions, and the current measures will not achieve a proportionate share of the fossil fuel emission reduction and violate the plaintiffs' right to protection from environmental degradation under the public trust doctrine.[111] The petitioners relied on the doctrine of public trust which protects the rights of present and future generations to the enjoyment of those essential natural resources that are of public concern to the citizens and the overarching public trust resource included the climate system, encompassing atmosphere, water and oceans, and the government has a responsibility to maintain the atmosphere free of 'substantial impairment.[112] In its order dated 10 November 2016, the Court observed that 'exercising my "reasoned judgment," I have no iota of qualm or doubt that the right to a climate system capable and competent of sustaining human life is fundamental and central to a free and ordered society … a stable climate system is quite literally the foundation "of the society, without which there would be neither the civilization nor the progress" … (without "a balanced and healthful ecology," the posterity "stand to inherit nothing but desiccated and scorched earth incapable of sustaining the existence of life".[113] The

---

[106] Ibid.

[107] Ibid.

[108] In Re: National Inquiry On The Impact Of Climate Change On The human Rights Of The Filipino People CHR-NI-2016-0001 (19 March 2018) www.ciel.org/wp-content/uploads/2018/03/Joint-Summary-Amicus-submitted.pdf accessed 23 March 2018 (Joint Summary of the Amicus Curiae).

[109] Ibid.

[110] *Juliana v United States* Case No 6:15-cv-01517-TC.

[111] Ibid, 2.

[112] Ibid.

[113] Ibid, 32.

Court designated the public trust obligations of government as 'inherent attributes of sovereignty' and stated that 'the sovereign's public trust obligations prevent it from "depriving a future parliament to decide the contours of the natural resources and reserves that would be necessary to provide for the human well-being, existence, and survival of its citizens"'.[114] The date for trial was fixed for February 2018 and in July 2017,[115] the administration filed a writ of *mandamus* seeking to bar the proceedings. The arguments on the matter took place in December 2017, and the case awaits the judges' determination.[116]

The above-cited cases are representative of the climate change litigation in general and can be employed and extended to situations of climate-induced displacement. New Zealand has been faced with matters directly about climate-induced displacement. In an extraordinary case, Mr. Ioane Teitiota and his wife moved to New Zealand from their homeland in Kiribati in 2007 and remained after their permits expired in October 2010.[117] To avoid deportation for the extended illegal stay in New Zealand, Mr. Tettionta and his family applied for the refugee status under the New Zealand Immigration Act, which incorporates the 1951 Refugee Convention within the national law.[118] The core argument raised by Mr. Teitiota was that the extreme climate change situation in Kiribati makes it impractical and impossible for them to return. It was claimed in the application that their right to the environment could no longer be protected in Kiribati due to rising sea level, and hence, they should be allowed refugee status on the grounds of climate change. The refugee officer denied this claim and subsequently by the Immigration and Protection Tribunal (IPT).[119]Although recognizing the future scope of protection of climate refugees, the tribunal held that the present case dealt with voluntary adaptive migration and was not recognized as a ground for refugee protection under the New Zealand legal regime. The IPT also took into account the human rights aspect under international law and said that the present nature of climate change in Kiribati was not so parlous as to put the applicant's life in jeopardy. An appeal to the High Court and later to the Court of Appeals was dismissed on similar grounds, though the courts expressed sympathy towards the people of Kiribati,[120] and in July 2015, the Supreme Court of New Zealand endorsed the decisions of the lower courts and rejected Mr. Teitiota's application for leave to appeal[121] citing similar cases where the plea of refugee status

[114]Ibid, 49.

[115]Joel Stronberg, 'Julaina versus US: For Children of all ages' (Resilience) www.resilience.org/stories/2017-12-14/juliana-vs-u-s-children-ages2/ accessed 23 March 2018.

[116]Ibid.

[117]*Teitiota v Chief Executive of the Ministry of Business Innovation and Employment* [2015] NZSC 107, 4.

[118]Ibid 5.

[119]Ibid 6.

[120]*Teitiota v The Chief Executive of the Ministry of Business, Innovation and Employment* [2013] NZHC3125; *Teitiota v The Chief Executive of the Ministry of Business, Innovation and Employment* [2014]NZCA 173.

[121]Teitiota (n 117).

on this ground had been refused.[122] It was held that the case did not raise arguable questions of law or public importance and the petition could not establish the presence of severe harm in Kiribati, or the fact that the government of Kiribati was unable to provide adequate protection to citizens affected by environmental degradation.

The fundamental question raised by Mr. Teitiota was whether the international law considers 'environmental refugee' as qualifying for protection under article 1A(2) of the Refugee Convention.[123] The Court, while rejecting the contention advanced by Mr. Teitiota that for the Refugee Convention the term 'being persecuted' does not always require a human agency, observed that the jurisprudence developed around the term persecution requires human agency depicted through failure of state protection necessitating the systematic and perennial violation of fundamental human rights.[124] McAdams observes that even if it could be argued that climate change and its devastating consequences do amount to persecution, the grounds of persecution have to be in consonance with the 1951 UN Convention grounds including race, religion, nationality and membership of a particular social group or political opinion.[125]

The Court held that the Refugee Convention has to be assessed in the background of its historical development and purpose.[126] The Court remarked that there is no evidence to demonstrate that the government of Kiribati has plummeted to take steps to protect its citizens from the environmental degradation and repatriating Mr. Teitiota does not expose him to serious harm.[127] However, the Supreme Court did not categorically negate the contention that 'environmental degradation stemming from climate change or the other environmental or natural disasters could never formulate a gateway into the Convention or protected person jurisdiction'.[128] Thus, a positive aspect of the case is the observation of the ITP, the High Court and the Supreme Court that the outcome of this case did not rule out the future possibility

---

[122]Refugee Appeal No 72185 [2000] NZRSAA 335 (10 August 2000); Refugee Appeal No 72189 [2000] NZRSAA 355 (17 August 2000); Refugee Appeal No 72315 [2000] NZRSAA 493 (19 October 2000) (refusal by New Zealand Refugee Status Appeals Authority to grant refugee status to Tuvalu citizens on grounds of environmental and economic hardship); *Minister Mohammed Motahir Ali v Minister of Immigration, Local Government and Ethnic Affairs* [1994] FCA 887 (4 February 1994) (refusal by Australian Supreme Court to grant refugee status to Bangladeshi immigrant on grounds of hardship due to climate change).

[123]*Teitiota* (n 117) 11.

[124]Ibid 8; Mark Baker Jones and Melanie Baker-Jones, 'Teitiota v The Chief Executive Of Ministry Of Business, Innovation And Employment—A Person Displaced' (2015) 15(2) QUT L. Rev. 102–121, 114.

[125]Harriet Farquhar, 'Migration with Dignity: Towards a New Zealand Response to Climate Change Displacement in the Pacific' (LLB (Honours) Degree, Faculty of Law, Victoria University of Wellington 2014) 4–5 http://researcharchive.vuw.ac.nz/xmlui/bitstream/handle/10063/4302/thesis.pdf?sequence=2.

[126]Christel Cournil, 'The inadequacy of International Refugee Law in Response to Environmental Migration' in Benoît Maye and François Crépeau (Eds), *Research Handbook on Climate Change, Migration and the Law* (Edwar Elgar 2017) 85–107, 99.

[127]Teitiota (n 117) 12.

[128]Ibid, 13.

of granting recognition to climate refugees and such protection might be accorded in an appropriate case.

In the case of the AD (Tuvalu), a humanitarian approach was taken by the Immigration Tribunal.[129] The petitioner and his wife arrived in New Zealand in 2007 and decided to seek refuge as they felt the consequences of climate change makes it difficult for them to stay in Tuvalu.[130] They have already endured and suffered as the lack of adequate medical services in Tuvalu has caused the petitioners to lose two babies at the last stages of pregnancy.[131] Since they have failed to obtain protection under the Refugee Convention, an application on humanitarian grounds under the Immigration Act, 2009 was filed claiming that they 'would be deprived of their ability to have "a safe and fulfilling life" if forced back to Tuvalu because of the effects of climate change.'[132]

For a successful appeal on the grounds of humanitarian consideration, the petitioner was obligated to satisfy that (a) there are extraordinary scenarios of a humanitarian nature that would make it deleterious or biased for the appellant to be involuntarily exiled from New Zealand and (b) it would not in all the conditions be contradictory to the public interest to allow the appellant to remain in New Zealand.[133]

The tribunal while considering the fact that his children's are born in New Zealand and their young age makes them vulnerable to the negative impact of climate change. His close family bonds including his aged mother in New Zealand have referred to climate change not as a decisive ground for granting protection.[134] Media and academics have celebrated the decision as a positive assertion of climate refugee status. However, McAdam opines that the decision does not represent the first successful climate change refugee case since the tribunal explicitly endorsed that the applicant is not a refugee within the meaning of the UNCSR; (b) are not protected persons within the meaning of the Convention Against Torture; (c) are not protected people within the meaning of the Covenant on Civil and Political Rights.[135]

At the domestic level, climate litigation has met with varying degree of results. However, the recent experiences suggest a proactive role on the part of the judiciary to lay down guidelines and obligations on the governments to reduce greenhouse emissions. At the same time except in New Zealand, litigations dealing directly with climate displacement/migration have been absent. In the case of South Asia, the judiciary has been a potent force in the acceptance and expansion of environ-

[129]*AD (Tuvalu)* [2014] NZIPT 501370-371 4.

[130]Ibid 28.

[131]Ibid 31.

[132]Hélène Ragheboom, *The International Legal Status and Protection of Environmentally-Displaced Persons: A European Perspective* (Brill Publishers 2017) 405.

[133]Immigration Act 2009, s 207.

[134]Ibid; Christel Cournil (n 126), 94; see Elisa Formale, Jeremie Guelat and Etienne Piguet, 'Framing Labour Mobility Options in Small Island States Affected by Environmental Changes' in Robert McLeman, Jeanette Schade and Thomas Faist (Eds), *Environmental Migration and Social Inequality* (Springer 2015) 9, 167–188, 179.

[135]Jane McAdam, 'The Emerging New Zealand Jurisprudence on Climate Change, Disasters and Displacement' (2015) 3(1) Migration Studies 131–142, 137.

mental protection. The enforcement and fulfilment of environmental protection have been implemented within a constitutional framework of fundamental rights mainly through the expanded notion of a right to life and incorporation of international environmental principles.[136] The following section will analyse the evolving environmental jurisprudence and climate jurisprudence if any and discuss whether the judicial decisions can form the basis of protective measures for climate displacement.

6.4 Bangladesh

In Bangladesh, right to a clean environment has been enunciated by the judiciary as part of the broader interpretation of the constitutional right to life in the landmark case of *M. Farooque v. Bangladesh and Others*.[137] The Court acknowledged the *locus standi of* a public-spirited person, civil society institution or organization to file cases on behalf of others who are unable to do so because of their inability to gain access to the justice system.[138] The Court took the assistance of article 31 and article 32 of the Constitution of Bangladesh whereunder 'every citizen has the right to protection from action detrimental to the life liberty, reputation, body or property' unless these are taken by law.[139] However, the article 32 states: 'No person shall be deprived of his life or personal liberty except by law'.[140] These two articles jointly constitute the fundamental right to life, and the right to life also includes the right to a healthy environment that is capable of supporting the growth of a meaningful existence of life.[141] Thus, the right to life was construed as including a right to a decent environment. The case was followed by some PILs, which played an instrumental role in the development of environmental jurisprudence in Bangladesh.[142] However, despite repeated interventions by the Court, the government on many occasions has been reluctant to implement decisions which do not bring long-term results.[143]

In the case of *Dr. Mohiuddin Farooque v. Bangladesh and Others*,[144] the Court was seized with the matter of Flood Action Plan and examined the need for incorporation of the precautionary approach. However, the threshold of the seriousness of such damage was not examined, and the court did not accept it as entrenched in the

[136]Parvez Hassan and Azim Azfar, 'Securing Environmental Rights through Public Interest Litigation in South Asia' (2004) 22(3) Va. Envt.l L. J 215–247, 219.

[137]*M Farooque v Bangladesh and Others* (1997) 49 DLR (AD) 1.

[138]Ibid.

[139]Constitution of People's Republic of Bangladesh 1972, art 31.

[140]Ibid, art 32.

[141]Ibid.

[142]Md Saiful Karim and others, 'Legal Activism For Ensuring Environmental Justice' (2012) 7(1) ASJCL 1–46, 14 https://eprints.qut.edu.au/61471/4/61471.pdf accessed 23 March 2018. See generally Saiful Karim, Shipbreaking in Developing Countries: A Requiem for Environmental Justice from the Perspective of Bangladesh (Routledge 2017) 112.

[143]Ibid, 37–38.

[144]*Dr. Mohiuddin Farooque v Bangladesh and Others* WP No 92 of 1996.

domestic law.[145] Yet, the Court incorporated the concept of sustainable development as a concept of balancing and integrating the quality of life that is economically and ecologically sustainable concept albeit in an indirect way.[146]

In the year 2000, Bangladesh enacted new legislation to establish a separate court to adjudicate environmental pollution cases.[147] In 2003 and 2005 at Dhaka, Chittagong and Sylhet three Environment Courts (ECs) were established.[148] Though environmental problems were multiplying yearly, the number of institution and disposal of cases by these courts were severely handicapped. Eventually, the Courts established under this Act failed to fulfil the purpose. While taking it into consideration the dismal performance of the Courts, the government of Bangladesh repealed the Act and passed the Environment Court Act (ECA) 2010.[149] Under ECA 2010 Courts can adjudicate offense and claims of compensation arising under 'environmental law'.[150] According to ECA 2010, 'environmental law' is defined as Bangladesh Environment Conservation Act, 1995 and other laws subsequently specified by the gazette notification. However, in spite of the multitude of environmental legislations, no gazette notification has been issued to include any other Act within the jurisdiction of EC as 'environment law'. Further access to environmental justice refers to effective judicial and administrative remedies and procedures available to a person who is aggrieved or likely to be aggrieved by environmental harm.[151] ECA 2010 has made an aggrieved person's right to access to justice subject to the inspection report of the department of environment (DoE). This restrictive approach is a hindrance to environmental justice.[152] After all access to justice is the foundation of human rights, without which the positive realization of other human rights becomes arduous.[153] In terms of climate change, no cases have been brought before the judicial forum in Bangladesh.

---

[145]Saiful Karim and others (n 142) 24.

[146]Zona Razaque, 'Access to Environmental Justice Role of the Judiciary in Bangladesh' (2000) 4(1) BJL 1, 25, 12.

[147]Environment Court Act 2000, Preamble (Bangladesh).

[148]Abu Bakar Siddique, 'Environment Court a Failure' Dhaka Tribune (2 April 2016) www.dhakatribune.com/bangladesh/2016/apr/02/environment-court-failure accessed 15 April 2016.

[149]'Environment Court in Every District' Bangladesh News (04 October 2010) www.independent-bangladesh.com/2010100412013/country/environment-court-in-every-district.html accessed 15 April 2016.

[150]Environment Court Act 2000 (n 148) s 6 & 7.

[151]Robert Kibugi, 'Enhanced Access to Environmental Justice Kenya' in Jamie Benidickson and others (eds), *Environmental Law and Sustainability after Rio* (Edward Elger 2011) 158–178, 159.

[152]Brian J Preston, 'Characteristics of Successful Environmental Courts and Tribunals' (2014) 26(3) J 365–393, 368.

[153]Robert Kibugi (n 151) 158.

6.5 India

Indian judiciary has been hailed as one of the most powerful judiciaries in the world.[154] According to Gadbois, 'The power of 'Judicial Review' enabling the courts to set limits, not only on executive actions but also on legislative power, both in theory and practice makes the Indian Supreme court wield a political power unknown elsewhere'.[155] Judicial review has been recognized as one of the basic features of the Constitution of India.[156]

In the aftermath of independence till the Declaration of emergency phase, the Supreme Court of India adopted a positivist approach and followed restrictive interpretation.[157] Bhopal gas disaster provided an occasion for the judiciary to reinvent, and the judiciary was eager to play an activist institutional role.[158] The Supreme Court adopted an ambitious interpretation and innovatively brought in the principle of absolute liability.[159]

The principle of 'absolute liability'[160] was truly an indigenous innovation and a unique contribution made by the judiciary when international legal systems and domestic legal systems are grappling with fixation of corporate responsibility. The Constitution of India did not explicitly provide for the right to 'clean environment'. Starting with the landmark case of *Maneka Gandhi v Union of India,*[161] the Courts deviated from the narrow and prismatic approach to wider and sketchier interpretation to 'right to life' and progressively expanded the ambit of the right to life under article

---

[154]PTI News, 'Indian judiciary one of most powerful in world: CJI Altamas Kabir' *Economic Times* (17 June 2013) https://economictimes.indiatimes.com/news/politics-and-nation/indian-judiciary-one-of-most-powerful-in-world-cji-altamas-kabir/articleshow/20634022.cms accessed 21 April 2018.

[155]Quoted in CM Abraham, '*Environmental jurisprudence in India*' (Martinus Nijhoff Publishers 1999) 33.

[156]Stellina Jolly, 'Ninth Schedule, Basic Structure, and Constitutionalism: Issues and Perspectives' (2007) 48 PULR 51 (The basic structure doctrine was propounded in the case of *Kesavannda Bharati v State of Kerala* (1973) 4 SCC 225. The doctrine puts limitation on the amending powers of Parliament and states that there are certain basic features of the Constitution which cannot be amended. In *Raj Narain v Indira Gandhi* (1975) AIR 865 judicial review was recognized a constituting basic structure to the Constitution).

[157]Manoj Mate, 'Globalisation, Rights and Judicial Activism in The Supreme Court of India' (2016) 25 Pac Rim L & Pol'y J 643–671, 647; Armin Rosencranz and Michael Jackson, 'The Delhi Pollution Case: The Supreme Court of India and the Limits of Judicial Power' (2003) 28 Colum J Envtl L 223–254, 229; Lennon Banks Haas, 'Saving the Trees One Constitutional Provision at the time: Judicial Activism and Deforestation in India' (2011) 40 Ga J Int'l & Comp L 751–779,764; see also Ayesha Dias, 'Judicial Activism in the Development and Enforcement of Environmental Law: Some Comparative Insights from the Indian Experience' (1994) 6(2) JEL 243–62.

[158]Bharat H Desai, 'The Bhopal Gas Leak Disaster Litigation: An Overview' (1993) 3 Asian Y B Int'l L 163–179, 168.

[159](1868) LR 3 HL 330; See Dr SK Kapoor, *Law of Torts and Consumer Protection* (9th edn, Central Law Agency 2013) 223.

[160]The 'absolute liability' principle was reiterated in the case of Bhopal Judgment *Union Carbide Corporations v Union of India* (1990) AIR SC 273.

[161]1978 AIR SC 597.

21[162] to include the 'right to a clean environment' as a fundamental right.[163] Judiciary furthered the environmental jurisprudence by adopting various environmental principles identified at the multitude of international Declarations and Conventions.

One of the most fascinating features of environmental jurisprudence in India has been the liberal use of Public Interest Litigation (PIL) and expansion of locus standi.[164] PIL, considered as the incarnation of people oriented and focused judicial litigation, became the vehicle through which substantive justice was brought to millions of marginalized individuals and communities affected by developmental projects.[165]

Further, in India, the relevance and significance of environmental issues were considered,[166] and specialized environmental courts were recommended by judicial decisions[167] as well as by the 186 Law Commission of India recommendation.[168] Based on these interventions, National Green Tribunal (NGT) was set up by the Parliament of India to hear issues involving substantial environmental problems.[169] NGT has the jurisdiction to review government decisions on projects affecting the environment, including environmental clearances (ECs) and the grant of consent or licence to operate an industry. The NGT is empowered to enforce any legal right relating to the environment and is enjoined to follow internationally recognized envi-

162See Arun K Thiruvengadam, 'In Pursuit of "The Common Illumination of Our House"; Trans-Judicial Influence and The origins of PIL Jurisprudence in South Asia' (2008) 2 IJCL 66–103, 94.

163Shubhankar Dam and Vivek Tiwari 'Polluting Environment, Polluting Constitution: Is a 'Polluted' Constitution Worse than a Polluted Environment?' (2005) 17 JEL 383–393; see Kuldip Singh, 'Environmental Protection—The Role of Judiciary' (2004) CULR 15–24; Michael R Anderson, 'Individual Rights to Environmental Protection in India' in Alan Boyle and Michael Anderson (Eds), *Human Right Approaches to Environmental Protection* (Oxford: Clarendon University Press 1998) 199–225, 201.

164Origin of PIL could be traced to the historic case of the *Hussainara Khatoon v State of Bihar* 1979 AIR 1819, decided in 1979, which lead to case led to the release of more than 40,000 under trial prisoners in the Indian state of Bihar. The concept got a filling with the decision in *S.P. Gupta v Union of India*, 1981. This is in stark contrast to the European Union, where Public Interest Litigation by non-governmental organisations ('NGOs') is discouraged by adopting restrictive rules on standing: See Bilun Muller, 'Access to the Courts of the Member States for NGOs in Environmental Matters under European Union Law' (2011) 23(3) JEL 505–516. For a discussion on *locus standi*, see Susan D Susman, 'Distant Voices in the Courts of India: Transformation of Standing in Public Interest Litigation' (1994–95) 13 Wis Int'l LJ 57–101.

165J Mijin Cha, 'A Critical Examination of the Environmental Jurisprudence of the Courts of India' (2005) 10 Alb L Envtl Outlook J 197, 199–204.

166See generally, Stuart Bell, Donald Mcgillivary and Ole W Pederesen, *Environmental Law* (8th edn, Oxford University Press 2008) 350 (for a discussion on specialized environmental courts and its significance).

167*MC Mehta v Union of India* 1986 (2) SCC 176; *Indian Council for Environmental-Legal Action v Union of India* 1996(3) SCC 212; *AP Pollution Control Board v MV Nayudu* 1999(2) SCC 718.

168Law Commission of India, *Proposal to Constitute Environmental Courts* (186th report 2003).

169Gitanjali Nain Gill, 'Environmental Justice in India: The National Green Tribunal and Expert Members' (2016) 5(1) TEL 175–205 (established under the 2010 National Green Tribunal Act, the NGT is a specialized body equipped with the necessary expertise to effective and expeditious dispose environmental disputes involving multidisciplinary issues).

ronmental principles such as sustainable development, the precautionary principle and the 'polluter pays' principle while issuing an order, decision or award.[170]

In case of climate litigation, various legal avenues are open for the petitioners, including actions for nuisance, negligence, writ petition, redressal under statutory provisions.[171] However, the section will not deliberate on the possible options, but the focus will be on the jurisprudence which has evolved and or evolving with regard to climate change and environmental protection which is relevant in the context of climate change refugee protection. About climate change, several cases have highlighted the significance and the necessity to initiate legislative and policy measures. As Rajmani points out, the issue of climate change was underscored in various cases by petitioners and respondents.[172]

In the case of *Goa Foundation v. Goa State Coastal Zone,*[173] *the court observed that 'As such, there cannot be any doubt that there should be sustainable development and there should be a (sic) and harmonious co-existence of nature's bounty and development of the humankind. One should not grieve or be sacrificed at the cost of the others. It should never be overlooked that man only holds leasehold rights on this earth and posterity should not be divested of the primeval glory of the earth. The State of Goa and its entire coastline is full of dunes. the sand dunes do play a central role in the protection of the vicinity, in as much as the sand dunes as a sentinel against any destructive cyclones, rising water level of the sea due to a global increase in temperature. Sand dunes also play an essential role in protecting the agricultural and the potable water supply of the hinterland coming from springs and wells. Dunes are thus purely nature's gift to the coastal regions, Their protection is, therefore, necessary and they have rightly placed in CRZ I category. At the same time, it cannot be neglected that we have come a long way since the Stockholm conference of 1972. Tourism all over the world has been being a flourishing industry and a major source of revenue for the State. The balancing interest, therefore, between the preservation of ecology and environment and development of tourism has to be congenially maintained'.*[174]

In *B. B. Nalwade v. Ministry of Environment and Forests*,[175] the issue of climate change issue was considered indirectly. The NGT upheld the grant of environmental clearance for a coal-based thermal plant on the ground that the project operated within an ecolegal framework and contributed significantly to sustainable industrial development.[176] The Court referred to the doctrine of sustainable development as a

---

[170] National Green Tribunal Act 2010, s 20 (India).

[171] Arindam Basu, 'Climate Change Litigation In India: Seeking A New Approach Through The Application Of Common Law Principles' (2011) 1 Env. L. & Practice Rev. 34–50, 38.

[172] Lavanya Rajamani, 'Rights Based Climate Litigation in the Indian Courts: Potential, Prospects & Potential Problems' (2013) Centre for Policy Research, Climate Initiative Working Paper 2013/1(May) https://ssrn.com/abstract=2464927 accessed 24 April 27, 2018.

[173] *Goa Foundation v Goa State Coastal Zone Management* 2001 (4) Bom CR 226.

[174] Ibid [39].

[175] Appeal No 21 of 2011, Judgment dated 29 November 2011.

[176] Gitanjali Nain Gill, 'The National Green Tribunal Of India: A Sustainable Future Through The Principles Of International Environmental Law' (2014) 16(3) Environmental Law Review 183–202.

balancing concept aimed at ensuring a better living, social and economic conditions for human beings and on the other hand providing the environmental protection.[177]

The Court observed, 'production of electricity is very essential for industrial growth apart from domestic need. In the light of the existing power scenario in the country, the project under consideration when operated within the eco-legal frame work may contribute significantly to sustainable industrial development in the area under consideration'.[178]

In *Indian Council for Enviro-legal Action (ICELA) v. Ministry of Environment, Forest and Climate and Others*,[179] the applicant, a registered voluntary organization working for the protection of the environment, had applied to have directions issued to the respondents to check the production of HCF-23.[180] As per the applicant, such guidelines and directions need to be issued as HCF-23 is being generated as a by-product while manufacturing of HCF-22 has serious climatic impacts. However, the Ministry of Environment and Forest (MoEF), one of the respondents, submitted that the subject matter of the application is a matter of global concern and a policy that is continuously evolving in commensurate with the needs of the hour required to be framed as there is no domestic law/rules regulating this aspect.[181] The ministry also stated that HCF-23 is not a pollutant or a toxic gas which is harmful to human health, and it is just one of the greenhouse gases identified under the UNFCCC.[182] The Montreal Protocol already regulates HCFC-22 and HFC-23 internationally. Therefore, it did not come under the purview of the NGT, and thus, the application should be dismissed.[183]

Rejecting the contention of the MoEF, the NGT held that HFC-23 may or may not be a pollutant per se. However, it will not be correct to contend that the tribunal would have no jurisdiction. Once it is a part of the greenhouse gases, which admittedly are responsible for causing global warming there appears to be no reason as to why the tribunal will not exercise its jurisdiction on the principles stated in Section 20 of the NGT Act, 2010 read with Section 2(a) of the Environment Protection Act of 1986.[184]

Coming to the merits of the case, the tribunal stated that there is no domestic law regulation in place to regulate any of the facets of HFC-23 in the country and it is entirely governed internationally by the UNFCCC and Kyoto Protocol.[185] The tribunal held that the contents of the application are a matter of global policy, and therefore, there would be a minimal role for the statutory authorities within the country to take appropriate measures. The international Convention and treaties have

---

[177] *BB Nalawade v Ministry of Environment and Forests*, NGT Appeal No 21 of 2011 [22].

[178] Ibid.

[179] *Indian Council for Enviro-legal Action (ICELA) v Ministry of Environment, Forest and Climate and Others* NGT Order dated 10 December 2015.

[180] Ibid [2].

[181] Ibid [7].

[182] Ibid [8].

[183] Ibid [7].

[184] Ibid [21].

[185] Ibid [23].

to provide a path for domestic legislation, and in any case, it has failed, and it has to be regulated without further delay.[186]

An application was filed by Ratandeep Rangari alleging inaction by environmental regulatory authorities of the Ministry of Environment and Forest (MoEF) and Maharashtra Pollution Control Board (MPCB) regarding violation of the terms and conditions of consent granted to the Maharashtra State Power Generation Company Limited (MAHAGENCO) and the resultant environmental pollution.[187] He claimed that his village is affected by the air pollution resulting from the nearby coal-based power stations. The applicant stated that the Vidarbha region already has the most significant power generation capacity in Maharashtra, and such a large-scale power generation concentrated over the Vidarbha region would cause severe environmental degradation if the environmental norms are not strictly enforced.[188] He claimed that the coal supplied by these coal companies violates Environment Protection Rules, 1986 that has stipulated maximum coal ash content of 34% on the annual average basis for such coal-based thermal power plants.[189] The tribunal held that the rules brought by the MoEF are a pragmatic and proactive initiative to ensure generation of 'clean energy regarding less ash handling/disposal and stack air emissions.[190] The important co-benefit of such an initiative would be lesser GHG emissions, i.e. lesser carbon footprint in thermal power generation.[191]

In *Tamil Nadu Newsprint And Papers Ltd. v. Tamil Nadu Electricity Regulatory Commission,*[192] the appellant, in this case, has highlighted issues relating to power purchase and allied issues linked with non-conventional energy source-based generation and co-generation plants.[193] The tribunal instructed the commissions to give proper direction to the power sector by framing regulations to increase the generational capacity of these plants without causing harm to the environment.[194] The tribunal referred to the 2007 assessment report of the IPCC that revealed that the glaciers and the snow cover have receded, level of oceans has risen, and sea water has expanded due to the global warming contributed by the use of fossil fuels.[195] The tribunal while highlighting the environmental friendly facets of the Electricity Act, 2003, observed that small steps need to be taken to reduce dependence on fossil fuels to the extent possible without hampering the progress of electricity sector.[196]

---

[186]Ibid [23].

[187]Application No 19/2014 (WZ)AND MA No66/2015 (The National Green Tribunal (Western Zone) Bench Pune).

[188]Ibid [1].

[189]Ibid [2].

[190]Ibid [34].

[191]Ibid [34].

[192]*Tamil Nadu Newsprint And Papers Ltd v Tamil Nadu Electricity Regulatory Commission* 2007 ELR (APTEL) 157.

[193]Ibid [1].

[194]Ibid [17].

[195]Ibid [11].

[196]Ibid [11].

In *Narmada Bachao Andolan v. Union of India*,[197] the petitioners had sought the Supreme Court to review the entire Sardar Sarovar Dam project and to examine the overall costs including the environmental costs of this project.[198] The Court, in this case, gave a green signal to the government to continue with the construction of the Sardar Sarovar Dam as per the award of the Narmada Water Disputes Tribunal.[199] Thermal power projects that are being relied upon have been contributing towards global warming by introducing green house gases into the atmosphere. The emission of greenhouse gases has become a cause of global concern that has also resulted in the restriction of sanction of funds for thermal power projects by the World Bank.[200] Due to the hydel power project's negligible contribution to the greenhouse effect, the Court terms it as ecologically friendly and even economically more viable.[201] Keeping in mind these considerations, the Court upholds the tribunal's award even though the hydel projects cause comparatively more displacement. The issue of climate change and environmental friendliness of hydroelectricity was given prominence even over the possible displacement of population brought by the dam construction.[202]

In *Nar Bahadur Bhandari and others v. The State of Sikkim and ors*,[203] the petitioner had filed the petition challenging the construction of a hydroelectric power plant on the River Teesta.[204] Though, the Court dismissed the petition as devoid of merits and want of legal contentions, the court referred to an affidavit filed by the Ministry of Environment and Forests wherein it highlighted the severe peak power deficit facing the nation and advocated the promotion of hydropower projects as these power projects unlike the thermal power projects do not depend upon the combustion of fossil fuels that in turn contribute enormously towards atmospheric pollution and global warming and thus are environmentally more viable.[205]

In *Mohd. Salim v. Uttarakhand and or,*[206] the High Court of Uttarakhand exercised its *parens patrie* jurisdiction to declare rivers Ganga and the Yamuna as juristic/legal persons/living entities having the status of a legal person with all corresponding rights, duties and liabilities of a living person to preserve and conserve rivers Ganga and Yamuna.[207] The Director NAMAMI Ganga, the Chief Secretary of the State of Uttarakhand and the Advocate General of the State of Uttarakhand were declared *persons in loco parentis* as the human face to protect, conserve and preserve these rivers.[208] The Uttarakhand High Court directed the central government to constitute a

[197]*Narmada Bachao Andolan v. Union of India AIR 2000 SC 3751.*

[198]Ibid.

[199]Ibid.

[200]Ibid.

[201]Ibid.

[202]Ibid.

[203]Writ Petition (C) No 40 of 2005 (Sikkim High Court).

[204]Ibid.

[205]Ibid.

[206]*Mohd Salim v Uttarakhand and ors* 2017 (2) RCR (Civil) 636 (Uttarakhand High Court).

[207]Ibid [19].

[208]Ibid [9].

Ganga Management Board and make it functional within 3 months.[209] The judgment could prove potentially useful in future climate litigations as climate change will have a significant consequence on the quality as well as the quality of the availability of water and water bodies.

Ridhima Pandey v. UOI and CPC is the first case directly about climate change.[210] This application was filed under Section 14 of the National Green Tribunal Act, 2010 broaching substantial questions of the environmental problems where the community at large is adversely affected and is likely to be influenced by the adverse environmental consequences.[211] It was submitted that the applicant is part of a societal class that among all Indians, is most vulnerable to climate changes in India and yet is not the part of the decision-making process having regard to the nature of the climate change threat, children, and their caregivers have no pragmatic way of protecting themselves from the dangerous situation in which states have placed them.[212]

The application asserted that the government of India has failed in fulfilling its duty of mitigating climate change impacts. The applicant sought the intervention of the National Green Tribunal to direct the respondents to take effective, science-based action to reduce and minimize the adverse impacts of climate change in the country.[213] The applicant also pointed out the vast gap in the implementation of the environmental legislation. The applicant submitted that the state and its machinery are a trustee of vital natural resources that are central to the human survival and welfare dependency and are, therefore, bound in a fiduciary duty under the public trust doctrine to mitigate the sufferings caused by the climate change-induced displacement and migration to protect such resources for the utilization of the present and future generations.[214] As per the applicant, the decisions taken by the government that impact the essential natural reserves and resources must be based on the best contemporary climate science as it offers a road map for climate recovery that requires states to decrease atmospheric CO_2 levels below the 350 particles per million (ppm) by the year 2100 and stabilize global average temperature increase at not higher than the 1 °C for a long-term strategy.[215]

The arguments draw support from the fundamental rights granted under the Constitution of India, the public trust doctrine,[216] intergenerational equity[217] and alleges non-implementation of environmental laws in India. The application stated that the

---

[209]Ibid [3].

[210]*Ridhima Pandey v Union of India and Ors* OA No 187 of 2017 (Principal Bench NGT, New Delhi).

[211]Ibid [3].

[212]Ibid [1].

[213]Ibid [1].

[214]Ibid [2].

[215]Ibid [9].

[216]*MC Mehta v Kamal Nath* (1997) 1 SCC 388 (first mention of the doctrine by the Indian judiciary); see also, *MI Builders Pvt Ltd v Radhey Shyam Sahu* AIR 1999 SC 2468; *Th Majra Singh v Indian Oil Corporation* AIR 1999 J&K 81 (Jammu & Kashmir High Court, India).

[217]*MC Mehta v Union of India and others* (1997) 11 SCC 312; *ND Jayal v Union of India* (2004) 9 SCC 362.

impacts of climate change include augmented harm to human health, inferior agricultural yields, droughts, the rise in sea level and reduced availability of fresh water.[218] However what stands out is the mention of the climate refugees, as a result of environmental degradation, in the petition. The applicant stated that India already has climate refugees and supports her argument by referring to various newspaper articles.[219] The application related to a Times of India December 2009 article titled 'Hungry Tide, Homeless People' which stated that since the 1970s the area of Ghoramara Island in the Sunderbans area has shrunk from 9 km^2 to less than 3.7 km^2, while two smaller islands nearby have already sunk beneath the sea.[220] The Guardian reported that two-thirds of the Ghoramara Island's population have been forced to leave. The Bay of Bengal region has been witnessing the creation of tens of thousands of climate refugees due to the rapid disappearance of other islands.[221] The petition stated that rising sea levels have created numerous climate refugees in the Bay of Bengal region, primarily affecting the people from Bangladesh, who have taken shelter in India as a result. It is argued that such movement not only has social and human rights implications but also causes stress to economy and damage to national property.[222]

The application discussed India's obligations under the Paris Agreements, 2015, India's policy on climate change. The Paris Agreement, 2015 recognizes that climate change is a common concern of humankind and identifies the substance of the conservation and expansion of carbon sinks and reservoirs of the greenhouse gases depicted.[223] The agreement directed to make stronger the global response to the menace of climate change by maintaining the increase in the global normal temperature to comfortably below 2 °C beyond pre-industrial levels and following the efforts to restrict the temperature increase to 1.5 °C above pre-degree celsius.[224] The applicant has criticized these limits fixed in the Paris Agreement as to be a result of political negotiations and compromise and to be not based on science.[225] The applicant submitted that even though India ratified this agreement on 2.10.2016 and the country has committed to follow a low-carbon path, still India has not made any attempt to integrate its international obligations and commitments into the national framework of the law.[226]

The applicant is of the view that India's National Environmental Policy adopted in 2006 recognized that anthropogenic climate change would likely to have adverse

[218]Ridhima Pandey (210) 6.
[219]Ibid.
[220]Ibid.
[221]Ibid.
[222]Ibid.
[223]Ibid [12].
[224]Ibid.
[225]Ibid.
[226]Ibid [12].

impacts on India and found the following to be essential to the response of India to climate change[227]:

(a) Adherence to the principle of CBDR and respective capabilities of different countries in respect of both mitigations of GHGs and adaptation steps[228];
(b) Dependence on multilateral approaches contrary to bilateral or plurilateral or unilateral measures[229];
(c) Equal per-capita entitlements of global environmental resources to all countries;
(d) Embracing the over-riding priority of the right to development;
(e) Identify key vulnerabilities of India to climate change, in particular impacts on water resources, forests, coastal areas, agriculture and health;
(f) Assessing the need for adaptation to future climate change scenarios, and evaluating the scope for integrating these measures in relevant programmes, including coastal zone planning, watershed management, forestry management, health programmes, scientific advancement in agriculture and practice and regulation;
(g) Encouraging the Indian industry to partake in the Clean Development Mechanism (CDM) through the capacity building for identifying and planning the CDM projects, including the financial sector of the region;
(h) Participating in the voluntary partnerships with the countries of Global South and Global North while addressing the challenges of the sustainable development and climate change-induced displacement consistent with the provisions of the UNFCCC.[230]

The petition highlighted that a high-level advisory group on climate change, namely the Prime Minister's council on Climate Change, was constituted in June 2007 and reconstituted in November 2014 with the following objectives[231]:

(i) Coordinate national action Plans for assessment, adaptation, and mitigation of climate change.
(ii) Advise government on proactive measures that can be taken by India to deal with the challenge of climate change.
(iii) Facilitate inter-ministerial coordination and guide policy in relevant areas.[232]

The petition elaborately pointed to the National Action Plan on Climate Change, which has been implemented through eight national missions.[233] The applicant submits that even though the Union of India has proclaimed several initiatives to handle climate change crisis, however, none of them have been translated into effective measures.[234] There are various environmental legislations and rules/notifications

[227] Ibid [13].
[228] Ibid [13].
[229] Ibid [13].
[230] Ibid [13].
[231] Ibid [14].
[232] Ibid [14].
[233] Ibid [15].
[234] Ibid.

made thereunder existing in India. Therefore, if these measures are effectively implemented in their true spirit, it would aid in handling the problem of negative climate change impacts as mentioned hereinabove.[235] These include the Acts mentioned in the Schedule-I of the NGT Act, 2010. The applicant stated that the Union of India is bound under article 48A to take effective and science-based measures to ensure that climate change impacts are mitigated and prevented. The applicant states that the right to life under article 21 would necessarily include the right to a healthy environment.[236] The petition is pending before the tribunal, and the issue has not been decided yet. However, going by the humanitarian stance of Indian courts towards the plight of refugees as well as environmental issues and given the fact that country has already incorporated and elevated environmental concerns to the paradigm of fundamental rights, one can hope that the tribunal shall take note of the matter and make provisions in favour of climate refugees.

6.6 Nepal

In Nepal, the judiciary has been progressively interpreting the constitutional and statutory provisions as including the protection of the environment. In *Suraya Prasad Sharma Dhungel v. Godavari Marble Industries and others*,[237] the petitioners filed a writ petition in 2001 against the decision of the government to grant permission for mining in the rich biodiversity Godavari region that has caused severe environmental degradation.[238] The petitioners contended that the impugned decision is inconsistent with the constitutional mandate of the protection of the environment.[239] The government contended that the development and destruction were interdependent issues. The course of development brings certain degree of destruction and the proper course of action would be to envisage proper management mechanism.[240] The main legal question was whether the right to clean environment is included as part of the right to life and whether there was locus standi of NGOs or individuals working for the protection of the environment. The Court interpreted that a clean and healthy environment was part of the right to life under article 11 (1) of the Constitution of the

---

[235]Ibid [21].

[236]Ibid.

[237]*Suray Prasad Sharma Dhungel v Godavari Marble Industries and others* WP 35/1992 (Nepal Supreme Court).

[238]Ibid.

[239]Ibid.

[240]Ibid; see, David R Boyd, *The Environmental Rights Revolution: A Global Study of Constitutions, Human Rights, and the Environment* (UBC Press 2011) 17–171.

Kingdom of Nepal 1990.[241] With regard to the *locus standi* of the petitioners, the Court stressed the public interest as a protectable fundamental right.[242]

Regarding climate change, the Court observed

> On account of the impacts caused by climatic changes, the whole earth and air have become heated more than two degree Celsius. Due to yearly increasing temperature, the life of some densely populated cities is becoming affected. Likewise, due to gradual melting, there has emerged increasing danger of white Himalayan ranges getting converted into black and ugly Mountains. On account of early increase in the sea level, there appears a probability of even habitation being exposed to risks. Ever increasing are the effects caused by untimely floods, landslides, and desertification resulting from drought, excessive rain or lack of rain. The sources of water are getting dried up whereas the available sources of water are becoming unusable having become polluted. The existence of marsh area has been exposed to risk, and air pollution and noise pollution have transgressed the limit of natural tolerance.[243]

The Court also highlighted the effort of the international community to fight climate change through the Convention on Climate Change and Kyoto Protocol. The Court also interpreted the emerging concepts/principles of international environmental law including sustainable development, precautionary Principle, polluter pays principles.[244] In the case of *Advocate Prakash Mani Sharma v. Cabinet Secretariat*, it was observed by the Court 'Private interest must yield to the public interest'. It is the constitutional duty of government, which is entrusted with the constitutional duty of the government of the state to implement every letter of the law to protect against the harms caused to public health by a polluted environment.[245] In *Yogi Narahari Nath and other v. Honourable Prime Minister Girija*,[246] an agreement was signed between Ministry of Education, Culture and Social Welfare and the International Society for Medical Education (USA) to establish a medical college in Nepal.[247] Due to the public interest associated with the sacred Devghat area, Supreme Court quashed the decision of the government. The doctrine of public trust was invoked and declared that protection of the environment is a primary responsibility of government.[248] With regard to climate change, PIL was filed in the case of *Prabindra*

---

[241] Shayami Puyimanasinghe, 'From a Divided Heritage to a Common Future, International Investment Law, Human Rights and Sustainable Development' in Shawkat Alam, Sumudu Atapattu, Carmen G Gonzalez and Jona Razzaque (Eds), *International Environmental Law and the Global South* (Cambridge University press 2015) 321.

[242] Alice Palmer and Cairo AR Robb, *International Environmental Law reports* (4th Volume, Cambridge University Press 1999) 321–330, 322.

[243] Ibid.

[244] Ibid.

[245] *Advocate Prakash Mani Sharma v Cabinet Secretariat* Writ No 3027/2059 (Supreme Court of Nepal).

[246] *Yogi Narahari Nath and other v Hon Prime Minister Girija Prasad Koirala and other* 33 NLR 1955 (Spreme Court Of Nepal).

[247] Ibid; see also, Stellina Jolly, 'A Legal Analysis of Linking Human Right Approach to Access to Water and Sharing of Trans Boundary Rivers in South Asia' in J L Kaul and Anupama Jha (Eds), *Shifting Horizons of Public International law in South Asia* (Springer, 2018) 135–158.

[248] United Nations Environment Programme, *Compendium of Summaries of Judicial Decisions in Environment Related Cases* (UNEP 2005) 162.

Shakya, Chairperson, Community Empowerment and Social Justice Foundation v. the State of Nepal requesting an interim order to not to proceed with the REDD mechanism violating the rights of communities till the time proper guidelines have been framed.[249]

The petitioner has pointed to the complex linkage between population growth and development and resultant climate change. The petition highlighted the major impact of climate change in Nepal including bursting of glaciers, the occurrence of floods and landslides.[250] The petitioner has pointed to the increasing role of REDD mechanism adopted to mitigate and adapt to climate change. The petitioner has highlighted incidents where REDD mechanism has been implemented without proper safeguard mechanisms in favour of indigenous and forest communities resulting in the violation of human rights.[251] The petition has highlighted that strategy report of Nepal has identified serious environmental risks of REDD especially in the process of lack of consultation and participation of the stakeholders in the decision-making. The petition has contended that the situation violates the constitutional guarantees of the right to live with dignity under article 16 (1), the right to preservation of cultural identity under article 32 and the right to environment guaranteed under article 30 of the Constitution of Nepal.[252]

6.7 Pakistan

Following the Indian example, Pakistan judiciary has been proactive in the case of environmental protection. In the case of *Shehla Zia v. WAPDA,*[253] judiciary in Pakistan interpreted the constitutional provisions dealing with 'right to life' under article 9 of the Constitution innovatively. In the case, some citizens of Islamabad forwarded a petition to the Supreme Court of Pakistan complaining about the construction of a grid station in their locality.[254] The Court formulated two questions for resolution.

- Whether any government agency has a right to imperil the lives of citizens by its actions without the consent of such citizens?
- Whether zoning laws vest rights in citizens, which could not be withdrawn or altered without the citizen's consent?[255]

---

[249] *Prabindra Shakya, Chairperson of Community Empowerment and Social Justice Foundation v State of Nepal* https://cemsoj.files.wordpress.com/2015/09/redd-writ-eng.pdf accessed 25 April 2018.

[250] Ibid.

[251] Ibid.

[252] Ibid.

[253] *Shehla Zia and Others vs. WAPDA* 1994 PLD SC 693 (Supreme Court of Pakistan).

[254] Ibid.

[255] Ibid.

The Court observed that 'article 14 of the constitution provides that the dignity of man subject to law, and the privacy of shall be inviolable.'[256] Article 9 guarantees the fundamental right to preserve and protect the dignity of man and right to 'life'.[257] Court held that 'if both are read together, the question will arise whether a person can be said to have the dignity of a man if his right to life is below bare necessity line without proper food, clothing, shelter education, health care, clean atmosphere and unpolluted environment'.[258]

In *Sindh Institute of Urology and Transplantation v. Nestle Milkpak Limited*,[259] the question concerned was the extent of exploitation of aquifer and its consequence on the environment. The Court while granting relief to the petitioner relied on the doctrine of public trust while drawing elaborately from the jurisprudence developed in India.[260] In the Lahore Canal Bank Road case,[261] the Supreme Court considered a challenge to a canal-widening project in Lahore, which the petitioner said threatened to cause environmental degradation to a green belt.[262] Supreme Court declared the Bambawali-Ravi-Bedian (BRB) Canal and the Canal Bank Road green belt from Jallo Park till Thokar Niaz Beg, a public trust. The court also instructed the provincial government to pass an Act through the Punjab Assembly to declare them a public trust.[263]

However, the cases which have relied on other international environment principles like sustainable development, precautionary principle is negligible in Pakistan.[264] *Asghar Leghari v. Federation of Pakistan* etc. marks the first case where climate concerns were agitated directly.[265] The petitioner challenged the indecisiveness, delay and mere procrastination on the part of government of Pakistan to cope with the challenges posed by climate change on the present and future generations.[266] Petitioner relied on National Climate Change Policy, 2013 and Framework for Implementation of Climate Change Policy (2014–2030) and contended that the respective authorities have failed to develop adaptive capacity and resilience to address climate change.[267]

---

256 Ibid.

257 Ibid.

258 Ibid.

259 2005 CLC 424 (Karachi).

260 Ibid; see also Waris Husain, 'Pakistan's Patch Work of High Court Justice', in Aparna Pande (Eds), *Routledge Handbook of Contemporary Pakistan* (Routledge 2017).

261 Suo Moto Case No. 25 of 2009 (Cutting of Trees for Canal Widening Project Lahore), 2011 SCMR 1743.

262 Wen-Chen Chang, Li-ann Thio, Kevin YL Tan and Jiunn-rong Yeh, *Constitutionalism in Asia: Cases and Materials* (Bloomsbury Publishing 2014) 1015.

263 2011 SCMR 1743.

264 Irum Ahsan and Saima Amin Khawaja, *Development of Environmental Laws and Jurisprudence in Pakistan* (Asian Development Bank 2013) 17.

265 Asghar Leghari (n 87).

266 Ibid; see James R May and Erin Daly, *Judicial handbook of Environmental Constitutionalism* (United Nations Environment Programme 2017) 119–120.

267 *Asghar Leghari v Federation of Pakistan and others* WP No 25501/2015 [4].

Petitioner highlighted the devastating impact of climate change on Pakistan and the floods faced by the nation.[268] He contended that the situation offends the fundamental rights under article 14 dealing with human dignity and right to life under article 9 of the Constitution of Pakistan. He submitted that international environmental principles like the doctrine of public trust, sustainable development, the precautionary principle and intergenerational equity form part of the fundamental rights.[269] The representatives of various ministries made brief submissions but could not satisfactorily show that adaptation measures as listed in the framework were seriously considered.[270] The Court observed that the interruption and lassitude of the State in executing the framework transgress the fundamental rights of the citizens that include the right to life and the right to human dignity, read with constitutional principles of democracy, equality, socio-economic and political justice that include within the idea of constitutionalism ambit and constitutional commitment.[271] The international environmental principles of sustainable development, the precautionary principle, environmental impact assessment, inter- and intra-generational equity and public trust doctrine were accentuated upon by the Court for a 'move to climate change justice in the country.[272]

The Court observed '*environment and its protection have taken center stage in the scheme of our constitutional rights. It seems that we have to move on. The present environmental jurisprudence has to be shaped to cater the needs of something more urgent and overpowering, i.e., Climate Change. From environmental justice, who was largely localized and confined to our ecosystems and biodiversity, we require to move to climate change justice. The Fundamental rights in the Constitution of Pakistan set at the foundation of these two coinciding justice systems. The Right to life, the right to human dignity, the right to property and the right to information under articles 9, 14, 23 and 19A of the Constitution of Pakistan read with the constitutional values of political, economic and social justice that offer the necessary judicial toolkit to address and monitor the Government's response to climate change*'.[273]

The Court noted that Pakistan's contribution to the global greenhouse gas emissions is very small, but as a responsible member state of the international community, it should give importance to mitigation efforts in multiple sectors like agriculture, livestock, forestry, industries, transport, energy and urban planning.[274] The Court ordered that the relevant ministries, departments and authorities should nominate a climate change focal person within their institution to closely work with the Ministry of Climate Change to ensure the implementation of the framework for climate change

[268]Ibid [3].

[269]Ibid [4].

[270]Ibid [11].

[271]Ibid [7].

[272]Ibid [7].

[273]Ibid [7]; JessicaVents, (26 September 2015). *Lahore Court orders Pakistan to act on Climate Change*. Retrieved May 4, 2016, from http://blogs.law.columbia.edu/climatechange/2015/09/26/lahore-high-court-orders-pakistan-to-act-on-climate-change/.

[274]*Asghar Leghari v Federation of Pakistan and others* WP No 25501/2015 [8].

policy implementation.[275] To expedite the matter and to effectively implement the fundamental rights of the people, the Court constituted Climate Change Commission under order 26 Civil Procedure Code.[276] It comprises (a) representatives of the key ministries/departments, (b) NGOs, (c) technical experts, and two members are from Ministry of Climate Change.[277]

As per the direction of the Court, Chairman of the Climate Change Commission has submitted the report of the Climate Change Commission in 2016.[278] The report pointed out the lopsided policy measures, lack of coordination, inadequate budget allocation, the absence of reliable data among departments and lack of implementation that the National Climate Change Policy appears not to have fully enforced.[279]

Based on the assessments, the Court recommended the establishment of Climate Change Authority, as a joint statutory body between different departments of governments.[280] The Court also recommended coordination between governments and emphasized on creating public awareness and participation at the local level.[281] The case is a momentous decision in the history of evolving climate justice from the position of environmental justice. The multiple orders and establishment of institutions have helped in steering the course of climate implementation in the country. From the procedural level, the case is significant, as it has passed recurrent orders for the prompt action on the part of the executive.

However, the case is silent as far as the issue of climate displacement is concerned. During the preceding of this case, Senate of Pakistan has passed Climate Change Act on 17 March 2017.[282] Pakistan became a handful of countries, which enacted legislation to combat climate change. Using a human rights-centric approach that is based on the fundamental rights and constitutional duties, the judiciary had enforced within months what others, including civil society and legislators, had failed to do for years.[283]

---

[275] Ibid [8].

[276] Ibid [8(3)].

[277] Ibid [8(3)].

[278] Ibid [11].

[279] Ibid [9]; see Parvez Hassan, 'Judicial Commission on Climate Change in Pakistan', (APCEL Climate Change Adaptation Platform, A paper for the 20th APCEL Anniversary conference Panel on Climate Change Adaptation) https://law.nus.edu.sg/apcel/cca/4.%20JudicialCommissiononClimateChangeinPakistan.pdf.

[280] Ibid.

[281] Ibid.

[282] Pakistan Climate Change Act 2017.

[283] Jacqueline Peel and Hari M Osofsky, 'A Rights Turn in Climate Change Litigation?' (2018) 7(1) TEL 37–67, 38.

6.8 Sri Lanka

In Sri Lanka, the constitutional provisions explicitly do not guarantee the right to life.[284] Hence, unlike other South Asian jurisdictions, right to clean environment has not adopted out of the expanded notion of right to life.[285] A large body of jurisprudence has been interpreted and innovated the specific rights contained in the Constitution of Sri Lanka that elaborated the right to environmental protection through the emphasis on the right to equality.[286] In *the case of Environmental Foundation Ltd. and Others v. The Attorney General and Others*,[287] a petition was filed against the *blasting of* rock quarry operation and the severe injury to their environment and health of the communities. In the absence of explicit provisions about environmental protection, the petitioners relied on various constitutional provisions including article 3,[288] article 11,[289] article 14(1)(g)[290] and article 12 (1) of the Constitution.[291] The Court held that right to environment and the norms of intergenerational equity pertaining to the protection, promotion and preservation of the environment are intrinsic in a pragmatic reading of the constitutional provisions.[292] The rules of the court were broadened to provide for PIL in the case of *Environmental Foundation Ltd. v. Ratnasiri Wickremanayake, Minister of Public Administration* et al.[293]

Similar to other South Asian jurisdiction, international environmental principles were adopted as part of Srilankan domestic jurisprudence. In *Gunarathne v. Homagama Pradeshiya Sabha* et al., the concept of sustainable development was brought in emphasizing the publicity, transparency and fairness.[294] In *Bulankulama & Others v. Secretary, Ministry of Industrial Development and Others (Eppawala*

---

[284] Sumudu Atapattu, 'A Commentary on the Draft Fundamental Rights Chapter' in *Sri Lanka: State of Human Rights 1998* (Colombo: Law & Society Trust, 1998) 173–191.

[285] Ibid.

[286] Camena Guneratne, 'Using Constitutional Provisions to Advance Environmental Justice—Some Reflections on Sri Lanka' (2015) 11(2) Law, Environment and Development Journal 3–22, 7.

[287] *Environmental Foundation Ltd and Others v The Attorney General and Others* SC Application No 128/91.

[288] Jane E Schukoske, 'Enforcing Environmental Laws in Sri Lanka through Fundamental Rights Litigation' (1996) 8(2) International Legal Perspectives 155–172, 162.

[289] The Constitution of the Democratic Socialist Republic of Srilanka, art 11 (No person shall be subjected to torture or to cruel, inhuman or degrading treatment or punishment).

[290] Ibid, art 14 (g) (the freedom to engage by himself or in association with others in any lawful occupation, profession, trade, business or enterprise).

[291] Ibid, art 12(1) (all persons are equal before the law and are entitled to the equal protection of the law).

[292] Ibid.

[293] *Environmental Foundation Ltd v Ratnasiri Wickremanayake, Minister of Public Administration and others* 3(4) S Asian Envtl L Rep 103 (1996).

[294] *Gunaratne v The Homagama Pradeshiya Sabha* App No 210/97(FR) of 4 March 1998; see Shyami Fernando Puvimanasinghe, 'Towards a Jurisprudence of Sustainable Development in South Asia: Litigation in the Public Interest' (2009) 10(1) Sustainable Development Law & Policy 41–49, 46.

Mining),[295] the case concerned the process of mining, extracting and processing the phosphate and the serious harmful impacts on the environment, the people of the area and the country as a whole. The Court observed, 'Human beings are the main concerns for sustainable development as they are entitled to enjoy a healthy and productive life in the lap of nature. To achieve the sustainable development, environmental protection shall form an inalienable part of the development apparatus and cannot be considered in isolation therefrom'.[296] In *Wattegedera Wijebanda v. Conservator General of Forest and Eight Others*,[297] the Court observed that 'the Courts in Sri Lanka, have long since recognized that the branches of the State are guardians to whom the people have committed themselves for the care and preservation of the natural resources.'[298] This recognition of the doctrine of 'public trust' accords a mammoth responsibility upon the government to preserve, promote and protect the environment and its resources.

This chapter sought to articulate the common features emerging from a range of very different cases decided in South Asian jurisdictions about refugee protection and environment protection. The basic objective was to highlight and analyse the role played by the judiciaries in South Asia in the matters of climate displacement. The cases reveal the humanitarian approach exhibited by the judiciary while dealing with the political refugees. The discussion also explains that though regarding environmental protection the role of the South Asian judiciary has been praiseworthy, there has not been a single occasion for South Asian judiciaries to consider the issue of climate displacement.

The cases also highlight the expanding horizons of environmental jurisprudence. The chapter had addressed how national judiciaries in South Asia have assessed the responsibility of governments in environmental matters. In particular, the chapter described how judiciary has played a proactive role in identifying the international environmental law principles to environmental protections. Enforcement and fulfilment of environmental protection were most effectively achieved within a constitutional framework of human rights.[299] If the global climate litigations are any indications for South Asia, the cases reveal that in climate litigations courts attempt to balance the competing interests through the doctrine of sustainable development.[300] In certain cases, courts have found climate concerns to be negligent in comparison with other developmental concerns and in other climate concerns prevailed over competing interests.[301]

---

[295] *Environmental Foundation Ltd and Others v The Attorney General and Others* SC Application No 128/91.

[296] Ibid.

[297] *Wattegedera Wijebanda v Conservator General of Forest and Eight Others* SC Application No 118/2004.

[298] Ibid.

[299] S Puvimanasinghe(n 294) 42.

[300] Lisa Vanhala, 'The Comparative Politics of Courts and Climate Change' (2013) 22(3) Env. Politics 447–474.

[301] Meredith Wilensky, 'Climate Change in the Courts, An Assessment of Non-U.S. Climate Litigation' (2015) 26 Duke Envt'l. L.& Pol. F 131–179, 175–177.

Cases also reveal that constitutional rights provided both the substantive and procedural foundation on which environmental rights were conceptualized and interpreted. Most of the South Asian jurisdictions have provided for substantial fundamental rights through which a catena of environmental rights was interpreted. Procedural rights were expanded through the PIL with a focus on equality, non-discrimination and due process.[302]

Despite climate migration being an issue of concern for the South Asian nations, not much has been contributed by the judiciaries of these nations regarding climate refugees and a climate change jurisprudence has not been evolved.[303] One reason for this could be the national policy frameworks of the states, which do not consider the climate migrants as a distinct category of displaced population. In fact, the Indian government has refused entry of climate migrants in the country when faced with such situation.[304] Another reason is that the refugee problem is usually masked under the garb of socio-economic or political persecution, and hence, the matter of climate refugees does not come to the fore. Also, most of the climate displacements in South Asian nations happen intra-State.[305] These movements fail to garner adequate attention and thus stay out of judicial purview and scrutiny. All these factors combine to cause the lack of development of climate refugee jurisprudence.

What is noteworthy is the fact that not only in the South Asian region, preferably in most parts of the worlds the judiciary has not had the opportunity to deal with the issue of climate refugees. Moreover, in fact, the small numbers of issues that have been brought to judicial scrutiny have not received a favourable response.[306] However, given the adoption of international environmental principles to domestic jurisprudence and expanded notion of constitutional rights, there are enough hooks in the South Asian legal structure to frame reliefs in matters of climate impacts and adaptation including climate displacement. Principles like sustainable development and public trust doctrine which are already part of the South Asian jurisprudences are a potentially promising argument, which could be used to bring state liability regarding protection of climate-displaced population. It has been submitted and argued that the concept of Atmospheric Trust Litigation (ATL) as an extension of public trust doctrine could be useful in climate litigation to impose a duty on the state to

---

[302] Ibid.

[303] David Markell and JB Ruhl, 'An Empirical Assessment of Climate Change In The Courts: A New Jurisprudence or Business As Usual?' (2012) 64(1) Fla L Rev, 17–86, 77.

[304] Hans Gunter Brauch and others, *Facing Global Environmental Change: Environmental, Human, Energy, Food* (Springer 2009) 303.

[305] Neeta Lal, 'Climate Migrants Lead Mass Migration to India's Cities' *Relief World* (26 July 2016) https://reliefweb.int/report/india/climate-migrants-lead-mass-migration-india-s-cities accessed 17 December 2017 (migration from drought hit Maharashtra and Andhra Pradesh and migration from flood prone Uttarakhand to Delhi are examples of intra-state migration in India).

[306] Paulina Tandiono, 'Legal Conundrum in the Plight of Pacific Island Countries: Climate Change, Displacement and Human Rights' (M.Sc. Thesis, London School of Economics and Political Science 2016) 16.

restore and improve the climate and impose liability on the part of states to protect climate-displaced populations.[307]

While considering the lack of cases pertaining to climate refugees and also the novelty of the concept, it is crucial that any jurisprudence on the issue is developed with utmost care. The climatic conditions of South Asian nations differ considerably, and every one of them has faced the problems of refugees in general and is likely to face an influx of climate migrants in the future. With no proper statutory laws enforced in this region, the burden of refugee protection can immediately fall on the judiciary, in cases where the executive is unwilling to provide an adequate safeguard to such people. In such a situation, it would be a welcome step if intra-judicial interactions could be organized between the members of the judiciary of the South Asian states. Such collaboration would lead to the development of new ideas and would also help the judges of one nation in understanding the problems faced by the other countries, regarding climate change and its impacts. It would also provide the judicial bodies an opportunity to develop a standard structure, which can be applied in cases of climate refugees and their protection. Lack of judicial decisions, coupled with a refusal to accept climate refugees, should not become a reason for the judicial failure to accord protection to those in dire need of shelter. Hence, an intra-judicial interactive mechanism can be a positive step forward in both environmental and human rights jurisprudence.

6.9 Making a Case for Intra-South Asian Judicial Interactions on Climate Refugees

The refugee experience of South Asia dates back to antiquity, and every perspective on the refugee crisis in South Asia is likely to be innovative in the absence of legally binding refugee regime in the region.[308] The CDPs or climate refugees in the SAARC region must enjoy the same array of rights that are available to political and humanitarian refugees.[309] But, unfortunately, the South Asian national governments have disenchanted their people and have not been successful in addressing the climate change-induced displacement and migration that has produced and will produce a new class of migrants called climate refugees.[310] The climate migrants are invisible in South Asia for policymakers and administrative authorities even though

---

[307] Mary Christina Wood and Charles W Woodward IV, 'Atmospheric Trust Litigation and the Constitutional Right to a Healthy Climate System: Judicial Recognition at Last' (2016) 6(2) Wash. J. Envt'l L & pol'y 634–683.

[308] See generally Jeff Handmaker and others (Eds), *Advancing Refugee Protection in South Africa* (1st edn, Berghahn Books 2011) 1–8.

[309] Sebastien Malo, 'Climate refugees in South Asia need Protection, Advocates Say' *Thomson Reuters* (9 December 2016) https://in.reuters.com/article/asia-climate-migrants/climate-refugees-in-south-asia-need-protection-advocates-say-idINKBN13X2PK accessed 28 March 2018.

[310] Ibid.

South Asia is most climate disaster-prone region of the world.[311] The climate refugee protection remains a political problem, and it is not considered social and humanitarian issue impregnated with grave violations of human rights. Therefore, in South Asia, the concept of comparative judicial understanding at regional and international levels creates leverage for interstate exchange and reception of constitutional, ethical and moral ideas to migrate, integrate and assimilate in different judicial jurisdictions confronted with the problems of refugees, climate change migration and displacement.[312] Further, such leverage promotes the judicial integration to address the plight of refugees and climate refugees who are getting unequal treatment in different jurisdictions.[313]

However, there is no conspicuously formal arrangement in South Asia to have judicial interactions among all the SAARC judiciaries regarding the legal protection of climate refugees or CDPs and other issues. In South Asia, refugees, asylum seekers, climate refugees and other migrants are subjected to a multitude of procedures where under their plight of migration and displacement from their homelands or country of origin do not get humane treatment consistent with global standards refugee law and human rights. Even South Asian states have adopted administrative and executive policies and practice that are deviant to international legal conformity, global constitutional uniformity and anthropological homogeneity. Their status as refugees is determined callously that violates fundamental principles of human rights. Moreover, the legal protection of climate refugees in the SAARC jurisdictions is the most neglected area of human rights and human dignity. In South Asia, even refugee status determination procedures varied from country to country and based on personal history of the refugees, political predilections of the country of reception, religious choices of the refugee-receiving states, the emergence of ultra-right politics and national security narratives. Therefore, there is no single model of refugee treatment, and there is no harmony among the national governments as to how to address refugee exoduses in the region. Thus, refugee status determination (RSD) mechanism is discriminatory, intolerant and prejudiced for some classes of refugees and humanitarian and dignified for other refugees.

In this context, the judiciary is perceived as a viable institution of hope for refugees in South Asia that requires evolving a mechanism for Intra-South Asian Judicial Interactions (ISAJI) to deal with the problem of refugees in the region. The SAARC judiciaries can play a pragmatic role in the climate refugee governance by introducing judicial and the rule of law principles in the national laws of South Asian

---

[311]UN reports, 'As 'most disaster-prone region,' Asia-Pacific needs risk-sensitive development' *UN News* (10 March 2016) https://news.un.org/en/story/2016/03/523982-most-disaster-prone-region-asia-pacific-needs-risk-sensitive-development-un accessed 26 March 2018.

[312]See generally for comparative approach of judical statement on asylum practice Andrew I Schoenholtz, Philip G Schrag and Jaya Ramji-Nogales, 'Refugee Roulette: Disparities in Asylum Adjudication' (2007) 61 Stan. L. Rev 295–412.

[313]Sean Rehaag, 'Troubling Patterns in Canadian Refugee Adjudication' (2008) 39 Ottawa L. Rev 335–365.

countries.[314] The critical position of the judiciary in environmental matters was recognized in 2002 at the Global Judges Symposium at Johannesburg in South Africa.[315] The Johannesburg principles were agreed to by the judges to resort to (a) utilizing the judicial mandate for sustainable development and adhering to the rule of law principle and democratic processes[316]; (b) recognizing an crucial need for regional and national initiatives and incentives to educate and train judges on international environmental law and (c) cooperating within and across the regions to improve environmental law compliance and enforcement.[317] Again, the UNEP had organized regional conferences[318] of judges around the world including South Asia to address the problems of climate refugees. In June 2010, the largest assemblage of judges around the world along with other stakeholders convened in Manila, Philippines, at the Asian Judges Symposium on Environmental Decision-Making, Rule of Law and Environmental Justice.[319] The symbosium discussed the environmental principles adhered to and introduced by the judiciary in South Asia to conserve cultural legacies and natural heritages like Taj Mahal,[320] River Ganges,[321] maintaining the perennial public benefit of natural resources,[322] human dignity,[323] issuance of writ of continuing *mandamus* on pollution cleaning[324] and the doctrine of public trust in preserving natural resources.[325]

In South Asia, there has been a rich tradition of borrowing of judicial principles. For instance, Bangladesh has developed environmental jurisprudence where

---

[314]UNEP, 'The Role of the Judiciary in Promoting Environmental Governance and the Rule of Law, Global Environmental Governance: the Post-Johannesburg Agenda' (New Haven 23–25 October 2003) 3–12 www.academia.edu/4624916/The_Role_of_the_Judiciary_in_Promoting_Environmental_Governance_and_the_Rule_of_Law_Lal_Kurukulasuriya_Chief_Environmental_Law_Programme_UNEP_Prepared_for_Global_Environmental_Governance_the_Post-Johannesburg_Agenda accessed 25 April 2018.

[315]Report of the Global Judges Symposium on Sustainable Development and the Role of Law, 2002 UNEP/GC/22/INF/24, https://www.eufje.org/images/DocDivers/Johannesburg%20Principles.pdf accessed 10 April 2018.

[316]Ibid; see also Brian Preston, 'The Role of the Judiciary in Promoting Sustainable Development: The Experience of Asia and the Pacific' (2005) 9 Asia Pac. J. Env. L 109, 113–14.

[317]Ibid.

[318]The meeting for South Asian countries was organized in collaboration with the South Asia Co-operative Environment Programme (SACEP) at Colombo in Sri Lanka in July 1997.

[319]Asian Judges Network to Strengthen Environmental Law Enforcement, https://www.adb.org/news/asian-judges-network-strengthen-environmental-law-enforcement accessed 16 April 2018.

[320]*MC Mehta v Union of India (Taj Trapezium case)* AIR 1988 SC 1037.

[321]*MC Mehta v Union of India (Ganges Pollution case)* AIR 1988 SC 1115.

[322]*Concerned Residents of Manila Bay v MMDA* GR Nos 171947-48 (8 December 2008) (Philippines).

[323]*Bandhua Mukti Morcha v Union of India* AIR 1984 SC 802; *Concerned Residents of Manila Bay v MMDA* GR Nos 171947-48 (8 December 2008) (Philippines).

[324]*Vineet Narain v Union of India* AIR 1996 SC 3386.

[325]*Bulankulama v Secretary, Ministry of Industrial Development (Eppawela case)* Application No 884/99 (Supreme Court of Sri Lanka).

under 'right to life'[326] was interpreted to include 'right to protection and preservation of the ecology'[327] and the 'right to have a pollution-free environment'[328] while emulating similar principles from Indian and Pakistani[329] jurisprudence. Though the protection of climate refugees is still absent in this jurisprudence, the historical cross-fertilisation of judicial norms offer invaluable lessons in forging cooperative interaction. In this regard two developments are momentous. In European Union, national judges rely on EU environmetal law and contribute to a more effective enforcement of environmental law.[330] In addition, EU Forum of Judges emphasis on shared experience in the area of training of judiciary in environmental laws and cases thus contributing to a better knowledge of environmental law. Secondly, the judges from 15 countries under a Charter have established the Global Judicial Institute for the Environment (GJIE) on 23–24 April 2016 in Brazil where under its constitution, structure, objective and mission have been formulated to address the plight of planet earth and seeks to develop and enhance the capacity of judges to exercise their role in environmental matters through the effective implementation, compliance and enforcement of the law. Therefore, such institution is equally important for South Asia and SAARC jurisdictions should also ponder over to have South Asian Judicial Institute for the Environment (SAJIE) to exchange information, to collaborate in partnerships and to provide research opportunities on the issues of climate change migration, displacement and climate refugees.

There are other networks of judges with a similar mandate like European Union Forum,[331] LAWASIA[332] and Asian Judges Network on the Environment (AJNE).[333] Similarly, at the regional level in South Asia, judiciaries require an ISAJI mechanism[334] to develop regional jurisprudence on climate change migration and displacement and climate refugees. On AJNE model, South Asian Judges Network of Environment (SAJNE) must be contemplated and established for the SAARC region. Therefore, in South Asia, Intra-South Asian Judicial Interactions model of climate change displacement governance must establish:

---

[326]Constitution of People's Republic of Bangladesh (141) art 31.

[327]Jona Razzaque, (n 146).

[328]*Mohiuddin Farooque v Bangladesh* 1997 BLD (AD) 1.

[329]*Shehla* Zia (n 253).

[330]The European Union network for the implementation and enforcement of environmental law is a network of national inspectors, formed in 1992. For more information, see: http://ec.europa.eu/environment/impel/index.htm.

[331]See 'European Union Forum of Judges for the Environment', www.eufje.org/ accessed 28 March 2018.

[332]See Lawasia, 'The Law Association for Asia and the Pacific' http://lawasia.asn.au/ accessed 26 March 2018.

[333]News Release, 'Asian Development Bank, Asian Chief Justices, Judges Propose Network to Promote Environment Justice' *Asian Development Bank* (30 July 2010) www.adb.org/news/asian-chief-justices-judges-propose-network-promote-environment-justice accessed 27 March 2018.

[334]See George Pring and Catherine Pring, *Greening Justice: Creating and Improving Environmental Courts and Tribunals* (The Access Initiative 2009) 14–16.

(a) A South Asian Judges Network on Environment (SAJNE) must also be formed to create a South Asian Comparative Judicial Integration to appreciate and understand the common problems of climate refugees.
(b) The SAJNE would promote the integrity,[335] human dignity and climate justice to climate refugees with global standards of human rights.
(c) The SAJNE would facilitate Intra-South Asian Judicial Interactions on collective judicial ideas sharing, intra-sub-continental exchanging of judicial experiences, transnational judicial capacity building and cooperation in cases of the climate refugee by promoting judicial regionalization as other trans-governmental alliances[336] failed to deliver desired outcomes.
(d) The evolution of South Asian Regional Climate Refugee Justice programme on access to justice for climate refugees, strengthening the rule of law and climate decision-making, and alternative dispute resolution[337] (ADR).
(e) The Intra-South Asian Communication on improving the delivery of benefits from ISAJI protection measures through building confidence among the judges, better knowledge about climate refugees and lego-institutional framework responsiveness on climate refugee rights must also be emplaced.
(f) The SAJNE ought to develop a comprehensive compendium of all international laws, global pacts, Declarations, principles and understandings relating to climate change and migration and displacement and their introduction in the national adjudicatory systems through the judges.
(g) The SAJNE should institutionalize the holding of annual meet of South Asian Judges to better understand the patterns, trends and issues relating to climate change migration and human displacement and needs of climate refugees, CDPs and other citizens of the region.
(h) The Kathmandu Resolution 2015[338] should be taken forward to institutionalize the judicial framework of methodologies to access climate justice to climate refugees, judicial collaboration for sustainable development, maintaining the climate change resilience, identifying measures for information sharing, preserving the environment, promoting regional judicial understandings in South Asia.

---

[335] Anne-Marie Slaughter and David Zaring, 'Networking Goes Global: an Update' (2006) 2 Ann. Renn. Rev. L & Soc. Sci 211–229.

[336] Anne-Marie Slaughter, 'Judicial Globalization' (2000) 40 Va J. Int'l. L 1104–1124.

[337] Chief Justice Corona, 'Asian Judges and the Environment, Capacity Needs and the Potential for a Network' Asian Judges Symposium (28–29 July 2010) www.scribd.com/doc/37133068/Renato-Corona-Talking-Points-AsianJudges-and-the-Environment-Capacity-Needs-and-Potential-for-a-Network accessed 28 March 2018.

[338] Irum Ahsan and Gregorio Rafael P Bueta, 'Proceedings of The Fourth South Asia Judicial Roundtable on Environmental Justice' (Kathmandu 28–29 November 2015) (Asian Development Bank 2016) 53.

(i) The SAJNE would work as a transnational judicial network[339] for regional cooperation for the protection, integration, rehabilitation and resettlement of climate refugees.

It is now axiomatic from the ongoing deliberations that these institutional measures would ensure the accelerated protection of climate refugees in South Asia in a manner that has not been resorted to earlier. Although the climate change impacts a range of human rights that have been recognized under core Conventions and treaties, these negative climate change impacts upon the peaceful enjoyment of human rights could be mitigated by having a climate justice mechanism as envisaged hereinabove. The inception of Intra-South Asian Interactions on Climate Refugees would enhance the accountability, human rights-based development, regional cooperation, ensuring equity in climate action, equality and non-discrimination, informed participation, maximum resource creation, capacity building measures, speedy climate justice environment and generate the regional and national audacity to attend the climate change justice challenges in South Asia and above all adhering to and promoting 'the Geneva Pledge'[340] within the framework of international law by utilizing judicial creativities and adopting additional regional Protocols on climate refugees.

The Intra-South Asian Judicial Interactions could play a pivotal role in not only highlighting the climate displacement concerns but also gradually evolving a consensus on the mechanisms for adapting and responding to the climate displacement. Such consensuses are decisive for South Asia to develop regional cooperative arrangement in the wake of the growing and potential security concerns. These mechanisms will fit into the more substantial south–south cooperation acknowledged by the international community in Lima for the first South-South Climate Change Cooperation (SSCCC) forum.[341]

South Asian nations are not parties to the international refugee protection frameworks and neither do they have any specific national laws to deal with refugee issues.[342] In such situation, most decisions regarding the status of refugees are taken either by acts dealing with foreigners, the constitutional rights or by judicial interference. This does not allow holistic protection of refugees, who are forced into sub-human lives due to non-recognition of their status as refugees despite claims that adequate safeguard has been provided to them.[343] Thus, there is a greater need for protection of refugees, who have to leave their homes due to mass human rights

[339] Kal Raustiala, 'The Architecture of International Cooperation: Trans-governmental Networks and the Future of International Law' (2002) 43 Va. J. Int'l. L 1–103.

[340] Promoting the Geneva Pledge for Human Rights in Climate Action, http://climaterights.org/our-work/unfccc/geneva-pledge/a accessed 15 march 2018.

[341] Silvia Casetta, 'South-South Cooperation An Emerging Dimension of the Global Response to Climate Change' in Ed Couzens, Tuula Honkonen and Melissa Lewis (eds), *International Environmental Law Making and Diplomacy Review* (University of Finland-UNEP Series 2017) 117–125.

[342] V Suryanarayan, 'The need for National Refugee Law' (2001) 5 ISIL Year Book of International Humanitarian and Refugee Law www.worldlii.org/int/journals/ISILYBIHRL/2001/15.html accessed 20 December 2017.

[343] Sidharth Kaushik and Stuti Bhatnagar, 'Need for Codification of Refugee Laws in India' (2014) 2(1) IJRA 390–401, 400.

abuses.[344] One of the means by which it can be achieved is the development of regional cooperative mechanisms between SAARC nations dealing with the legal provisions for displacement issues.

Two significant views are present among scholars regarding the need for a regional structure relating to refugees. One view has been expressed by scholars like B.S. Chimni who has advised against the formulation of any multilateral agreement on refugee problem and speaks in favour of the passage of national legislation which would permit states to recognize and discuss individual concerns, both at the level of security and resources, thereby bringing to the forefront the conflicting insights into the problem.[345] The other opinion speaks in favour of a regional framework for refugee protection. These scholars suggest that the coordination and collaboration of concerned states are essential for the promotion of the safety of refugees.[346] They argue that despite the accession to international human rights instruments and the positive action taken by the judiciary of these South Asian jurisdictions, refugee protection has been jeopardized on multiple occasion in these states due to the absence of legal provisions and the prevalence of foreign policy and political consideration have put refugees under vulnerable situations.[347] It has also been stated that the observance to the policy of finding political resolutions through bilateral negotiation among the host state and the country of origin (emphasizing on sovereign jurisdiction) is the most significant hindrance in evolving a formal refugee regime in South Asia.[348]

The most common arguments in advocating the regional approach are[349] that the South Asian states have sufficient commonalities regarding the problems, policies and practice, thus making it the perfect platform for developing a regional approach. Secondly, the large movement of population and the political complexity of the region make it unsuitable for *ad hoc* measures to be employed. Thirdly, a regional approach would permit South Asian states to improve cooperation and solidarity among countries and increase the prospects for a solution. It has been suggested that a regional framework would depoliticize the act of granting asylum, increase the liability of administrative actions and promote burden sharing within the region by creating a stable approach to the difficulties of refugee flows.[350] The need for depoliticizing can be understood by looking at the example of Bhutan and Nepal,

---

[344] BC Nirmal, 'Refugees and Human Rights' (2001) 1 ISIL Year Book of International Humanitarian and Refugee Law 94, http://www.worldlii.org/int/journals/ISILYBIHRL/2001/6.html accessed 15 April 2018.

[345] B.S. Chimni, 'The Law and Politics of Regional Solution of the Refugee Problem: The Case of South Asia' (1998) 4 RCSS Policy Studies, 1–16, 6.

[346] Abrar R, Chaudhary, 'Legal Position of Refugees in South Asia', FMR 10, 21–13.

[347] Ibid.

[348] Abrar R, Chaudhary (n 346) 22.

[349] Susan Kneebone, 'Comparative Regional Protection Frameworks for Refugees: Norms and Norm Entrepreneurs' (2016) 20(2) The Int'l J. Hum. Rts 153–172; Narayan Sharma, 'Refugee Situation in South Asia: Need of A Regional Mechanism' (2012) 1(1) Kathmandu Law Review 103, 120.

[350] Smrithi Talwar, 'Building a Regional Consensus on Asylum: The Indian Perspectives' (2000) 1(2) Bulletin on IHL and Refugee Law 251.

where Bhutan has on multiple occasions, accused Nepal of sheltering stateless and willingly migrated people to generate international antipathy against Bhutan.[351]

A study by the UNHCR in 2009 exhibited that approximately 22.5 million had to leave their homeland or country of nationality due to adverse impacts of climate change.[352] With growing climate threat, this number is set to increase in the coming years and might emerge as the cause of violent conflicts.[353] In such a scenario, nations can no longer be immune to the climate refugees and would be required to take positive actions for their protection. The current process followed by South Asian nation-states borders on humanitarianism and extension of protection to any refugee group (be it those facing persecution due to sociopolitical threat or the newly emerging climate refugees) is dependent upon the nation's willingness. However, scholars feel that policy solutions must be guided by human rights, not humanitarianism and climate change adaptation, socio-economic vulnerability, and cross-border legal responses must be resorted to within a legal human rights framework.[354]

It has become increasingly important for nations, especially those from South Asia to take action for dealing with climate migration crisis. In a study by Subramanian and Urpelainen, the game theory analysis was used to show that regional collaboration and treaty formation are easier under conditions of mutual vulnerability and hence are a feasible method to address cross-border displacement caused by environmental degradation.[355] The South Asian states have similar climatic conditions, and, therefore, the shared vulnerabilities exist in the region. To address these dimensions, a regional agreement would be of great importance.

One suggestion that can be extended regarding a probable refugee framework is the formation of a model similar to the Bangkok Principles propounded by Asian–African Legal Consultative Organization (AALCO).[356] The Bangkok principle covered areas like incorporating a definition for the term refugee, the concept of asylum and standard of treatment to refugees, durable solutions and burden sharing,.[357] The definition provided in this framework is a forward-looking one, stating

---

[351]Narayan Sharma (n 349) 121.

[352]UNHCR, 'Climate Change and Disasters' (14 December 2009) www.unhcr.org/climate-change-and-disasters.html accessed 28 March 2018.

[353]UN Office of the High Commissioner for Human Rights report on the Relationship Between Climate Change and Human Rights UN Doc A/HRC/10/61 (15 January 2009); Oli Brown and Elec Crawford, *Rising Temperatures, Rising Tensions: Climate Change and the Risk of Violent Conflict in the Middle East* (International Institute for Sustainable Development 2009) www.iisd.org/pdf/2009/rising_temps_middle_east.pdf accessed 23 December 2017.

[354]Shweta Jayawardhan, 'Vulnerability and Climate Change Induced Human Displacement Consilience' (2017) 17(1) JSD 103–142, 134.

[355]Narayan Subramanian and Johannes Urpelainen, 'Addressing Cross-Border Environmental Displacement: When Can International Treaties Help?' (2013) 14(1) Int. Environ Agreem. P, 25–46, 25–46.

[356]Asian-African Legal Consultative Organization (AALCO), Bangkok Principles on the Status and Treatment of Refugees (31 December 1966) www.refworld.org/docid/3de5f2d52.html accessed 24 December 2017 (Bangkok Principles).

[357]See Pia Oberoi, 'Regional Initiatives on Refugee Protection in South Asia' (2001) 3(1) Bulletin on IHL and Refugee Law -193-201.

that refugee includes every person who is compelled to leave his habitual place of residence owing to events seriously disturbing the public order partially or wholly in his country.[358] An important factor about such expansive definition is that it can be extended to incorporate climate refugees as well. Climate change is undoubtedly an event, which is likely to cause extreme disorder in the region of occurrence. In such a scenario, a definition like the one provided under the principle is precisely what is required to tackle the immense threat posed due to climate migration. Even though climate migrants have not been explicitly recognized thereunder, the definition is still far broader than what is present under the 1951 Refugee Convention.[359]

Another model that can be followed is the Kampala Convention, adopted by the African Union (AU), 'addressing a diverse variety of cases for dislocation including natural or human-made environmental disasters, climate change-induced displacement and migration'.[360] Adoption of a similar model in South Asia and discussions over the looming climate migrant issue are essential for the protection of the numerous displaced people and those facing the threat of displacement due to environmental damage. A regional instrument can serve as collective insurance against climate migration.

Therefore, the SAARC nation-states must strive to evolve a SAARC regional legal responces on the lines of UN Global Compact on Migration (GCM) to protect the rights of climate refugees in conformity with global constitutionalism. However, such a mechanism can be explored as a long-term solution. For achieving such a regional protection arrangement, regional governments must shed their diplomatic, legal and political prejudices and predilections and start the SAARC consultations on developing a regional[361] all-encompassing protection system, exploring a viable refugee protection agenda, promoting technical know-how and institutionalizing the uniform judicial approach and overall cooperation[362] that would emplace and shape

[358]Refugee includes every person who owing to external aggression, occupation, foreign domination or events seriously disturbing public order in either part or in whole of his country of origin or nationality, is compelled to leave his place of habitual residence in order to seek refuge in another place outside his country of origin or nationality under article 1 (2) of the AALCO agreement. article 1 (1) of the AALCO definition is the same as that of the 1951 Refugee Convention definition.

[359]Refugee under 1951 Convention is a person who owing to well-founded fear of being persecuted for reasons of race, religion, nationality, membership of a particular social group or political opinion, is outside the country of his nationality and is unable or, owing to such fear, is unwilling to avail himself of the protection of that country; or who, not having a nationality and being outside the country of his former habitual residence as a result of such events, is unable or, owing to such fear, is unwilling to return to it.

[360]Subramanian and Urpelainen (n 355) 25–46.

[361]Bali Process, 'Fourth Bali Regional Ministerial conference on People Smuggling, Trafficking in Persons and Related Transnational Crime: Co-Chairs' Statement' (March 2011) [16] www.baliprocess.net/UserFiles/baliprocess/File/110330_FINAL_Ministerial_Co-chairs%20statement%20BRMC%20IV(1).pdf accessed 20 March 2018.

[362]Madeline Gleeson, 'Where to From Here? report from the Expert Roundtable on Regional Cooperation and Refugee Protection in the Asia-Pacific' Andrew and Renata Kaldor Centre for International Refugee Law (2016) 37 www.kaldorcentre.unsw.edu.au/sites/default/files/Where_to_from_here.pdf accessed 20 March 2018.

the contours of institutional constitutionalism. The constitutionalism narrative pervades the global human rights and domestic constitutional rights that destined to protect the multidimensional facets of human displacement with a regional thrust.

6.10 Evolving Climate Change Regional Constitutionalism

The climate change displacement will transcend national boundaries and has been evolving as a problem of global ramification and repercussion.[363] The protection of climate refugees and their reception, resettlement and reintegration have posed governance challenges[364] to existing international legal frameworks, regional arrangements and national mechanisms.[365] The absence of an international legal framework, uneven regional initiatives and inconsistent national legal responses to climate displacement have brought the attention of international community to the role of climate constitutionalism founded upon the principles of equality, justice[366] dignity[367] and ecological integrity.[368]

Such a basis of climate change constitutionalism approach[369] must be allowed to supervene the sovereign narratives, individual state interests and foreign policy concerns to protect the climate refugees in a democratic, transparent, responsive and participatory lego-institutional framework.[370] Therefore, the climate change constitutionalism approach must be the ethical foundation and normative process[371] for initiating an action in favour of the global community of climate refugees. Climate change constitutionalism will be helpful in dismantling the political clout of international organizations, powerful non-state actors and hegemonic nation-states while

---

[363]Roger Zetter, *Protection in Crisis: Forced Migration and Protection in a Global Era* (Washington DC, Migration Policy Institute 2015) 3.

[364]Koko Warner, 'Global Environmental Change and Migration: Governance Challenges' (2010) 20(3) Glob. Environ. Chang. 402–414, 413.

[365]Koko Warner and others, 'Integrating Human Mobility Issues within National Adaptation Plans, United Nations University' (2014) Nansen Initiative Joint Policy Brief #2, UNU-EHS Publication Series Policy Brief No 9, 16–18.

[366]Erica De Wet, 'The International Constitutional Order' (2006) 55 ICLQ 51–76.

[367]Douglas Kysar 'Global Environmental Constitutionalism: Getting There from Here' (2012) 1(1) TEL 83–94.

[368]Tim Stephens, *International Courts and Environmental Protection* (Cambridge University Press 2009) 8.

[369]Aileen McHarg, 'Climate Change Constitutionalism? Lessons from the United Kingdom' (2011) 2(4) Climate Law 469–484.

[370]Robert V Percival, 'Environmental Law in the Twenty-First Century' (2007) 25(1) Va. Envt'l. L. J., 19.

[371]Louis J Kotzé, 'Arguing Global Environmental Constitutionalism' (2012) 1(1) TEL 199–233, 200.

promoting the principles of sovereign equality[372] for a concerted action to accomplish climate change constitutionalism.[373]

The SAARC nation-states share a common past and are to have a common future in the post-colonial era wedded with the democratic, plural and progressive constitutions. The idea of constitutionalism has been perceived as a test to determine the legality, legitimacy and validity of the law and the governmental action by the spirit of the constitution.[374] There are four dimensions of the constitutionalism such as (1) social constitutionalism, (2) institutional constitutionalism, (3) normative constitutionalism and (4) analogical constitutionalism.[375] This classification gives us the understanding to evolve climate change constitutionalism under a compendium of approaches called global constitutionalism in the regional legal framework.[376] Many regional organizations have been proactively protecting and adhering to the principles of constitutionalism like *inter alia* the rule of law and the human rights in the national governance models.[377] Therefore, an independent and strong judicial dispensation in all the SAARC jurisdictions is fundamental to transform the constitutionalism on the ground.

Therefore, evolving Climate Change Regional Constitutionalism has become indispensable and inalienable norms of refugee law framework to protect the lives and rights of the climate refugees in the SAARC jurisdictions that would also open new avenues for collaborative legal protection engagements in this region.[378] The modern expositions of global constitutionalism[379] that subscribe to global participation based on democratic systems are uniform and ubiquitous in their application but lack a concerted approach to address the climate change migration and displacement.[380] However, in protecting the climate refugees in the SAARC region and beyond, we contemplate a paradigmatic shift from global constitutionalism to social

---

[372] Bardo Fassbender, *The United Nations Charter as Constitution of the International Community* (Martinus Nijhoff Publishers 2009) 12–26.

[373] Hilal Elver, 'New Constitutionalism and the Environment: A Quest for Global Law', http://www.global.ucsb.edu/climateproject/publications/pdf/Hilal%20Elver_New%20Constitutionalism%20and%20the%20Environment_A%20Quest%20for%20Global%20Law.pdf accessed 16 April 2018.

[374] Jeffrey Sachs, 'Globalization and the Rule of Law' (1998) Occasional Papers 2, 1–11.

[375] See generally.Christine EJ Schwöbel, 'Organic Global Constitutionalism' (2010) 23 LJIL 529–553, 530.

[376] Jeffrey L Dunoff and Joel P Trachtman, 'A Functional Approach to International Constitutionalization' in Jeffrey L Dunoff and Joel P Trachtman (eds), *Ruling the World: Constitutionalism, International Law, and Global Governance* (Cambridge University Press 2009) 3–35.

[377] Micha Wiebusch, 'The Role of Regional Organizations in the Protection of Constitutionalism, International' IDEA Discussion Paper 17/2016 8–10.

[378] Koko Warner, Walter Kälin, Susan Martin and Youssef Nassef, 'National Adaptation Plans and Human Mobility: Disasters and displacement in a changing climate' (2015) 49 FMR 8–9.

[379] Anne Peters and Klaus Armingeon, 'Introduction: Global Constitutionalism from an Interdisciplinary Perspective' (2009) 16(2) Ind. J. Global Legal Studies 385–395, 387.

[380] Emily Wilkinson, Amy Kirbyshire, Leigh Mayhew, Pandora Batra and Andrea Milan, *Climate-Induced Migration And Displacement: Closing The Policy Gap* (Overseas Development Institute, London 2016) 1–12.

constitutionalism[381] that envisions international sphere as an order of coexistence wherefrom we cull the idea of 'assorted asylum'[382] to address the problem of human migration and displacement due to climate change, natural disasters and man-made environmental changes.

The climate change constitutionalism is akin to international legal order that is equally applicable to all the nation-states of the world.[383] Similarly, climate change has an international character that makes it a part of global constitutionalism where under nation-states yield their sovereignty to international legal order[384] to protect the rights of climate refugees with collective machinery.[385] But there is a need for constitutionalization of international climate law that takes up the issues of climate refugees.[386] The climate change nemesis would wipe out the human existence if the problem of climate refugees is not attended in commensurate with climate change constitutionalism. The climate change displacement and migration are not a manifestation of human intrusions or incursions in violation of the concept of sovereign rights of states.[387] However, the climate change constitutionalism that has been evolved and developed in different parts of the world has been attempting for its recognition, reception and respect[388] within the climate change trajectory of mitigation, adaptation and migration agenda to protect the climate refugees. But, unfortunately, that appears to be unachievable in South Asia for addressing the plight of climate refugees. Thus, the South Asian judiciary must strive to develop inalienable principles of South Asian climate change migration law and jurisprudence. The concept of climate change constitutionalism ought to be a fundamental law and must create a binding pre-emptory norm that transcends all other laws in national and global contexts to protect the climate refugees beyond the configuration of international legal

[381] Christine EJ Schwöbel, *Global Constitutionalism in International Legal Perspective* (Martinus Nijhoff Publishers 2011) 11–14.

[382] As per the authors, the idea of 'assorted asylum' conceives and includes all reasons of persecution, migration and forced human displacement on grounds of age, ethnicity, gender, hate, race, religion, nationality, membership of a particular social group, political opinion, external aggression, occupation, foreign domination, events seriously disturbing public order in either part or the whole of his country of origin, war, climate change, natural disaster, human-made environmental changes and individual or group of persons or communities displaced thereunder and unable or unwilling to avail the national protection of their country of origin or grounds incidental thereto or connected therewith will be entitled to 'assorted asylum' in another country and 'assorted asylum' is not available to those who violate the principles and purposes of the United Nations Charter.

[383] E Murugaesan, 'Role Of International Constitutionalism For Mitigating Earth's Climate Change—Legal Perspective' (2016) 7(7) Int'l J. Scient. & Eng. Research 153–157.

[384] Margaret Rosso Grossman, 'Climate Change and the Law' (2010) 58 Am. J. Comp. L 223–256.

[385] Siobhan McInerney-Lankford, Mac Darrow and Lavanya Rajamani, *World Bank Studies: Human Rights and Climate Change: A Review of the International Legal Dimensions* (World Bank Publications 2011) 28.

[386] Jan Klabbers, Anne Peters and Geir Ulfstein, *The Constitutionalization of International Law* (Oxford University Press 2009) 12–15.

[387] Anne Peters and Klaus Armingeon (n 379) 387.

[388] Douglas A Kysar, *Regulating from Nowhere: Environmental Law and the Search for Objectivity* (Yale University Press 2010) 245.

frameworks.[389] The concept of climate change constitutionalism could operate as a normative mode of South Asian climate refugee protection regime within a process of climate justice framework. Therefore, climate change constitutionalism approach must be the common ideology to protect the rights of climate refugees in South Asia and beyond.

6.11 Conclusion

The sluggish and slacken headway to address the climate change impact under international and national legal frameworks has pandered to the development of an alternative generation of climate change policy based on judicial redressal system. The climate change litigation is being presented as a legal apparatus that fills the gaps in the law and provides climate justice to the climate refugees while introducing and beginning legislative reactions. The judicial process has an intrinsic competence to treat the narrative of sufferings from theoretical to practical levels by galvanizing the climate justice debate to appreciate the substantial lurking alarms and apprehensions set in the perception of the victims. The unique victim-oriented approach of the judicial process augments and amplifies the significance of adaptation to climate change, compensation to victims, mitigating the extreme climate change impacts in future. Therefore, the rulings pronounced by the municipal legal systems across the world are encouraging trend that nation-states and private actors cannot take refuge in the absence of connectedness, and they could be made accountable and liable for their failure for not adequately contributing in resolving the problem of climate change.

The instant chapter analysed the present legal mechanisms for providing and affording protection to refugees in the SAARC jurisdictions. In the absence of statutory protection in SAARC countries, ad hoc apparatus, administrative bodies and the judicial establishments have been advancing and offering protection to refugees by having recourse to the relevant municipal constitutional provisions and principles of global constitutionalism. Further, the South Asian judiciaries are permeated with a progressive attitude, humanitarianism and constitutionalism towards the refugees that have secured fundamental human freedoms, global human rights and domestic constitutional rights to refugees and asylum seekers while striking a balance between refugee rights and state rights. However, it is conspicuously evident that existing refugee protection architecture in South Asia is flagrantly absent and there is an urgent requirement to emplace a responsive lego-institutional refugee protection framework.

The present chapter examined the evolving concept of climate refugees and judicial treatment meted out to them in the contemporary circumstances. The judicial establishments in South Asia have not had to address the plight of climate refugees in their traditional trajectory of adjudication. However, other judiciaries across the

[389] Robert S Summers, 'The Principles of the Rule of Law' (1999) 74(5) Notre Dame. L. 1691–1712, 1692.

world have not yet acknowledged and recognized the climate change as a ground to determine the refugee status as has been the practice in case of conventional refugees in many domestic legal systems. The SAARC region has been confronting severe climate change and damage, but climate change cases have not been raised in a befitting manner. Thus, the lack of cases is astounding and indicates that South Asian countries have not recognized the problem of climate refugees and could not generate faith and trust in the minds of affected people. Therefore, the countries in South Asia are under an unprecedented obligation to have a comprehensive refugee law covering all dimensions of human displacement and the judicial bodies must also need to appreciate, understand and develop climate refugee jurisprudence.

The South Asian Judges Network on Environment must be conceived, formulated and developed for appreciating, understanding and conetxtalusing the climate concerns and for tackling the deficit that is being faced by the conventional refugee law and climate refugee law. The nation-states in South Asia collectively owe a constitutional, regional and international responsibility to protect the climate refugees based on social humanitarianism, global constitutionalism and asylum institutionalism.

Chapter 7
The Role of the SAARC: Way Forward

Abstract South Asia has been witnessing an earth-shattering humanitarian crisis of migration and forced displacement due to unparalleled global, regional and internal disturbances resulting in systematic and gross transgressions of the global human rights mandate, international humanitarian law obligations and climate change law norms. There are more than 3 million refugees in South Asia, and 90% out of them are victims and product of intra-regional migration. The SAARC jurisdictions are both refugee-producing and refugee-hosting nation-states. Pakistan has been hosting the most massive refugees of Afghan origin; India is home to the diverse groups of intra- and extra-regional refugees including latest addition of Rohingya refugees, and rest of the SAARC nations are also bracing the refugee crises in the region, and the crisis is further compounded by the returnees from the Global North countries. Among the SAARC jurisdictions, there is a problem of negative attitude towards refugees based on preposterous political indoctrination. The instant chapter examines the relevance of refugee crisis to regional collaboration and advocates for a regional institution to address the crisis while critically evaluating the role of the SAARC in protecting the rights of climate change-induced displaced persons. The needs of the SAARC countries have galvanized an understanding to address the complexity of the climate change migration by adopting a SAARC Climate Declaration and an Action Plan on Climate Change. Therefore, the SAARC is gradually ushering in the climate change field. However, the institution is structurally and politically weak. The matters get further dimmer as the position of countries in SAARC varies regarding climate change refugees with Maldives and Bangladesh expressing opinion in favour of their recognition and protection. Primarily, the chapter tries to identify the climate change consciousness and challenges in South Asia and climate law responses of the SAARC by espousing the hybrid integration of international legal norms with regional aspirations. It recognizes the need for regional trans-boundary cooperation to address the climate change displacement and migration and makes a case for the advocacy for an alternative regional regime on climate refugees.

The authors acknowledge the research assistance rendered by Amit Kumar, Tanaya Thakur, Sumedha Chaudhury, Manini Syali, Ravneet Sandhu, LLM students at SAU and Kanika Jamwal, an LLB student at RGNUL, Patiala.

S. Jolly and N. Ahmad, *Climate Refugees in South Asia*, International Law and the Global South, https://doi.org/10.1007/978-981-13-3137-4_7

7.1 Introduction

The prediction about the scope and extent of climate-induced migration involves a certain amount of uncertainty and complexity. South Asian region characterized by its fragile economy, low-lying islands, impoverished and vulnerable population and economy is reported to be highly susceptible to climate change and the consequent climate-induced migration/displacement. The book does not subscribe to an alarmist orientation, advocacy or conservative approach with regard to the scope and scale of the climate change-induced displacement and migration. Though, there are indications to the point that climate change with all its complexities will result in a mass of population fleeing from extreme disasters and areas of uninhabitable places. There is need to conceive and foster a pragmatic discourse to address climate change human displacement in international and national policy responses adequately. South Asian countries with its limited adaptive capacity have a particular concern for climate change as highlighted by the fifth IPCC assessment report.[1] International law has paid scant and inconsistent attention to the new trajectory of climate change-induced displacement contributing to its lack of recognition. The inaction of international community is fuelled by a host of complicated and multifaceted factors including the convoluted interplay of economic, environmental and social factors in climate displacement, the sheer diversity of situations and profile of climate migrants, lack of empirically precise data and multipronged approaches to climate displacement. Recently, one of the Louisiana tribe has been granted official climate refugee status.[2] The government has granted them support in the form of developing resilience and adaptation.[3] The question arises whether the international community can follow the direction in this regard.

This chapter critically evaluates the role of SAARC regarding protecting the rights of the climate-displaced population. Energy concerns motivated the initial focus of SAARC on climate change. In 2007, SAARC initiated an Action Plan for Climate Change and later signed the SAARC Climate Declaration. In 2010, an Intergovernmental Expert Group on Climate Change was constituted to oversee the implementation of the Plan. Though the SAARC is slowly entering into the climate area, the institution is structurally and politically weak. The matters get further murkier as the position of countries in SAARC varies about climate change refugees with Maldives and Bangladesh expressing stronger opinion in favour of their recognition and protection. In 1994, the former United Nations High Commissioner for Refugees (UNHCR) Mrs. Sadako Ogata unveiled 'Regional Consultation on Refugee and

[1]Intergovernmental Panel on Climate Change, 'Climate Change 2014: Synthesis report' (Contribution of Working Groups I, II and III to the Fifth Assessment report of the IPCC, Geneva, 2014) http://www.ipcc.ch/pdf/assessment-report/ar5/syr/SYR_AR5_FINAL_full_wcover.pdf accessed 16 December 2018.

[2]Debra Utacia Krol, 'In Louisiana, A Plan to Relocate the Country's First "Climate Refugees" Hits a Roadblock' *Huffington Post* (23 March 2018) https://www.huffingtonpost.in/entry/louisiana-climate-refugees-plan-roadblock_us_5ab402ade4b008c9e5f55c1b accessed 21 April 2018.

[3]Ibid.

Migratory Movements in South Asia' (RCRMMSA); therefore, having completed a long-drawn process of regional consultations, SAARC grouping came out with a regional Declaration and National Model Law on Refugees in 2004 whose fate is still in limbo but, unfortunately, the initiative does not address the problems arising out of the climate change and climate refugees in South Asia. Therefore, recognizing the emerging climate displacement scenario, there has been a demand to frame a SAARC Charter or cooperative SAARC climate change responces whereunder 'climate change refugees' or 'climate refugees' must have the right to free movement in South Asia. Other suggestions include an exposition about a comprehensive development Plan and an implementation mechanism thereunder for the SAARC member states. The instant chapter will critically examine and evaluate these vital legal developments. Though the climate displacement is a global phenomenon, its impacts are felt locally and more at a regional level and a global treaty to tackle the climate displacement may mask the regional and local dimensions of the issue.

7.2 Climate Change Migration Consciousness and Challenges in South Asia

As the climate change crisis looms large across, South Asia is one of the hardest hit regions in the world. Given the vulnerability of the region, climate change-induced migration is an important environmental, social and political challenge for the region's peoples and governments.[4] In 2018, the World Bank published its report stating therein that climate change could force around 140 million people across the globe.[5] The question is whether the issue has been elevated to the level of popular and policy consciousness level to elicit legal and policy responses.

---

[4]Arabinda Mishra et al., 'Climate Risks in the SAARC Region: Ways to Address the Social, Economic & Environmental Challenges' (The Energy and Resources Institute, New Delhi, 2012, Project report No. 2012GW03) 60 https://www.researchgate.net/publication/283436395_Climate_risks_in_the_SAARC_region_ways_to_address_the_social_economic_environmental_challenges accessed 23 June 2018.

[5]Kanta Kumari Rigaud, Alex de Sherbinin, Bryan Jones, Jonas Bergmann, Viviane Clement, Kayly Ober, Jacob Schewe, Susana Adamo, Brent McCusker, Silke Heuser, Amelia Midgley, 'Groundswell: Preparing for Internal Climate Migration' (World Bank, Washington, DC, 2018) 181 https://openknowledge.worldbank.org/handle/10986/29461 accessed 20 May 2018.

The governments, international organizations, academics[6] and civil society organizations have directed their focus to climate change-induced migration in the region and highlighted the manifold consequences including conflict and violence.[7] Especially, international institutions like Asian Development Bank have highlighted the extent, vulnerability and the possible solution for climate-induced migration and displacement in the region.[8] However, a dedicated study eliciting the extent of public awareness about climate as a primary factor in migration is missing at the regional level, though various pilot studies have attempted to trace the migration and the adaptive capacity of people and communities.[9] The studies have revealed the complex array of pull and push factors causing climate displacement which operating in the local context give rise to diverse local consequences and circumstances.[10] The studies have also pointed to the high vulnerability characterized by the high susceptibility to risk and inability to recover and respond to a disaster situation in South Asia owing to lack of coherent institutional and governance structure, and the prevalent social and economic inequalities.[11] The situation needs further analysis of the process and patterns of climate displacement. There is a dire need to increase public awareness and focus on capacity building and resilience, improve governmental coordination at all

---

[6]Frank Biermann and Ingrid Boas, 'Preparing for a Warmer World: Towards a Global Governance System to Protect Climate Refugees' (2010) 10(1) Global Envt'l Politics 60–88 https://www.researchgate.net/profile/Frank_Biermann/publication/227627225_Preparing_for_a_Warmer_World_Towards_a_Global_Governance_System_to_Protect_Climate_Refugees/links/5422bc260cf238c6ea6b866f/Preparing-for-a-Warmer-World-Towards-a-Global-Governance-System-to-Protect-Climate-Refugees.pdf accessed on 25 May 2018; Esmaralda Colombo, 'Land or Utopia? Seeking a New Climate for Migration' (2015) Green European Journal http://www.greeneuropeanjournal.eu/land-utopia-seeking-new-climatemigration/ accessed 19 April 2018; Astrid Epiney, '"Environmental Refugees": Aspects of International State Responsibility' in Jane McAdam (ed.), *Climate Change and Displacement: Multidisciplinary Perspectives* (Bloomsbury Publishing 151–75; International Organization for Migration, 'Migration, Climate Change and the Environment' (2009) IOM Policy Brief 4 https://www.iom.int/sites/default/files/our_work/ICP/IDM/iom_policybrief_may09_en.pdf accessed on 25 June 2018.

[7]Ibid., see also Clionadh Raleigh, Lisa Jordan and Idean Salehyan, Assessing the Impact of Climate Change on Migration and Conflict, http://siteresources.worldbank.org/EXTSOCIALDEVELOPMENT/Resources/SDCCWorkingPaper_MigrationandConflict.pdf accessed on 25 June 2018.

[8]Asian Development Bank, 'Addressing Climate Change and Migration in Asia and the Pacific' (ADB, 2012) https://www.adb.org/sites/default/files/publication/29662/addressing-climate-change-migration.pdf accessed 18 April, 2018.

[9]Ibid, 41, Nansen Initiative Regional Consultation, 'Climate Change, Disasters and Human Mobility in South Asia' (Nansen Initiative South Asia Regional Civil Society Meeting, Kathmandu, Nepal, 2–3 February 2015) 17–18 http://www.nanseninitiative.org/wpcontent/uploads/2015/07/150715_FINAL_OUTCOME_REPORT_SOUTH_ASIA_REGIONAL_CIVIL_SOCIETY_MEETING_screen.pdf accessed 25 June 2018.

[10]'Addressing Climate Change and Migration in Asia and the Pacific' (n 8) see also, Gunvor Jónsson, 'The Environmental Factor in migration Dynamics—A Review of African Case Studies', International Migration Institute, Working Paper 21, 2010. https://www.oxfordmartin.ox.ac.uk/downloads/WP21%20The%20Environmental%20Factor%20in%20Migration%20Dynamics.pdf accessed 10 June 2018.

[11]Nansen Initiative Regional Consultation (n 9).

levels, increase international partnerships and collaboration. Article 6 of UNFCCC and article 12 of the Paris Agreement states that the state parties should cooperate in facilitating climate change education, training programmes, public awareness, public participation and public access to information on climate change. Thus, climate change education and consciousness become an important aspect along with other mechanisms adopted in response to climate change. In spite of the clear exposition, except Nepal, none of the climate change policies of the South Asian countries talks about extending adaptation policy with transboundary application.

South Asian countries have responded to climate change challenge through national policy and legal initiatives. In addition, they have also developed specific legal stances on refugee protection at national levels. However, similar to international law, lack of synergy exists between both the regimes as both the regimes operate independently without having any coordination with each other. Furthermore, most of the climate change national programmes adopted by individual countries identify the climate change-induced displacement but do not address the same in a significant manner, and the challenge is to look at climate-induced migration comprehensively. Historically, given the porous border, dividing South Asian countries[12] and the migration patterns and climate change-induced migration in South Asia will not be confined to being limited to national borders, but it will have the tremendous regional impact as well. The situation calls for greater cooperation and grander role for SAARC as a regional organization. The following section will address the role played by SAARC in elevating climate consciousness to the policy and implementation level for regional collaboration.

7.3 SAARC and Climate Law Responses

The SAARC is a consolidation of sovereign jurisdictions of nation-states such as Afghanistan, Bangladesh, Bhutan, India, Maldives, Nepal, Sri Lanka and Pakistan. The fundamental objective of this regional body is to promote the overall economic, cultural and social welfare and growth of the South Asian people.[13] The nation-states and their behaviour have been changing at a pace that has not been seen before, and behaviour of states in South Asia is not an exception thereto. Presently, SAARC has been operating under the impact of wide-ranging dimensions of cooperation in all the fields including commerce, cultural interactions and exchange, conflict resolutions, human security and national defence, global terrorism. However initially SAARC as a permanent organization was established to accomplish limited objectives by

[12]Partha S. Ghosh, *Unwanted and Uprooted: A Political Study of Migrants, Refugees, Stateless and Displaced of South Asia* (Sanskriti, 2004), 40 &152.

[13]South Asian Association for Regional Cooperation, 'About SAARC' (*SAARC*) http://saarc-sec.org/about-saarc accessed 01 March 2018.

coordination and cooperation among its member states.[14] The SAARC subscribes to the purposes and principles[15] of the UN Charter while functioning as a regional body in South Asia milieu. It upholds the principle of sovereign equality,[16] territorial integrity,[17] political independence,[18] non-interference[19] in the internal affairs of member states,[20] peaceful settlement of all disputes[21] and mutual benefits. The SAARC Charter is not a substitute for bilateral and multilateral cooperation in the region, but it works as an element of complementarity for the South Asian region. SAARC regional body was established in 1985 at Dhaka[22] with the seven member states of South Asia except Afghanistan that joined later[23] in 2007. Initially, SAARC focused on non-controversial fields of cooperation[24] and understandings on agriculture and rural development,[25] health and population interests, and postal services and communications. Subsequently, SAARC has also started attending other pressing issues of the region relating to economy finance, trade, transport, connectivity, energy and climate change.[26] Presently, SAARC has been addressing all areas of mutual cooperation and benefits, and it has also created and established a number of intergovernmental organizations that have been guiding and shaping regional cooperation in these areas. SAARC Charter stipulates the structural and institutional framework[27] for managing the administration and functions of the SAARC organization. It provides a hierarchical arrangement with summit at the top backed by the Council of Ministers[28] comprising of Foreign Ministers from the SAARC member states and the Standing Committee[29] consisting of the Foreign Secretaries of the

---

[14]Lawrence Saez, *The South Asian Association for Regional Cooperation (SAARC): An Emerging Collaboration Architecture*, Routledge (2012) 8. See generally for Organisations, Robert Presthus, *The Organisational Society: An Analysis and A Theory* (Virtage Books, New York, 1962).

[15]Charter of the South Asian Association for Regional Cooperation (adopted 8 December 1985) 4 ASYIL 473 (SAARC Charter) art II; and Charter of the United Nations (adopted 24 October 1945) 1 UNTS XVI (UN Charter) art 1.

[16]UN Charter (n 15) art 2(1).

[17]SAARC Charter (n 15); UN Charter (n 15) art 2(4).

[18]Ibid.

[19]SAARC Charter (n 15); UN Charter (n 15) art 2(4).

[20]SAARC Charter (n 15); UN Charter (n 15) art 2(7).

[21]SAARC Charter (n 15); UN Charter (n 15) art 33(1).

[22]South Asian Association for Regional Cooperation (n 13).

[23]Afghanistan joined the SAARC as member in April 2007.

[24]Sobhan Rehman, 'Regional Economic Cooperation in South Asia' in Pradip Ghosh, *A Modernization Perspective* (Greenwood Press, London, 1984) 268–271.

[25]Moazzen Hussain, Iyanlur Islam, Reza Kibra, *South Asian Economic Development: Transformation, Opportunities and Challenges* (Routledge Publishers, London, New York, 1999).

[26]Kishore Dash, 'The Political Economy of Regional Cooperation in South Asia' (1996) 69(2) Pacific Affairs 185–209, 205.

[27]SAARC Charter (n 15) art III.

[28]SAARC Charter (n 15) art IV.

[29]SAARC Charter (n 15) art V.

SAARC member states to devise and prepare the policies,[30] to assess the progress of collaboration,[31] understanding on new areas of collaboration,[32] to establish additional mechanisms[33] and to decide other matters of importance[34] among the member states. It is also having a system of Technical Committees[35] in the fields of collaboration that provides the basic structure to the administrative branch of the SAARC organization. The SAARC Council of Ministers operates as a cabinet that sends matters to the summit for its deliberation and decision and gets them implemented by the SAARC Standing Committee. However, the Standing Committee may also constitute the Action Committees[36] for the implementation of the projects of the SAARC. Therefore, the Standing Committee emerges as an executive branch of the SAARC institutional structure responsible for the decision reached by the SAARC Council of Ministers and the summit. There is also a Programme Committee that controls the performance and operation of the Technical Committee.[37] Thus, the SAARC Charter reaffirms the fundamental principles of the peace, progress and prosperity under international law with nuances of geopolitical and security interests for accomplishing amity, stability and tranquillity in South Asia as per the principles of non-aligned movement and UN Charter vision.

In the context of climate change issues, the SAARC paid attention to climate change in a manner that has a gradual transformation beginning with its focus on environment protection. The principle institutional response on environmental distresses and trepidations was initiated in the third SAARC summit held at Kathmandu.[38] A research study was commissioned on the 'Protection and Preservation of the Environment and the Causes and Consequences of Natural Disasters' by the SAARC summit under a well-organized comprehensive and consolidated framework.[39] The heads of the government expressed concern on the declining forest cover in South Asia along with the region's susceptibility to natural disasters like floods, earthquakes, cyclones etc.[40]

---

[30] SAARC Charter (n 15) art IV (a).

[31] SAARC Charter (n 15) art IV (b).

[32] SAARC Charter (n 15) art IV (c).

[33] SAARC Charter (n 15) art IV (d).

[34] SAARC Charter (n 15) art IV (e).

[35] SAARC Charter (n 15) art VI.

[36] SAARC Charter (n 15) art VII.

[37] Suman Sharma, *India and SAARC* (Gyan Publishing house, New Delhi, 2002) 147.

[38] Rekha Singh Yadav, 'Chapter IV: SAARC summits: India's Stand' (Shodhganga, 2012) 115–116. http://shodhganga.inflibnet.ac.in/bitstream/10603/86856/13/13_chapter%204.pdf accessed 01 March 2018.

[39] Suman Sharma, 'Climate Change in South Asia and Regional Initiatives: A Case Study of South Asian Association for Regional Cooperation (SAARC) Initiatives' (*All Academic Research*) 10 http://citation.allacademic.com//meta/p_mla_apa_research_citation/5/0/0/5/5/pages500558/p500558-10.php accessed 01 March 2018.

[40] South Asian Association for Regional Cooperation, 'The Kathmandu Declaration of the Heads of State or Government of the member countries of the South Asian Association for Regional Cooperation' (3rd SAARC summit, Kathmandu, 4 November 1987) (Kathmandu Decla-

The fourth summit, which was held in Islamabad, Pakistan in 1988, focused on the impact of carbon dioxide emissions on the region.[41] The regional discourse focused on sharing information on climate change technology transfer, discussing regional mitigation mechanisms and scientific technology.[42] The fifth summit held in Male, in 1990, adopted the Male Declaration.[43] Among other things, the Declaration endorsed the decision of the SAARC Council of Ministers to finish the regional study commissioned in the third summit within the time frame stipulated by the council.[44] The final report formulated in 1991, emphasized the establishment of SAARC network of environmental NGOs, and SAARC Fund for environment protection.[45] The report also advocated the establishment of the SAARC Committee on Environment to coordinate and implement its recommendations.[46] The summit also for the first time attended to the issue of climate change; the summit noted the special vulnerabilities of the region and emphasized financial assistance and technological transfer mechanisms.[47]

At the sixth summit held at Colombo, Sri Lanka, the emphasis was on the need for financial resources for pursuing and promoting the practice of sustainable development. The summit emphasized an equitable global partnership for the protection of environment.[48] The 7th SAARC summit concentrated and converged on intra-SAARC and south–south collaboration in environmental affairs and questions.[49]

ration) http://saarc-sec.org/uploads/digital_library_document/03-Kathmandu-3rdSummit1987.pdf accessed 01 March 2018.

[41] South Asian Association for Regional Cooperation, 'The Islamabad Declaration of the Heads of State or Government of the member countries of the South Asian Association for Regional Cooperation' (4th SAARC summit, Islamabad, 31 December 1988) (Islamabad Declaration) http://saarc-sec.org/uploads/digital_library_document/04-Islamabad-4thSummit1988.pdf accessed 01 March 2018; Manju Jain, Fourth 'SAARC summit Islamabad Declaration' (1989) 2(1) Indian J. Asian Affairs 56–64.

[42] Ibid.

[43] South Asian Association for Regional Cooperation, 'The Male Declaration of the Heads of State or Government of the member countries of the South Asian Association for Regional Cooperation' (5th SAARC summit, Male, 23 November 1990) (Male Declaration) http://saarc-sec.org/uploads/digital_library_document/05-Maldives-5thSummit1990.pdf accessed 01 March 2018.

[44] Ibid.

[45] South Asian Association for Regional Cooperation, *Regional study on the causes and consequences of natural disasters and the protection and preservation of the environment* (South Asian Association for Regional Cooperation, 1992).

[46] Down to earth, 'SAARC ratifies committee on environment' (*Indian Environment Portal*, 14 August 1992) http://www.indiaenvironmentportal.org.in/content/33045/saarc-ratifies-committee-on-environment/ accessed 03 March 2018.

[47] Male Declaration (n 43).

[48] South Asian Association for Regional Cooperation, 'The Colombo Declaration of the Heads of State or Government of the member countries of the South Asian Association for Regional Cooperation' (6th SAARC summit, Colombo, 21 December 1991) (Colombo Declaration) http://saarc-sec.org/uploads/digital_library_document/06-COLOMBO-6thSummit1991.pdf accessed 01 March 2018.

[49] South Asian Association for Regional Cooperation, 'The Dhaka Declaration of the Heads of State or Government of the member countries of the South Asian Association for Regional Cooperation'

In the 9th SAARC summit, the urgent need to curb the rising degradation of the environment was stressed. The SAARC Environment Ministers meet adopted the Environment Action Plan.[50]

The Plan purported to address environmental concerns, enhance environmental management capabilities and focus on concerted national and regional activities. With environmental protection at its axis, it aimed at adopting the following strategies:

1. The utilization and employment of 'environmental assessment' as a tool to help governments of SAARC member states to deal with environmental problems, natural calamities and climate change-induced displacement.[51]
2. Develop, improve and execute national and regional environmental action Plans to attend the climate apprehensions of the SAARC nation-states.[52]
3. Develop legal mechanisms to provide support for collaborative endeavours to protect and conserve the environment and promote the sustainable development.
4. Disseminate environmental awareness and climate consciousness amongst the general masses and hold, organize and raise funds for the execution of the Environment Action Plan.

The Plan stated that the SAARC nations are more susceptible to the adverse impacts of global climate change and sea-level rise. Member states reiterated the requirement of periodical analysis of the impact of climate change on the South Asian region and the establishment of a group of expert scientists who may do the same.[53] The Plan provided for the establishment of the two regional institutions, namely the SAARC Forestry Centre and the SAARC Coastal Zone Centre.[54] The action Plan also provided for evaluation of the viability of a Regional Treaty on Environment Law, coherent with the existing international Conventions.[55] For the implementation of this regional action Plan, the governments of member states were directed to formulate a National Action Plan on the overall environmental degradation and natural disasters and present it to the SAARC Secretariat.[56] Furthermore, member nations were to identify a nodal ministry, which would be responsible for monitoring the implementation and progress under this Plan, coordinate with the SAARC Secretariat and act as a communication link between the Secretariat and the government.[57] The focus of the 12th SAARC summit was on the drafting of a

(7th SAARC summit, Dhaka, 11 April 1993) (Dhaka Declaration) http://saarc-sec.org/uploads/digital_library_document/07-Dhaka-7thSummit1993.pdf accessed 02 March 2018.

[50] South Asian Association for Regional Cooperation, 'SAARC Environment Action Plan 1997' (3rd Meeting of the SAARC Environment Ministers, Dhaka, 3 July 2008) http://saarc-sec.org/assets/responsive_filemanager/source/Files%20for%20Areas%20of%20Cooperation/ENB/SAARC%20ENVIRONMENT%20ACTION%20PLAN.docx accessed 01 March 2018.

[51] Ibid 'Environment Assessment'.

[52] SAARC, 'Environment Action Plan' (n 50) 'Aims and Strategies'.

[53] SAARC, 'Environment Action Plan' (n 50) 'Environmental Assessment'.

[54] SAARC, 'Environment Action Plan' (n 50) 'Regional Institutions'.

[55] SAARC, 'Environment Action Plan' (n 50) 'Legal Framework'.

[56] SAARC, 'Environment Action Plan' (n 50) 'Implementation'.

[57] Ibid.

Regional Environment Treaty and its pragmatic execution along with the SAARC Environment Plan of Action.[58]

The 13th SAARC summit was held in Dhaka in 2005. The summit aimed at documenting the achievements at regional cooperation in the first two decades and planning an effective cooperation strategy for the years to come.[59] Expressing deep concern over the continuing environmental degradation, the member states accepted the decision of the SAARC council to establish a SAARC Forestry Centre in Bhutan. They also extended support towards formulating a Regional Environment Treaty in furthering environmental cooperation.[60] India particularly concentrated on the necessity to encourage and stimulate regional cooperation pertaining to the development of resilient disaster management system.[61]

Recognizing the development of a high-risk environment as one of the causes for increasing vulnerability of South Asian people, the SAARC Comprehensive Framework on Disaster Management (SAARC CFDM) was adopted to deal with the threat posed by recurring disasters on the development of South Asian nations. The SAARC Centre for Disaster Management and Preparedness in New Delhi, SAARC Coastal Zone Management Centre in Male and SAARC Meteorological Research Centre in Dhaka collectively formed the institutional support for the implementation of the CFDM.[62]

The 14th SAARC summit was held at New Delhi in 2007. The leaders recognized the need to prioritize environmental conservation and proposed to launch 2007 as the year of Green South Asia.[63] The Dhaka Declaration and SAARC Action Plan was a consequence of the 14th SAARC summit, held at New Delhi, which articulated the profound apprehension regarding global climate change menace and contended for the development of climate resilience structures in South Asia. Following this, Bangladesh proposed to arrange an expert meeting at Dhaka which the SAARC Council of Ministers welcomed. Thus, a SAARC Ministerial Meeting was held on climate change, and the same was preceded by an Expert Group Meeting on climate

---

[58] South Asian Association for Regional Cooperation, 'Islamabad Declaration' (12th SAARC summit, Islamabad, 06 January 2004) http://saarc-sec.org/uploads/digital_library_document/12_-_Islamabad_-12th_SAARC_Summit,_4-6_January_2004.pdf accessed 01 March 2018; Saleemul Huq, 'The South Asian For Regional Cooperation' in Michael H Glants (ed), The role of Regional Organisations in the context of Climate Change (NATO ASI Series1: (Springer 1994)) 28 Baniateiling Majaw, 'Climate Change and South Asian Association for Regional Cooperation: A Regional Response' (2012) 2(4) IJSCHS 71–80, 75 http://www.sobiad.org/ejournals/journal_ijss/arhieves/2012_2/baniateilang_majaw.pdf accessed 03 March 2018.

[59] South Asian Association for Regional Cooperation, 'The Dhaka Declaration' (13th SAARC summit, Dhaka, 13 November 2015) http://saarc-sec.org/uploads/digital_library_document/13_-_Dhaka_-_13th_Summit_12-13_Nov_2005.pdf accessed 02 March 2018.

[60] Yadav (n 38) 158.

[61] id, 159–160.

[62] SAARC, 'SAARC Comprehensive Disaster Management Framework 2006' (SAARC Expert Group Meeting, Dhaka, 9 February 2006) https://www.preventionweb.net/files/2637_SAARCRegionalframeworkforDM.doc accessed 01 March 2018.

[63] Yadav (n 38) 164.

change-induced displacement. The Expert Group proposed a draft SAARC Action Plan on Climate Change with the following objectives.

The objectives of the SAARC Action Plan was to identify and create opportunities for technology and knowledge transfer between member nations by enhancing regional cooperation.[64] It aimed at fostering regional and national action at mitigating climate change and facilitating the development, discussion implementation of the global Plans of the United Nations Framework for Climate Change (e.g. the Bali Plan), by fostering a common understanding of various issues in negotiation and cumulatively reflecting the concerns of the member states.[65]

The action Plan provided a non-exhaustive list of the possible thematic areas and their scope. These included the following:

a. Adaptation to climate change risks and impacts
b. Policy action for climate change mitigation by sharing best practices
c. Policies and action for technology transfer by promoting research and development and understanding the hurdles in technology development
d. Funding of National Action Plan (NAP) in accordance with SAARC norms and assessing the additional fiscal requirements
e. Creating education and awareness about climate change through various modes like mass media and incorporating the same in education curriculum
f. Supervision and regulation of adverse impacts and risks caused by climate change and facilitating capacity building measures for international climate change negotiations.[66]

The responsibility to implement the action Plan was laid on both the institutions under SAARC as well as the governments at the national level.[67] Regional cooperation should be fostered through a mechanism utilizing the SAARC arrangements. Each government through various means like technological interventions, stakeholder's participation, institutional framework should carry out national level implementation.[68]

Consequently, at the 15th SAARC summit which was held in Colombo in 2009 under the theme of 'Partnership for our people'. The member states signed the 41-point Colombo Declaration, which aimed at collective efforts for energy security,

---

[64]South Asian Association for Regional Cooperation, 'SAARC Action Plan on Climate Change 2008' (SAARC Ministerial Meeting on Climate Change, 3 July 2008) http://saarc-sec.org/assets/responsive_filemanager/source/Files%20for%20Areas%20of%20Cooperation/ENB/SAARC%20ACTION%20PLAN%20ON%20CLIMat.%20CHANGE.docx accessed 01 March 2018.

[65]South Asian Association for Regional Cooperation, 'Regional Cooperation on Climate Change Adaptation and Disaster Risk Reduction in South Asia: Road Map' (SAARC Workshop on Climate Change and Disasters—Emerging Trends and Future Strategies, Kathmandu, 21–22 August 2008) http://www.ifrc.org/docs/IDRL/Regional%20strategies%20on%20disaster%20management/Regional%20Cooperation%20on%20Climate%20Change%20Adaption%20on%20Disaster%20Risk%20Reduction%20in%20South%20Asia.pdf accessed 02 March 2018.

[66]SAARC, 'Action Plan' (n 64).

[67]Ibid.

[68]Ibid.

climate change, and terrorism, facilitation of trade, and increased economic growth as well as collective efforts for food security.[69] Focusing particularly on the risks and impacts of climate change, the leaders appealed for an painstaking study on 'Climate Justice: The Human Dimension of Climate Change' to propose a human rights-based approach. The member states advocated the assessment of per capita carbon emissions of Global South and Global North countries on an equitable criterion to cope with climate change.[70] They supported the idea that endeavours aimed at reducing climate change impacts must take into consideration the historical responsibility as per capita emissions and capacity of the nation-states.[71]

Further, a Declaration was adopted by the heads of the government at the 16th SAARC summit at Thimphu, Bhutan, in 2010, the purports to create varied intergovernmental panels supported by different SAARC bodies, to ensure preservation and sustainable utilization of natural resources.[72]

The member states agreed on the following points:

1. Decided to time-bound review and make sure the well-timed enforcement of Dhaka Declaration and Climate Change Action Plan.[73]
2. Resolved to institute an Intergovernmental Expert Group for policy direction and regional collaboration as mooted in the Climate Change Action Plan.[74]
3. Proposed to commission a study through SAARC Secretariat on 'Climate risks in the Region: ways to comprehensively address the related social, economic and environmental challenges'.[75]
4. Commission a research for the development of a mechanism which can provide low-carbon technology, renewable energy and establish a Low-Carbon Research and Development Institute in South Asian University.[76]
5. Member states should strive to establish regional linkages between institutions for exchange of knowledge, strategies and information relevant to climate change.[77]

[69] South Asian Association for Regional Cooperation, 'The Colombo Declaration Partnership for Growth for Our People' (15th SAARC summit, Colombo, 3 August 2008) (Colombo Declaration) http://saarc-sec.org/uploads/digital_library_document/15_-_Colombo,_15th_Summit_2-3_August_2008_-_for_printing.pdf accessed 02 March 2018.

[70] Ibid.

[71] SAARC, 'Colombo Declaration' (n 69).

[72] South Asian Association for Regional Cooperation, 'Thimpu Statement on Climate Change' (Heads of State or Government, 16th SAARC summit, Thimpu, 28–29 April 2010) SAARC/SUMMIT.16/15 (Thimpu Statement) http://saarc-sec.org/assets/responsive_filemanager/source/Files%20for%20Areas%20of%20Cooperation/ENB/THIMPHU%20STATEMENT%20ON%20CLIMat.%20CHANGE.docx accessed 01 March 2018.

[73] Ibid.

[74] SAARC, 'Thimpu Statement' (n 72).

[75] Ibid.

[76] Ibid.

[77] Ibid.

6. Commission an Intergovernmental Mountain supported by the SAARC Forestry Centre focused on the importance of glaciers, particularly on their contribution to sustainable development and livelihoods.[78]
7. Commission an Intergovernmental Monsoon Initiative to assess the pattern of rainfall and its erraticism and vulnerability owing to climate change. The panel should be supported by SAARC Meteorological Research Centre.[79]
8. Commissioning a Climate-Related Disasters Initiative on the integration of disaster risk reduction (DRR) and climate change adaptation (CCA), backed by the SAARC Disaster Management Centre.[80]
9. The accelerating the process of ratification of the SAARC Convention on Cooperation on Environment to facilitate its successful execution at the earliest.[81]

Similarly, it was agreed that the Intergovernmental Expert Group should meet at least twice a year to review and monitor the implementation of the statement and make recommendations to ensure its effective implementation and submit its report to the SAARC environment ministers.[82] The Addu Declaration was signed under the 17th SAARC summit, which was arranged under the theme of 'Building Bridges' in 2011 at the Maldives. The 20-point Addu Declaration focused on ensuring the timely implementation of the Thimpu Statement on Climate Change, making available an appropriate portion of national income for sustainable investment in renewable energy resources and formulating an operational framework to ensure access to clean drinking water.[83]

The summit also saw the adoption of the SAARC Agreement on Rapid Response to Natural Disasters in the region. The fundamental objective of the agreement was to strengthen the resilience of the South Asian nations to disasters and substantially reducing the damages and deprivation in lives and in the socioeconomic and environmental resources and reserves of the SAARC member nations.[84] Article III of the agreement lays down the principles for assistance and management of natural disasters.[85] The primary responsibility to assist in the scenario of a natural disaster rests with the state party so affected.[86] Only on request of assistance will another state party extend its aid to the requesting party. Extension of aid shall be in coherence with sovereignty, territorial integrity and national security of the requesting

---

[78] Ibid.

[79] Ibid.

[80] Ibid.

[81] Ibid.

[82] Ibid.

[83] South Asian Association for Regional Cooperation, 'The Addu Declaration: Building Bridges' (17th SAARC summit, Maldives, 11 November 2011) (Addu Declaration) http://saarc2014.ekantipur.com/pdf/17thSummit2011.pdf accessed 01 March 2018.

[84] South Asian Association for Regional Cooperation, 'Agreement on Rapid Response to Natural Disasters' (17th SAARC summit, Maldives, 11 November 2011) http://saarc-sec.org/uploads/digital_library_document/28_Rapid_response_to_Natural_disasters.pdf accessed 02 March 2018.

[85] Ibid art III.

[86] SAARC, 'Addu Declaration' (n 84) art III (1).

party.[87] Article IV lays down the general obligations of member nations. It obligates them to ensure immediate response to disasters within their territory, provide immediate assistance on request by another member state, share technical information and experiences relevant to mitigating disasters and disaster management, implement[88] Acknowledging the existential threat of climate change for the region, Kathmandu Declaration 2014 emphasized on regional cooperation.[89]

For the last more than one decade, climate change issues have occupied an important position in the SAARC summit from formulating norms and establishing institutions for taking concrete and pragmatic steps. However, effective information and data on the implementation status of Dhaka Declaration and the SAARC Action Plan on Climate Change are lacking despite the accentuation on project execution, cooperation, knowledge sharing, competence fostering and capacity building.[90]

In spite of the shared history and cultural similarity, unlike Association of South East Asian Nations (ASEAN) and other regional organizations, the regional collaboration between the SAARC nations have remained minimal, rhetoric and inconsequential and has not been able to put into practice many of its decisions and realize its commitments and promises.[91] The absence of dedicated leadership, fiscal impediments and lack of cooperation have posed irritants to achieve partnership.[92] The structure of the SAARC with its power asymmetry makes regional cooperation difficult and to form joint positions on common trans-boundary issues.[93] The political tensions and mistrust between India and Pakistan have stood in the way of regional collaboration and have limited impact in the area of environmental protection.[94]

The scenario becomes more complicated when we account for the fact that specific attention to climate migration/displacement has been almost absent at the SAARC policy level. Further at the national level, the migration from climatic and natural disasters concerns has not been conspicuous in the standards, strategies and policies in any of the South Asian countries. In this scenario, where countries at the South

[87] Ibid art III (4).

[88] Ibid art IV.

[89] The Declaration has decided to establish the SAARC Environment and Disaster Management Centre by merging the earlier institutions established the SAARC Forestry Centre (Bhutan), SAARC Disaster Management Centre (India), SAARC Coastal Zones Management Centre (Maldives) and SAARC Meteorological Centre (Bangladesh).

[90] 'The SAARC Declarations on Climate Change' (*Equity and Justice Working Group Bangladesh*) http://www.equitybd.net/wp-content/uploads/2015/10/The-SAARC-Declarations-on-Climate-Change.pdf accessed 05 March 2018.

[91] Dr. Manzoor Ahmad, 'SAARC summits 1985–2016: The Cancellation Phenomenon' (2017) XVII (1) IPRI Journal 43–71.

[92] Ibid, 48.

[93] Raghav Thapar, 'SAARC Ineffective in Promoting Economic Cooperation in South Asia' (2006) 7(1) SJIR https://web.stanford.edu/group/sjir/7.1.03_thapar.html accessed 04 March 2018.

[94] Ibid see also Parvez Hassan, 'Environmental Jurisprudence from Pakistan-Some Lessons for SAARC Region' (South Asia Conference on Environmental Justice, 24–25 March 2012, Bhurban, Pakistan). https://www.iucn.org/sites/dev/files/import/downloads/pk_1_environmental_jurisprudence_from_pakistan___some_lessons_for_the_saarc_region__d.pdf accessed 08 March 2018.

Asian region at national level and SAARC as an organization at the regional level have not taken practical and pre-emptive steps with regard to the recognition and in evolving protective mechanisms in the context of climate migration, we need to evaluate the multiple proposals suggested by the experts to address and rummage the climate-induced displacement at global level.

7.4 Hybrid Integration of International Legal Norms with Regional Aspirations

From the time, the critical and harsh perils of climate displacement and the absence of an effective international legal framework which recognizes climate refugees as an autonomous category have been highlighted by the international reports[95] and academics, multiple solutions and proposals have been suggested by academics and policy-makers to put in place legal norms. These Plans which are capricious in their scope, scale, purpose and background allude to the existence of a massive vacuum in the institutional systems of the international law in contemplating a climate change-induced protection apparatus.

The proposal by Docherty and Ginannini is to create an instrument as a part of the pre-existing legal framework for tackling climate change migration.[96] The instrument is intended to complement the larger existing framework like the UNFCCC rather than replacing it.[97] The proposed instrument does not under its scope cover the entire gamut of environmental refugees and limits the definition to cover only victims of climate change covering both sudden and gradual environmental disruption as well as temporary and permanent.[98] The Plan delineates the climate refugees on the anvil of environmental damage and migration eccentrics.[99] Limiting its scope, the proposal focuses only on refugees whose migration is forced and is forced to cross borders.[100] This distinction is in line with the Refugee Convention. Adopting this distinction prevents the breach of state sovereignty arguments.

The proposal identifies the core principle of protection of climate refugees. This includes laying standards for determination of status, human rights protections and humanitarian assistance. Therefore, refugee status determination of the climate refugees shall be carried out on the basis of legal precedents and unique features of

---

[95] International Organization for Migration; United Nations Refugee Agency; United Nations Office for the Coordination of Humanitarian Affairs, World Bank; Asian Development Bank; Internal Displacement Monitoring Centre; and Friends of the earth.

[96] Bonnie Docherty, Tyler Giannini 'Confronting A Rising Tide: A Proposal for A Convention On Climate Change Refugees' (2009) 33 Harv. Envt'l Rev 349-403.

[97] Ibid, 372–373.

[98] Ibid, 368.

[99] Ibid.

[100] Ibid 370.

climate change-induced migration; however, it is recommended that preference to be made on the group basis.[101]

The proposal guarantees civil, political, economic, social and cultural rights, drawing inspiration from the Refugee Convention.[102] At this juncture, the present instrument goes beyond the provisions of the Refugee Convention. Humanitarian assistance in the form of basic survival needs must be extended to refugees.[103] The instrument seeks to establish the responsibility of three parties, the host state, the home state and the international community. Host states must coordinate with CSOs and international agencies to provide humanitarian aid.[104] The responsibility of the home state is primarily to contribute to solving refugee problems. It is also obligated to facilitate the return of nationals to home state and foster family reunion, in cases where the return to the home State no longer threatens the survival of the refugees. Most importantly, the home state must either eliminate the causes of environmental disruptions causing climate refugees or if such situation arises, undertake adequate and reasonable steps to combat it.[105]

Finally, the responsibility is on the international community towards climate refugees since those persons migrating are mostly poor who contribute negligibly to climate change as compared to developed, sophisticated nations.[106] Thus, those who contribute most must take responsibility for their acts. States must work on principles of cooperation and assistance and discharge their responsibility by using finances from the global fund.[107]

The proposed Protocol by Frank Biermann and Ingrid Boas is called the Protocol for the Recognition, Protection, and Resettlement of Climate Refugees.[108] It will be an Additional Protocol to the UNFCCC implemented by the pre-existing international organizations like United Nations Environment Programme (UNEP), United Nations Development Programme (UNDP) and World Bank.[109] The protocol seeks to extend protection to climate refugees by linking them to the overall regime of climate change and protection. The protocol seeks to use the term 'environmentally displaced persons' as opposed to refugees. It seeks to develop the definition of environmentally displaced persons in line with UNHCR's definition of Internally Displaced Persons (or IDPs).[110] By adopting this approach, there shall be lesser responsibility on the international community. It is, indeed, a fallible approach as it does not address individual displacement and migration instances; however, the

---

101 Ibid, 374.

102 Ibid, 376.

103 Ibid.

104 Ibid, 379.

105 Ibid.

106 Ibid, 382.

107 Ibid.

108 Frank Biermann & Ingrid Boas, 'Protecting Climate Refugees: The Case for a Global Protocol, Environment' (2008) 50(6) Science & Policy for Sustainable Development 8–17.

109 Ibid 15.

110 Ibid, 13.

expression 'environmentally displaced persons' alludes to an individual also that requires to be attended in such situations.

The Protocol is based on five fundamental principles for its viable operation whereunder the core principle shall only be applied to the situations of resettlement and reintegration of population that steadily impacted by climate change as contrary to those situations where people seek disaster relief. Secondly, the Protocol makes an obvious difference on the basis of duration of migration, i.e. whether the migration is temporary or permanent.[111] Thirdly, unlike the UN refugee regime, the Protocol intends to cover groups of people as opposed to individuals.[112] It seeks to address the needs of villages, cities and even entire nations in cases of Small Island States.[113] The areas, which are in need of assistance, will be identified and provided as part of the instrument.[114] The underlying rationale being that environmental disruptions often affect larger sections of the population and have a community impact over a substantial geographical area or/and community. Fourthly, the Protocol shall focus at challenges faced in governance, i.e. lay down guidelines for assistance from international, regional and local community and national agencies, identifying funding resources and procedure, determine resettlement procedure, etc.[115] Finally, the Protocol seeks to urge states to recognize the climate refugees as a global problem and responsibility as opposed to the responsibility of the host state or the home state.[116] Since climate refugees are mostly poor persons from developing/underdeveloped nations who do not have a substantial contribution in global climate change, the Protocol seeks to oblige bigger nations of Global North by the CBDR norms in assisting the countries of Global South through the establishment of a fund.[117]

The proposed new protocols aim to establish new international structures within the confines of UNFCCC; however, in comparison to UNFCCC, UNHCR is the institutional framework well-equipped with expertise to handle refugee emergencies of all categories[118] of human displacements and has the best experiences to deal with the legal and technical requirements of new class of displaced people called the 'climate refugees'.[119] Moreover, incorporation of climate refugees into a comprehensive

[111] (Ibid) 12.

[112] Ibid.

[113] Ibid 12.

[114] Ibid, 14.

[115] (Ibid) 13.

[116] Ibid.

[117] Ibid, 13.

[118] UNHCR has been dealing with the political refugees, asylum seekers, returnees, stateless people and a portion of the world's Internally Displaced Persons (IDPs). See, Frank Biermann and Ingrid Boas, 'Climate Change and Human Migration: Towards a Global Governance System to Protect Climate Refugees' in Jürgen Scheffran, Michael Brzoska, Hans Günter Brauch, Peter Michael Link, Janpeter Schilling (eds), *Climate Change, Human Security and Violent Conflict: Challenges for Societal Stability* (Springer, Berlin, 2012) 291–300.

[119] Stephen Castles, 'Environmental change and forced migration: making sense of the debate: New Issues in Refugee Research' (2002) Oxford Refugee Studies Centre Working Paper 70, 8 http://tinyurl.com/cjrr4cy accessed 11 May 2018.

framework categorizes political refugees distinctly and separately with other interests of the displaced people. Therefore, such categorization might endanger the efficacious and pragmatic legal protection and financial support to the uprooted people who leave, flee, or migrate and relocate away from their roots. Thus, these probable conflicts and the holistic predictability of climate change-induced displacement require groundwork and Plans that must be parleyed, counteracted and manifested in the new Protocol.

The proposal by David Hodgkinson, Tess Burton, Heather Anderson and Lucy Young is for a more comprehensive stand-alone convection including within its ambit IDPs as well as refugees.[120] The proposal provides innovative definitions of the expressions like 'climate change events' and 'climate change displaced persons'.[121] The Convention defines climate change displaced persons (CCDPs) as 'groups of people whose habitual homes have become or will, on balance of probabilities, become temporarily or permanently uninhabitable as a consequence of a climate change event'.[122] The Convention proposes *en-masse* designation of refugee status as opposed to individual cases. A climate change event is defined as 'sudden or gradual environmental disruption that is consistent with climate change and to which humans very likely contributed'.[123]

The Convention has been proposed specifically focusing on the vulnerability of small island states, drawing upon the idea that these states might become inhabitable owing to climate change. As the proposed Convention covers as well as internal and external movement,[124] the matter of sovereignty is tackled by making available rights to all the stakeholders to seek aid and assistance. However, as per the Convention, merely fulfilling the criteria envisaged in the definition hereinabove will not *per se* grant status of 'climate refugees' to the group of displaced people in question who got displaced due to climate change. There is a formal requirement of recognition of status as climate refugees under the definition. Such recognition shall be founded on the scientific research studies, sharing of information by the affected communities and international organizations.[125]

The Plan also delineates governance architecture of the Convention and the institutional structure and obligations of the parties thereunder regarding aid and humanitarian assistance. The principle institution under this structure is called the Climate Change Displacement Organisation (CCDO).[126] The principle organs of the CCDO consist of an Assembly, a council, a Climate Change Displacement Environment

---

[120] David Hodgkinson, Tess Burton, Heather Anderson and Lucy Young, 'Convention for Persons Displaced by Climate Change', (2010) 36(1) Monash L. Rev. 69 http://www5.austlii.edu.au/au/journals/MonashULawRw/2010/4.html accessed 01 May 2018.

[121] Ibid.

[122] Ibid.

[123] Ibid.

[124] Kniveton et al., *Climate Change and Migration: Improving Methodologies to Estimate Flows* 30 (Research Series No 33, International Organisation for Migration, 2008).

[125] Hodgkinson, Burton, Anderson and Young (n 120).

[126] Ibid.

and Science Organization (CCDESO), a Climate Displacement Fund, and Climate Change Displacement Implementation Groups.[127] The high contracting parties to the Convention shall play a critical role and responsibilities by requesting the member states for resettlement assistance and *en-masse* designation.[128]

The Convention lays down for mandatory contribution from developed state parties under the Convention based on the principle of CBDR. The distinct level of contribution by such state party shall be gauged by the CCDESO by measuring the emissions.[129] The displacement agreement between the Small Island States and the host state must operate under the aegis of the CCDO and thrive on three principles, i.e. the proximity of resettlement (close to their cultural area), self-determination and preservation of intangible culture.

Despite the fact that it is across-the-board Convention its principal concentration is on small island nations. However, the issue of climate change-induced displacement is not limited to small island nations; rather, it is the issue of universal and multicausal ramifications that has been impacting the vulnerable populations of all nation-states. Moreover, the principle focus of the Convention is on aid and assistance and fiscal help constitutes a significant part of the protection of the climate refugees. The crystallization of main fundamental legal rights frames the central component of such protection instruments, but the instant Convention lacks the same in its representation. Furthermore, there is an apprehension of getting cases neglected since the ambit of the Convention is on group recognition. A regionally oriented programme was proposed by Angela Williams[130] to combat climate refugees and to build on the existing geopolitical and economic relationships, allowing states to develop responsive policies, which are localized and suited to their particular situation. Moreover, the regime thrives on the hypothesis that regional agreements ensure a greater level of commitment as opposed to that commitment achieved at the international level owing to inter-regional disparities.[131] For example, the Regional Seas Programme under the United Nations Law of the Seas Convention (UNCLOS) has been in operation for over two decades and obligates state parties to protect the shared marine environment by concluding various regional programs and implementing regional action Plans.[132]

The regional programme will develop under the existing framework of UNFCCC.[133] Article 4(1)(b) of UNFCCC states that 'parties will take measures to facilitate adequate adaptation to climate change'.[134] In the same way, the article 19 (b) of the Kyoto Protocol states that the parties shall 'formulate implementation,

---

[127]Ibid.

[128]Ibid.

[129]Ibid.

[130]Angela Williams, 'Turning the Tide: Recognizing Climate Change Refugees in International Law' (2008) 30(4) Law and Policy 502–529.

[131]Ibid, 517.

[132]Ibid, 518.

[133]Ibid, 519.

[134]Ibid.

publish and regularly improve national and where appropriate regional programs having measures to mitigate climate change and measure to facilitate adequate adaptation to climate change'.[135] These obligations to foster adaptation to climate change can reasonably be presumed to cover combating displacement coerced by climate change. The first suggestion that the proposal puts forth is that it must explicitly acknowledge the relationship between climate change and human migration.[136] It must also foster regional cooperation and provision of humanitarian assistance. As a third requirement, the proposal suggests that there must be a non-binding MOU or Plan of Action which consists of strategies and Plans to implement various provisions of the regional instruments and implement regional initiatives.[137]

This system will recognize the broader problem of climate refugees existing at the international level, leaving room for determination of details at the local level. Since the system thrives on pre-existing regional associations (e.g. the African Union, Organization of American States and SAARC) or strong regional bonds (e.g. the South Pacific), the operation of this instrument is comparatively smoother and viable as opposed to bilateral or multilateral instruments at the international level, establishing novel relationships.[138] A subsidiary body shall be placed within the UNFCCC which will be responsible for coordination between the local initiatives and also provide for inter-regional interaction, whenever possible.[139]

In attempting to define 'climate refugees', directorial and guidance-oriented definition shall be offered that shall be malleable to accommodate regional distinctions. Every regional grouping shall have the flexibility and liberty to delineate it as per their regional aspirations while tuning such definitions with the comprehensive definition. Such a threshold of flexibility in the definition would provide a leverage to address the perennial degradation and dilapidation in the environmental reserves and resources that might bring a lot of individuals within the purview of such definition. The definition consists of a 'graduating scale', i.e. the level of protection provided to a climate change refugee should directly proportional to the intensity of the environmental disruption. Hypothetically, a scale must be perceived, with one end as 'acute' cases and the opposite end as 'chronic' cases. Acute cases are those climate change refugees who are affected by severe problems, threatening their survival. These deserve the maximum amount of protection. Chronic cases include victims of gradual climate change, the impacts of which slowly make habitation treacherous. Furthermore, the definition must incorporate the causes of displacement. This will not only aid in recognizing the people who require the legal protection and ward off future recurrences of climate change-induced displacement in the regions that are prone to displacement due to such reasons. Moreover, a holistic understanding and a critical appreciation of the problem require the inclusion of displacement causes as the same will lead to pragmatic resolutions.

---

[135]Ibid 519.

[136]Ibid 520.

[137]Ibid.

[138]Ibid, 521.

[139]Ibid, 522.

Williams's alternative proposal for a 'regionally oriented regime'[140] is good and could be the basis of further deliberation, but there is a requirement to have the New Global Treaty/Additional Protocol both at regional and international levels on 'climate refugees' to address the current gaps in the protection of climate refugees, but it should operate under the UNHCR and not under UNFCCC and Kyoto Protocol as these two instruments are limited to the advocacy for regional cooperation for adaptation strategies.

These proposals represent the growing attention paid by international community to the all-pervading issues of climate displacement. These proposals have provided nuance to the concept of 'climate refugees'. These proposals have emerged in the background of a clear consensus on the term of 'climate refugees' and lack of an international institution or legal instrument to acknowledge, problematize and comprehensively address climate displacement facets. These proposals, though differ considerably in its scope, definition and coverage have highlighted the core principles of assistance and rely on the differential responsibility of the developed nations to assist the developing nations to address the issue of climate displacement. However, these proposals have taken a unidimensional approach and fail to capture the legal and policy implications of climate change-related migration expurgated across many different fields, including human rights, development, humanitarian assistance, asylum, immigration and the environment primarily touching upon environmental law, refugee law and human rights law. The need is to approach the issue with a holistic approach beyond the confines of subject-specific approach. Currently, climate change law addresses the question of states obligation in terms of mitigation and adaptation. Refugee law lays down the standard of protection afforded to refugees. International human rights law provides the minimum protection standards based on human dignity to all. The need is to utilize the hybrid integration of international legal norms in an inclusive and complementary manner.[141] The concept of hybrid international law represents an amalgamation of three branches of international law consisting of perspectives from environment, human rights and refugee law.[142] There are immense benefits flowing out of the use of hybrid system as it helps to capture the real complexity of the situation and develop holistic norms to address the climate interrelated and convoluted migration crisis.[143] In addition, the applications of hybrid approach can emphasis the human impact of climate issue, which the state-centric international law is unable to focus.[144] Though initially represented at the context of

---

[140]Ibid, 502, 519.

[141]Cosmin Corendea, 'Regionalism, Human Rights and Migration in Relation to Climate Change' (*The Conversation*, 1 March 2018) https://theconversation.com/regionalism-human-rights-and-migration-in-relation-to-climate-change-90129 accessed 10 April 2018.

[142]Ibid.

[143]Ibid.

[144]Ibid.

Pacific low-lying islands, there is enough research available to prove that the hybrid principle can be extended to other regions[145] including South Asia.

The traditional international law had Eurocentric predilections and developed initially with Eurocentric ethos and conviction with an imperial character sans the Third-World Alignment.[146] In the post-colonialism[147] era, there was a stratification of international human rights law, international environmental law, international refugee law, Investment law, and trade law with the specialized study and research of these different disciplines having the diverse participation of developing nations. However, international norms and principles claimed to have been developed in a Eurocentric tinge were absolute in nature and devoid of any observation for collaboration and interaction between different disciplines.[148] But such a stratification of international law individualistically was embellished with legal devices to address the climate change, human rights, and refugee catastrophe within the polemical gamut of Eurocentric development of international law.[149] However, when faced with multifaceted issues the stratification of these disciplines proved to be vulnerable in offering an acceptable solution. Climate change is such a defining moment of predicament for the humanity and challenges the foundation and understanding of international law. Therefore, law and societies must be vibrant and dynamic while responding vigorously to such challenges. Thus, climate change provides a crucial occasion for the international community to respond and react in an unprecedented and innovative manner by dismantling the traditional approach of law and adopting the concept of hybrid law. In the words of Ben Wisner[150] et al., 'Ironically, climate change offers humanity an opportunity for a quantum leap in sustainable development and peace-making' to stabilize the contemporary climate change-induced human mobility.

Therefore, the real challenge for international law is to scrutinise various fragments, break the barriers of that fragmentation and lack of synergy between different branches of international law while addressing the climate change-induced displacement and migration. In this scenario, where the term climate refugees do not have legal status, and an adequate legal protection apparatus is nonexistent at international law, the solutions can be sought from developing a bottom-up approach and

---

[145]Cosmin Corendea, 'Hybrid Legal Approaches Towards Climate Change: Concepts, Mechanisms and Implementation' (2016) 21(1) Ann. Surv. Int'l & Comp. L29-41.

[146]R. P. Anand, *Studies in International Law and History: An Asian Perspective* (Leiden, Boston and New Delhi: Martinus Nijhoff and Lancer's Book, 2004), 146 and 165–66; Adil Najam, 'Developing Countries and Global Environmental Governance: From Contestation to Participation to Engagement', *International Environmental Agreements: Politics, Law and Economics (Springer)*, 5/3: 303–321 (2005).

[147]Gurminder K Bhambra, 'Postcolonial and Decolonial dialogues' (2014) 17(2) Journal of Postcolonial Studies 115–121.

[148]Ram Prakash Anand, *Studies in International Law and History: An Asian Perspective* (Martinus Nijhoff Publishers, Leiden, 2014).

[149]María José Fernández, 'Refugees, climate change and international law' (2015) 49 FMR 42–43 http://www.fmreview.org/sites/fmr/files/FMRdownloads/en/climatechange-disasters/fernandez.pdf accessed 07 April 2018.

[150]Ben Wisner et all 'Climate Change and Human Security', 2007 1–116, 11. http://www.afespress.de/pdf/ClimateChangeandHumanSecurity.pdf accessed 10 June 2018.

envisaging cooperative mechanisms at regional level and national level through the integration of hybrid legal norms and gradually strive to evolve legal mechanisms promoting enhanced options for assuring dignified life to the displaced based on equitable burden sharing by the international community.[151]

The hybrid concept of law is a process of reformulating the existing legal norms rather than creating new ones. This is based on the assumption that in most of the complex scenario where the effects span to multiple branches of law, the legal gaps are essentially the result of lack of interconnectedness. Since the existing legal frameworks have miserably failed to address the climate change nemesis in a holistic manner there is a need to have a hybrid approach to deal with climate change-induced human sufferings based on the improvement and amalgamation of multiple branches of international law backed by the rule of law at the national, regional and global levels. The rule of law has to be central to the hybrid approach for ensuring the protection of rights, mitigation of risks and construction of resilience while empowering the impactees and enabling the pragmatic migration to construct the international hybrid climate refugee law. The hybrid approach contemplates that process of climate change migration must have requisite coordination and suppleness in conformity with the latest developments in international law.

Primarily, climate change violates human rights and decimates human dignity to the insurmountable extent that forces the people to flee, migrate and relocate themselves. Therefore, the hybrid approach is an interwoven proposition of law that represents a paradigmatic shift whereunder the duty of the nation-states has been reconceptualized for protecting the climate refugees by the climate refugee generating and hosting countries. The hybrid approach ensures the protection of human rights, and its application expands the gamut of state responsibility to rescue the people who got displaced due to the climate change-induced humanitarian crisis. For instance, even if the definition under the Refugee Convention is not expanded to include climate-displaced population, human rights-centred principles can be applied as part of the hybrid law to provide protection to climate displacement. Parties to the UNFCCC in the Bula Zone and Non-Party Stakeholders in the Bonn Zone participated in the COP 23 on Climate Talks in Germany at Bonn on 6–17 November 2017 to operationalize the Paris Agreement to 'halting the critical global warming levels and tackling the divisive issue of damages' to climate migrants with a human rights orientation. Similarly, the refugee law does not address the climate refugees, but the 1951 UNCSR has the principle of *non-refoulement* in its article 33 (1) that could be integrated into the international hybrid climate refugee law as it is equally legitimate and valid to protect the climate refugees. Thus, in the hostile scenario of climate change-induced displacement and migration coupled with anthropogenic actions inimical to international ecological stability, climate change crisis requires an international hybrid law at the national, regional and global stages while attending the situations of climate change migration, adaptation, and mitigation and climate change litigations. Moreover, it is considerably pertinent in case of the Refugee Con-

[151] Susan Martin, 'Climate Change, Migration, and Governance', (2010) Global Governance 16(3) 397–414.

vention that has not been signed by any SAARC member state except Afghanistan. SAARC states have regularly critiqued the failure of the normative and procedural substance of the Refugee Convention in addressing the grave anxieties of the Global South countries. In this context, the hybrid law can comparatively be the most appropriate response for principally two factors. Firstly, nation-states might not be in favour of amending the Refugee Convention to the extent of incorporating the climate refugees in its definition under article 1A (2) as it might jeopardize the purposes of the Convention. Secondly, the malleable nature of hybrid law accommodates the viewpoints of the Global South countries to be taken into consideration. There is no specific formula for the adoption of hybrid mechanisms and can be based on varied process and procedures of integration. Hybrid approach along with international hybrid climate refugee law can be developed as a groundbreaking and pioneering initiative through mechansims envisaged in the Paris Agreement like the progressive participation of non-state actors, determination of loss, liability, and damage and enactment of national climate change legislations with a focus on the interconnectedness between human rights and climate change. Ultimately, it is the human-driven solution that must be the focus of international hybrid climate refugee law to protect climate change refugees or climate refugees.

7.5 The Need for Regional Trans-boundary Cooperation to Address Climate Displacement and Migration

The first step towards building a hybrid international law approach to climate change migration is to envisage cooperative mechanisms at regional level built on the foundation of negotiation, dialogue conversation, cooperation, personal and community narrative and good will. Such an accommodating and obliging machinery makes realistic significance[152] as climate migration in the initial time is expected to be primarily internal and intra-regional predominantly for South Asia due to porous borders.

South Asian countries except Afghanistan are not parties to the Refugee Convention, and countries in the region have not elaborated on the legal framework. However, South Asia has been receiving and offering protective mechanisms to thousand of refugees from different parts of the world. Thus, given the historical experience of dealing with the refugee crisis, the countries in the region are well placed to evolve a cooperative regional mechanism to deal with climate refugee crisis.

For viable, effective and sustainable regional cooperative mechanisms, the SAARC needs to challenge its institutional and organizational structure. However, cooperation can be built by reinvigorating the existing institutional structure in the region.[153] There is a need of regional consultations based on consensus to develop a protection road map to address the requirements and concerns of people confronting

[152]Benoit Mayer, 'Environmental Migration: Prospects for a Regional Governance in the Asia-Pacific Region' (2013) 16 APJEL 77–103, 91.

[153]Mishra et al., (n 4) 111.

climate change-induced migration. Such consultation can be instrumental in shaping the response architecture amenable to local sensibilities, and regional consultations can be an occasion for victims of climate change-induced displacement to share their personal narratives to bring out their specific needs for legal protection based on the nature and pattern of displacement. For instance, people fleeing sudden environmental disaster may not have access to essential service, and as a result thereof displacement can be particularly harsh on women and children bringing a mass of human rights violations including trafficking and violence.[154]

SAARC Action Plan on Climate Change[155] and Thimbu statement on climate change[156] has already emphasized multiple cooperative strategies and thematic areas of regional cooperation to address climate change. These envisaged measures include the establishment of Intergovernmental Expert Group for policy direction and regional cooperation, institutional establishment like Disaster Management Centre, Coastal Zone Development, Forestry Centre and Meteorological Research Centre emphasis information sharing, cooperation and capacity building.[157] One of the successful examples of regional cooperation has been the development of the South Asian Disaster Knowledge Network (SADKN) as an access point to knowledge and information dissemination on disaster risk management in the SAARC jurisdictions.[158] In the context of climate displacement, the need is to focus on enhanced responsibility and impact of climate change.[159] SAARC summits have also emphasized research collaboration and information sharing in addressing knowledge gaps and capacity development, an active role for people's participation through civil society consultation, and the establishment of regional climate funds and south—south technology transfer.[160] Planned relocation is an initial strategy followed by the government when faced with the threat of displacement and especially in the wake of sudden disasters. The process is complicated and depends on a host of factors including availability of alternative safe lands and financial resources. In the context of Maldives and Sri Lanka, governments have successfully implemented many schemes of the planned relocation of people to other safe Islands in the aftermath of tsunami.[161] In Nepal, the severe water crisis has forced a planned relocation of

154 The Climate Change- Human Trafficking Nexus. 5 https://publications.iom.int/system/files/pdf/mecc_infosheet_climate_change_nexus.pdf accessed 02 March 2018.

155 SAARC, 'Action Plan' (n 64).

156 SAARC, 'Thimpu Statement' (n 72).

157 Ibid.

158 SAARC Disaster Management Centre, 'The South Asian Disaster Knowledge Network (SADKN)' http://saarc-sdmc.org/south-asian-disaster-knowledge-network-sadkn accessed 02 March 2018.

159 Subir Gokarn, 'Climate Change Imperatives for Regional Cooperation' in Brookings India (ed), *Reinvigorating SAARC India's Opportunities and Challenges* (Brookings India, 2014) 27–28 https://www.brookings.edu/wp-content/uploads/2014/11/saarc_briefing-book.pdf accessed 12 May 2018.

160 Ibid.

161 Nansen Initiative Regional Consultation (n 9) 17.

people from the village in Upper Mustang District.[162] Regional funds and technological cooperation can be utilized in planning for planned relocation in areas highly susceptible to climate change. A successfully planned relocation can often limit the displacement from crossing national boundaries.

In this context, the Temporary Protection Directive of European Union is significant. The directive aims to prescribe minimum standards for providing temporary protection in the wake of a mass influx situation of displaced people from third countries who are not able to return to their country of nationality or origin",[163] and the directive has not been put into practice much and has remained a dead letter directive.[164] Likewise, some instances can be quoted in the context of Africa where temporary protections were accorded. People escaping after the outbreak of Mount Nyiragongo in the Democratic Republic of Congo were allowed to remain and stay in Uganda.[165] Botswana and Tanzania have also admitted people are fleeing flooding in neighbouring states.[166] There is much for South Asia to adopt such a mechanism under the aegis of SAARC for promoting coordination between states, to improve upon existing laws, policies and practices in the SAARC region and to assist mass displaced population. However, as a precondition for temporary protection, countries should follow the respect for non-refoulement commitment flowing out of international law. An ADB study has specified the fundamental principles for regional cooperation in SAARC jurisdictions in the context of climate change. These principles consist of the development of climate resilience strategies, regional collaboration on adaptation to climate change, strengthening existing regional institutional framework and mandate, knowledge sharing through collaboration and capacity building and exposition of political position.[167] Though the study does not explicitly refer to climate displacement, the emphasis of knowledge sharing adaptation and institutional cooperation can be utilized in the context of evolving better understanding of the dynamics of climate displacement. Overall South Asia has a lot to benefit from a coordinated and collaborated action Plan on climate change-induced displacement. The norms originating from such cooperation could form the basis for a vibrant regional regime on climate refugees.

---

[162]Daniel Bernet, Fleeing Drought, https://www.dandc.eu/en/article/climate-change-nepal-entire-villages-must-relocate-because-water-scarcity-getting-worse, accessed 13 June, 2018.

[163]Giovanni Sciaccaluga, 'Sudden-Onset Disasters, Human Displacement, and the Temporary Protection Directive: Space for a Promising Relationship?' in Giovanni Carlo Bruno, Fulvio Maria Palombino and Valentina Rossi (eds), *Migration and the Environment: Some Reflections on Current Legal Issues and Possible Ways Forward* (CNR Edizioni, 2017) 75–100, 78 https://www.iriss.cnr.it/wp-content/uploads/2018/02/Migration_and_the_Environment_2017.pdf accessed 12 June 2018.

[164]Ibid, 82.

[165]Tamara Wood, 'Developing temporary protection in Africa' (2015) 49 FMR 23–25 http://www.fmreview.org/climatechange-disasters/wood.html accessed 10 May 2018.

[166]Ibid.

[167]Mishra et al., (n 4) 3.

7.6 Advocacy for an Alternative Regional Regime on Climate Refugees

In the third chapter, authors have made a call for the establishment of an Additional Protocol to the Geneva Convention which can take into account the plight of climate displacement. However, one of the prerequisites of such a Convention will be the participation and acceptance of refugee regime by the countries including South Asian nations. Such an initiative will depend on the reformulation of refugee framework by incorporating and addressing the concerns of developing nations. Given the political and structural apprehension, the probability of such a reformulation appears far-fetched. In this background envisaging regional cooperative mechanisms, by looking at the other regional initiatives could be paramount.

In the regional milieu, the Organization of African Unity (OAU) 'Convention Governing the Specific Aspects of Refugee Problems in Africa' has accepted a comprehensive definition of the term 'refugee'.[168] They have achieved this by interjecting an additional paragraph, to the definition of the expression 'refugee' which broaches that the term 'refugee' shall also be applicable to every person who, '*owing to external aggression, foreign domination, occupation, or events seriously disturbing public order in either part or the whole of his country of origin or nationality, is compelled to leave his place of habitual residence in order to seek refuge in another place outside his country of origin or nationality*'[169] The experts point out that the term 'events seriously disturbing public order' could include those caused by natural disasters.[170] The Lomé Declaration on Climate Change and Protection of Civilians in West Africa of the Economic Community of West African States (ECOWAS)[171] is immensely important. The Declaration addresses the requirements of the populations affected with the environmental disasters by establishing the fund. Thus, the recent adoption of Kampala Convention on IDPs has acknowledged the natural calamities and disasters as a cause of displacement.[172] The 'IDPs' are defined as 'persons or groups of

---

[168] Convention Governing the Specific Aspects of Refugee Problems in Africa (adopted 10 September 1969) 1001 UNTS 45 (OAU Convention) http://www.refworld.org/docid/3ae6b36018.html accessed 15 June 2018.

[169] Ibid.

[170] Koko Warner et al., 'Climate Change, Environmental Degradation, and Migration' (2010) 55(3) Natural Hazards: Journal of the International Society for the Prevention and Mitigation of Natural Hazards 689–715.

[171] 'Lomé Declaration on Climate Change and Protection of Civilians in West Africa' (The Regional Conference on Protection Challenges to Climate Change in West Africa, Lome, Togo, 16 September 2009). See generally Giovanni Carlo Bruno, Fulvio Maria Palombino and Valentina Rossi (eds), *Migration and the Environment: Some Reflections on Current Legal Issues and Possible Ways Forward* (CNR Edizioni, 2017) https://www.iriss.cnr.it/wp-content/uploads/2018/02/Migration_and_the_Environment_2017.pdf accessed 12 June 2018.

[172] African Union Convention for the Protection and Assistance of Internally Displaced Persons in Africa (adopted 23 October 2009) (Kampala Convention) https://au.int/sites/default/files/treaties/7796-treaty-0039_-_kampala_convention_african_union_convention_for_the_protection_and_assistance_of_internally_displaced_persons_in_africa_e.pdf accessed 25 May 2018.

persons who have been forced or obliged to flee or to leave their homes or places of habitual residence, in particular as a result of or in order to avoid the effects of armed conflict, situations of generalized violence, violations of human rights or natural or human-made disasters, and who have not crossed an internationally recognized state border'.[173] The Convention is a significant step in linking climate change to disasters and consequent displacement.

As pointed by Abebe that 'the Convention panoramically attempts to prevent and mitigate the climate change-induced displacement and offers protection to the displaced persons due to natural disasters and climate change. It exhorts the states to create conditions congenial to sustainable and permanent solution to the problem of IDPs. It also makes provisions for the establishment of national and regional infrastructure for advanced warning system, disaster risk reduction mechanism and coordination for humanitarian assistance. However, the core concentration of the Convention is on the protection of persons and groups of displaced within the national borders of a country. It strongly advocates the prevention of displacement and the exploration for durable solution methods. Consequently, many provisions of the Convention will cause essential but indirect impacts upon cross-border climate change displacement'.[174]

These regional instruments might not have accomplished the desired targets, but these Conventions would prove most fruitful to check the future cross-border migrations across the borders.[175] Compared to the truncated reception of global standards and the contestation, which still characterizes the global refugee regime, the participation of countries in the regional mechanisms needs to be appreciated.[176] Such a regional approach could also be instrumental in highlighting the unique model of governance representing Asian particularities advocated by Third-World Approaches of International Law (TWAIL) scholars.[177]

In the context of South Asian countries, they have adopted national policies legal frameworks in place to address climate change, disasters and migration which could be relevant in addressing the plight of climate-displaced population. However, South Asia has not initiated or evolved a regional or sub-regional protection mechanism.[178] There has been growing awareness and demand from the civil society institutions and organizations to develop and formulate a SAARC Charter for providing the rights of free movements to 'climate refugees' within the South Asian region.

The climate change-induced migration has been surpassing the national borders; therefore, genuine state-centric national response command machinery is inadequate

[173] Ibid art 1(k).

[174] Allehone Mulugeta Abebe, 'The Kampala Convention and environmentally induced displacement in Africa' (IOM Intersessional Workshop on Climate Change, Environmental Degradation and Migration, Geneva, Switzerland, 29–30 March 2011) 1https://www.iom.int/jahia/webdav/shared/shared/mainsite/microsites/IDM/workshops/climate-change-2011/SessionIII-Paper-Allehone-Mulugeta-Abebe.pdfaccessed 10 June 2018.

[175] Ferris (n 151).

[176] Mayer, 'Environmental Migration' (n 152) 92.

[177] Ibid, 95.

[178] Nansen Initiative Regional Consultation (n 9).

and unsatisfactory and requires an international response mechanism. Since climate change impacts intersect with local realities in complex ways to complicate migration patterns, a regional approach seems to be the starting point. As pointed out by Angela Williams, a regional approach has inherent advantages as it allows engagements at multiple levels based on the capacity and severity of the issue.[179] She also feels that a series of regional arrangement can be instrumental in the evolution of customary international law.[180] Thereafter, the CIL will play a formative role in the progressive development of regional regimes that will enrich the living standards of communities worldwide. Therefore, there is an emergent requirement to pursue a regional approach away from the traditional perceptions about the treatment of climate refugees produced by the climate change-induced displacement and migration. Historically, refugees have been recognized within the mandate of the UNCSR framework. However, the problems ascend when efforts are made to fragment a new class of refugees with legal obligation backed by the evolutionary CIL into pre-existing UNCSR adopted for a different category of refugees. Therefore, the current trends of refugee migration due to climate change become the determining ground for specific recognition and protection of climate refugees. The climate change phenomenon has created a new class of refugees requiring an autonomous status under the international law.

Therefore, in 1994, Former UNHCR Mrs. Sadako Ogata commenced the 'Regional Consultations on Refugees and Migratory Movements in South Asia' and completed the marathon procedure of regional consultations. Thus, SAARC assemblage concluded and adopted a Regional Declaration and National Model Law on Refugees in 2004 with strategies and common approaches for bolstering and consolidating refugee protection and deepening the quest for durable solutions, and its main features may be appreciated as under:

- That the 2004 SAARC Regional Declaration (SRD) and National Model Law on Refugees[181] have adopted a harmonized approach for reiterating the principles of refugee protection and reaffirming the requirement for SAARC states to promulgate national legislation on refugees;
- That the SAARC region has the similar reasons and causes of human displacement and migration such as intra-country conflict, human rights abuses and other forms of violence that has been affecting other regions of the world;[182]

---

[179] Williams (n 130) 518.

[180] Ibid, 522.

[181] Eminent Persons Group (EPG), 'The South Asia Declaration on Refugees' (Seventh Informal Regional Consultation of the EPG, New Delhi, 15–16 December 2002) http://shodhganga.inflibnet.ac.in/bitstream/10603/28291/17/17_appendices.pdf accessed 12 April 2018.

[182] Refugee crisis in South East Asia (Rohingya Refugees), South Asia, Syria, Africa and Far East and other regions of the world.

- That the SRD has adopted the definition under 4 (a) and (b) clause based on the definition under article 1 (A) (2) of the 1951 UNCSR[183] that has universal application for the refugee definition;
- That the SRD has identified the contemporary nature of conflicts and linked its definition to inter-ethnic violence and ethnic identity has also been included as per the definition of 1951 UNCSR;
- That the 1951 UNCSR has a ground of membership of a particular social group that includes gender-based persecution and the same has also been reflected upon in the SRD;
- That the SRD definition clause 4 (b) is grounded on a broader definition postulated under article 1 (2) of the 1969 OAU[184] that is in harmony with the advancement and safeguarding of human rights throughout the world resulting from the principles and purposes of the United Nations Charter;
- That the SRD definition clause 4 (b) also recognizes the massive violations of human rights across the world similar to the 1984 Cartagena Declaration on Refugees wherein the ground of massive violation of human rights has also been incorporated addressing the asylum seekers as refugees;
- That the SRD definition also respects and incorporates the principle of *non-refoulement* provided under article 33 (1) of the 1951 UNCSR and article 3 of the 1984 CAT[185];
- That the clause 5 of the SRD excludes people or refugees who have committed political crimes recognized and excepted under article 33 (2) of the 1951 UNCSR; a crime against peace, war crimes or crimes against humanity, as defined under international human rights and IHL instruments; and a profound non-political crimes beyond the country of reception or asylum prior to his/her entry or admission into this country or country of refuge or reception as a refugee.
- That the SRD clause 12 on the 'Determination of Refugee Status' is based on UNHCR ExCom Conclusion No. 8(XXVIII)-1977;
- That the SRD clause 15 grants and recognizes the rights and duties of the refugees and asylum seekers as per the 1951 UNCSR; and
- That the SRD clause 17 on voluntary repatriation derives its validity and legitimacy from UNHCR ExCom Conclusion 40 (XXXVI)-1985 at Para (b).

But, unfortunately, 2004 SRD does not implicitly or explicitly incorporate 'climate change-induced displacement and migration' as a ground to seek or claim asylum or refugee status in another country. Therefore, the impugned 2004 SRD must define and take climate change into account by organizing another around of regional consultation on climate change ground to recognize and protect climate refugees on the

[183]United Nations Convention Relating to the Status of Refugees (adopted 28 July 1951) 189 UNTS 137.

[184]OAU Convention (n 168), art. 1(2).

[185]United Nations Convention Against Torture and Other Cruel, Inhuman or Degrading Treatment or Punishment (adopted 10 December 1984) 1465 UNTS 85 http://www.refworld.org/docid/3ae6b3a94.html accessed 14 May 2018.

following considerations of Myers,[186] formulations that must be reflected upon while emphasizing the South Asian Regional Model Law on Refugees:

- That an immediate, rapid and sustainable response mechanism must be emplaced;
- That reinforcing respect for individual and collective rights must be central to protection narrative;
- That a global approach to climate change-induced displacement and migration must be the legal hallmark of their protection, recognition and settlement;
- That the principle of burden-sharing based on CBDR[187] must also be promoted;
- That the SAARC Joint Commission on Refugees must also be established as a part of implementation machinery while ensuring the participation and coordination of all the stakeholders of the climate change migration;
- That the implementation framework must also incorporate the accessibility of essential services to climate change displaced people that address livelihood opportunities, healthcare services, settlement, employment, education and lego-institutional access;
- That the dependence on global burden-sharing principle based on CBDR framework must be emphasized along with south–south fiscal cooperation;
- That the SAARC Development Fund[188] (SDF) could perform the task of routing the fund from the developed world to the SAARC region for climate change purposes; and
- That the support of agencies such as ADB with the mandate and institutional ability to route the funds must also be ensured.[189]

South Asian Model Law on Refugees offers a significant avenue for the elaboration of hybrid approaches. The Model law can incorporate the humanitarian principle of protection for climate-displaced persons including the grant of refugee status, relaxed visa rules and rules on stay. Similarly, it should invoke the human rights principles and focus on the rights of the displaced population on the anvil of human dignity. The core principle of non-refoulement which forbids returning a person to any country where his or her life is threatened should be part of the legal initiative. It should also embody international environmental law obligations and standards for states to managing climate risk in the country which points to taking measures of climate adaptation programs, by enhancing the resilience and adaptive capacity of the people. However, a complete focus on the regional initiative can have practical

[186]Benoit Mayer, 'The International Legal Challenges of Climate-Induced Migration: Proposal for an International Legal Framework' (2011) 22(3) Colo. J. Int'l Envt'l L. & Poli'y, 357–416 https://www.colorado.edu/law/sites/default/files/Mayer%20(Corrected)-S.pdf accessed 12 May 2018.

[187]Ibid, 394–396.

[188]In 1996, a first funding mechanism was created in SAARC, 'South Asian Development Fund (SADF), merging the SAARC Fund for Regional Projects (SFRP) and the SAARC Regional Fund. SADF objectives were to support industrial development, poverty alleviation, protection of the environment, institutional/human resource development and promotion of social and infrastructure development projects in the SAARC region. See, 'About SDF' (*SAARC Development Fund*) http://www.sdfsec.org/about-sdf accessed 11 May 2018.

[189]Mayer, 'The International Legal Challenges' (n 186) 372.

repercussions. Firstly international law shows a regrettable lack of norms language and principles guiding climate displacement. Secondly, a diversified and inflexible regional approach can generate troubles and bottlenecks in achieving an international consensus and acclimatizing to climate change-induced displacement scenario is an extremely costly affair, which requires generous international aid. Many of the regional bodies such as Organization of African Unity (OAU) and the SAARC are evolved between underdeveloped nations and operate within severe financial difficulties. A magnanimous and liberal enterprise for global aid and humanitarian assistance constitutes the basis for a successful and pragmatic regional initiative to address the climate change-induced displacement. A global fund with compulsory participation and financial contributions from the developed states that are responsible for climate change should be an inherent component of any legal initiative to address climate change. Such a system and procedure must accentuate on focusing their funds to the nation-states in vulnerable situations of addressing such low-lying islands[190] and must act as 'climate debt mechanisms born out of the need for climate justice and reparation'.

In this scenario, resort to international law principles to look for synergy and cooperation becomes imperative. International law has sufficient avenues to make nations of Global North accountable for assisting the countries of Global South while addressing the pernicious outcomes of climate change. These standards are based on the humanitarian assistance and international treaty obligation[191] and principle of CBDR. The Global North countries are obligated under the article 4 (4) of the UNFCCC to assist the developing country parties that are particularly vulnerable to the adverse effects of climate change.[192] The fundamental norms of CBDR are germane to climate change-induced displacement milieu because most of projected risk areas and affected people are those who have least contributed to climate change and they cannot adapt to climate change.

Thus, the regional mechanisms, which the chapter advocates, are envisaged only as part of a global framework of hybrid integration of legal and policy norms for addressing climate displacement. Chapter reiterates that a one-dimensional approach is inappropriate to address the complex interplay of climate displacement. Climate change displacement is an environmental and social crisis, and it is necessary that law reacts and responds. However, in addition to presenting the legal and policy actions *inter alia* the activism of civil society institutions, public awakening, general environment protection, inclusive development models that mitigate the climate change-

[190] Wanjala Dean George, '"Environmental Refugees": Exposing Their Protection Gaps in International Law' (Strathmore University Law School, 2017) 42 https://su-plus.strathmore.edu/bitstream/handle/11071/5209/Environmental%20refugees.pdf?sequence=1&isAllowed=y accessed 10 March 2018.

[191] Stephanie Dickson, Sophie Webber and Tim K. Takaro, 'Preparing BC for Climate Migration' (Canadian Centre for Policy Alternatives, 2014) 7 https://www.policyalternatives.ca/sites/default/files/uploads/publications/BC%20Office/2014/11/ccpa-bc_ClimateMigration_web.pdf accessed 10 March 2018.

[192] United Nations Framework Convention on Climate Change (adopted 29 May 1992) 1771 UNTS 107 (UNFCCC) art 4(4).

linked vulnerabilities and enhance the climate change adaptation and resilience are immensely critical to the protection rights of the climate refugees.

7.7 Conclusion

It may be submitted that the ramifications of climate change transcend the local, national, regional, continental and global geopolitical demarcations that have been adversely affecting the human existence in societies and nation-states. In South Asia, climate change vulnerability and vibrant interaction of socioeconomic dynamics will pander to climate change-induced displacement of people affecting their livelihoods, lands and homelands involving movements from temporary to permanent nature, from sudden to gradual mobility within the national borders, and from internal migration to international migration. Therefore, the contemporary international law requires more legal wherewithal to address the size and scale of the climate change-induced displacement crisis. The SAARC jurisdictions have subscribed to many legal Declarations and other international law *magnum opus,* but their commitment has not been up to the international standards in implementing the mandate of such frameworks.

SAARC, the product of a collective vision to harness regional cooperation, has placed priorities on building regional cooperative mechanisms in addressing climate risks. However, only scant attention has been paid to the emerging storm of climate displacement. Currently, the climate change displacement is mostly looked and addressed through the prism of adaptation, but this approach fails to evaluate the climate displacement in its multifarious nature. The SAARC region is direly in need of evolving credible basic norms of refugee protection with crystallization of roles of different stakeholders based on participation and inclusiveness. In the SAARC context, regional cooperation is exclusively state-sponsored with support from other regional and global organizations beyond South Asia such as Organization of American States, European Union, Bali Process, OAU, the IOM and the UNHCR. Civil Society Organizations in the SAARC region could provide one of the most effective leadership in the field of protecting the climate change-displaced populations. The SAARC nation-states collectively require acting decisively to perform their ethical and moral responsibility by engaging in cooperation, consultation, and good faith to evolve an inclusive, participatory and long-term legal approach to address the climate-induced displacement. A bottom-up participatory approach must be contemplated while accentuating on the multitude of vulnerabilities and adaptive competence at the local stage. Conclusively speaking, climate change has emerged as the shared responsibility, and nothing can be left to chance while recognizing the climate change as a common problem of the humankind. It is an issue that requires innovative and groundbreaking solutions and reconceiving, reformulating and redefining the legal stipulations and international responsibilities to endorse solutions for the climate change-induced displacement and migration. Therefore, there is a requirement to develop a hybrid legal mechanism that must have a fusion of the core principles of human rights, environment and refugee laws to implement and accomplish such a vision and values. Regional solutions on refugee protection and treatment with a hybrid approach will be the most pragmatic solution provided the core rule of law norms has been effectively implanted in the lego-institutional frameworks of the region. The international community must realize its ethical, moral and legal foundations deriving from humanitarianism[193] to address the pains and sufferings of climate change-induced displaced populations.

[193]Barry Munslow and Tim O Dempsey, 'From War on Terror to War on Weather? Rethinking Humanitarianism in a New Era of Chronic Emergencies' (2010) 31(8) Third World Q. 1223–1235.

MIX
Papier aus verantwortungsvollen Quellen
Paper from responsible sources
FSC® C105338

If you have any concerns about our products,
you can contact us on
ProductSafety@springernature.com

In case Publisher is established outside the EU,
the EU authorized representative is:
Springer Nature Customer Service Center GmbH
Europaplatz 3, 69115 Heidelberg, Germany

Printed by Libri Plureos GmbH
in Hamburg, Germany